WITHDRAWN
JUN 27 2024
DAVID O. McKAY LIBRARY
BYU-IDAHO

JOURNAL OF CHROMATOGRAPHY LIBRARY – volume 68

emerging technologies in protein and genomic material analysis

JOURNAL OF CHROMATOGRAPHY LIBRARY – volume 68

emerging technologies in protein and genomic material analysis

edited by

György A. Marko-Varga

Department of Analytical Chemistry, Lund University
Lund, Sweden

and

Peter L. Oroszlan

Zeptosens AG
Witterswil, Switzerland

2003

ELSEVIER
Amsterdam – Boston – Heidelberg – London – New York – Oxford – Paris
San Diego – San Francisco – Singapore – Sydney – Tokyo

ELSEVIER SCIENCE B.V.
Sara Burgerhartstraat 25
P.O. Box 211, 1000 AE Amsterdam, The Netherlands

First edition 2003

Library of Congress Cataloging in Publication Data
A catalog record from the Library of Congress has been applied for.

British Library Cataloguing in Publication Data
A catalogue record from the British Library has been applied for.

ISBN: 0-444-50964-X
ISBN: 0301-4770 (Series)

♾ The paper used in this publication meets the requirements of ANSI/NISO Z39.48-1992 (Permanence of Paper).
Printed in The Netherlands.

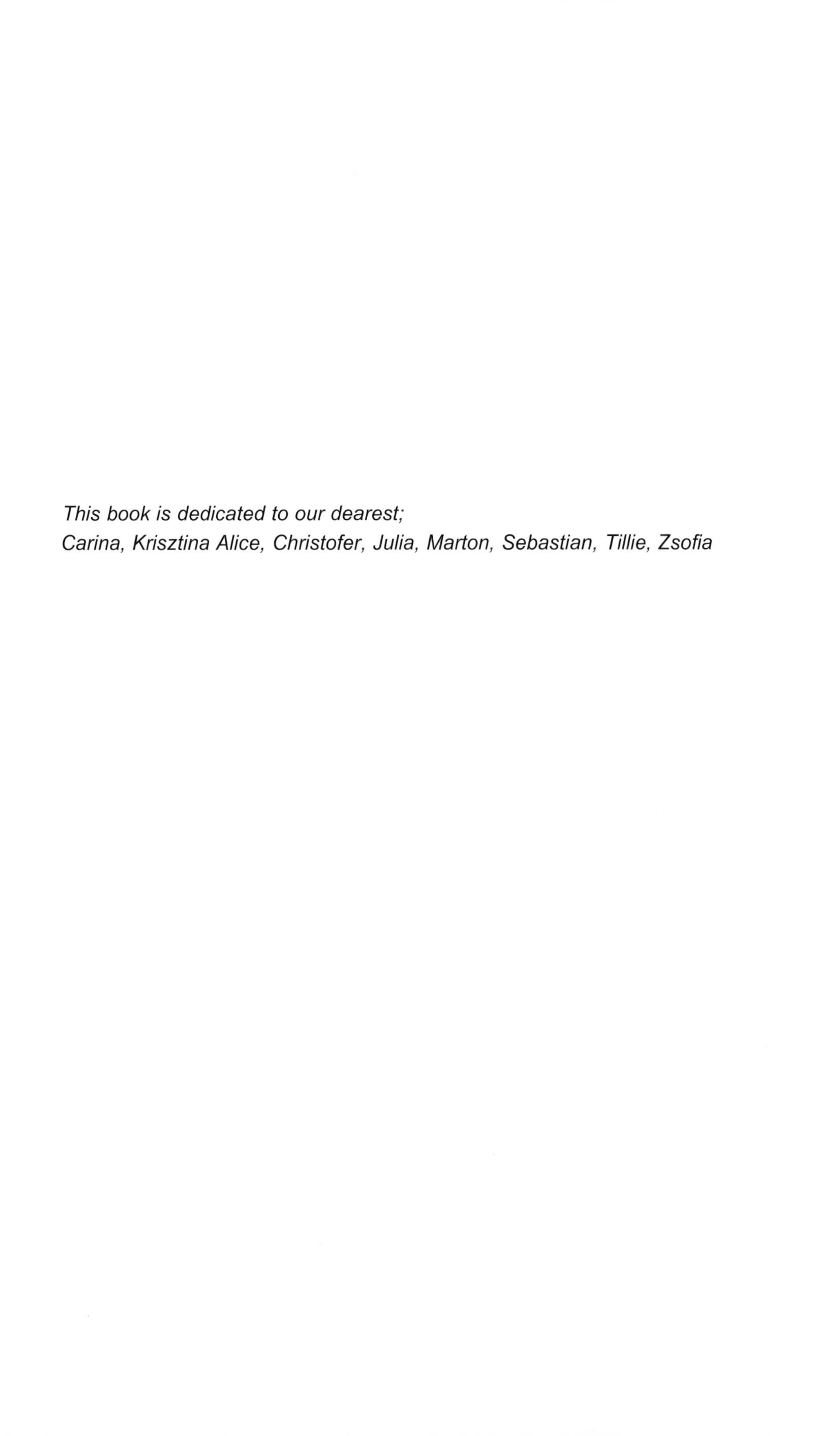

This book is dedicated to our dearest;
Carina, Krisztina Alice, Christofer, Julia, Marton, Sebastian, Tillie, Zsofia

Preface

The protein- and gene-expression research areas are developing very rapidly and it is fair to say that analytical chemistry plays a very important role in these developments. New challenges lie ahead, with new discoveries and tools opening up novel dimensions in life-science applications. One could argue that the developments in analytical technology are among the major drivers in the biological sciences of today. Examples include the ground-breaking work on DNA-sequencing using capillary electrophoresis, and mass spectrometers being standard instrumentation in research laboratories throughout the world for protein/DNA sequencing and structure work. The pioneering contribution of mass spectrometry to progress in biological sciences was also highlighted by the December 2002 Nobel prizes that were awarded for the invention of electrospray and laser desorption MS.

In parallel, new discoveries in biology and medicine also set new targets for the scientific community regarding performance of new analytical technologies, systems and methods. Our recognition of the pressing need for advanced technologies to fulfill the requirements of the demanding tasks of functional genomic initiatives is the underlying motivation for all of us working on the cutting edge of bioanalytical technologies. With this book, our intention is to

- compile state-of-the-art knowledge of the field and describe emerging technologies
- provide examples by relevant applications and other case studies
- provide insight into the new frontiers of the field that embrace relevant biological systems
- stimulate young scientists and newcomers to be a part of the ongoing biological revolution, where bioanalytical developments are a cornerstone of these successes

The book is intended to serve as a general reference for researchers and scientists within the bioanalytical field as well as for postgraduate students. Each chapter includes references to the corresponding literature to serve as valuable entry points to anyone wanting to move forward in this field, either as a practitioner or for acquiring state-of-the-art knowledge.

Finally, we would like to express our gratitude to the contributing authors for their time and effort in preparing the chapters. Without the engagement and lively interactions we have had over the last half year to get this project finalized, this book would not have been possible.

GYÖRGY A. MARKO-VARGA AND PETER L. OROSZLAN

List of Contributors

J. Abian
Structural and Biological Mass Spectrometry Unit, Department of Medical Bioanalysis, Instituto de Investigaciones Biomédicas de Barcelona-Consejo Superior de Investigaciones Científicas (IIBB-CSIC), Institut d'Investigacions Biomèdiques August Pi i Sunyer (IDIBAPS), Rosselló 161, 7ª Planta, 08036 Barcelona, Spain

P.E. Andren
Biological Mass Spectrometery Center and Department of Pharmaceutical Biosciences, Uppsala University, SE-75124 Uppsala, Sweden and Amersham Biosciences, Björkgatan 30, SE-75184 Uppsala, Sweden

Y. Arntz
Institute of Physics, NCCR Nanoscale Science, University of Basel, Klingelbergstrasse 82, CH-4056 Basel, Switzerland

J. Bergquist
Department of Analytical Chemistry, Institute of Chemistry, Uppsala University, P.O. Box 531, SE-751 21 Uppsala, Sweden

P. Bertoncini
Institute of Physics, NCCR Nanoscale Science, University of Basel, Klingelbergstrasse 82, CH-4056 Basel, Switzerland

M. Carrascal
Structural and Biological Mass Spectrometry Unit, Department of Medical Bioanalysis, Instituto de Investigaciones Biomédicas de Barcelona-Consejo Superior de Investigaciones Científicas (IIBB-CSIC), Institut d'Investigacions Biomèdiques August Pi i Sunyer (IDIBAPS), Rosselló 161, 7ª Planta, 08036 Barcelona, Spain

R.J.E. Derks
Department of Analytical Chemistry and Applied Spectroscopy, Division of Chemistry, Faculty of Sciences, vrije Universiteit amsterdam, De Boelelaan 1083, 1081 HV Amsterdam, The Netherlands

Ch. Gerber
Institute of Physics, NCCR Nanoscale Science, University of Basel, Klingelbergstrasse 82, CH-4056 Basel, Switzerland and IBM Zurich Research Laboratory, Säumerstrasse 4, CH-8803 Rüschlikon, Switzerland

W. Grange
Institute of Physics, NCCR Nanoscale Science, University of Basel, Klingelbergstrasse 82, CH-4056 Basel, Switzerland

M. Grønborg
Protein Research Group, Department of Biochemistry and Molecular Biology, Odense University/University of Southern Denmark, Odense, Denmark

A. Guttman
Torrey Mesa Research Institute, San Diego, CA 92121, USA

M. Hegner
Institute of Physics, NCCR Nanoscale Science, University of Basel, Klingelbergstrasse 82, CH-4056 Basel, Switzerland

A.C. Hogenboom
Department of Analytical Chemistry and Applied Spectroscopy, Division of Chemistry, Faculty of Sciences, vrije Universiteit amsterdam, De Boelelaan 1083, 1081 HV Amsterdam, The Netherlands

S. Husale
Institute of Physics, NCCR Nanoscale Science, University of Basel, Klingelbergstrasse 82, CH-4056 Basel, Switzerland

H. Irth
Department of Analytical Chemistry and Applied Spectroscopy, Division of Chemistry, Faculty of Sciences, vrije Universiteit amsterdam, De Boelelaan 1083, 1081 HV Amsterdam, The Netherlands

O. Nørregaard Jensen
Protein Research Group, Department of Biochemistry and Molecular Biology, Odense University/University of Southern Denmark, Odense, Denmark

H.P. Lang
Institute of Physics, NCCR Nanoscale Science, University of Basel, Klingelbergstrasse 82, CH-4056 Basel, Switzerland and IBM Zurich Research Laboratory, Säumerstrasse 4, CH-8803 Rüschlikon, Switzerland

T. Laurell
Department of Electrical Measurements, Lund Institute of Technology, Lund University, P.O. Box 118, S-221 00 Lund, Sweden

G.A. Marko-Varga
Department of Analytical Chemistry, Lund University, P.O. Box 124, SE-221 00 Lund, Sweden

J. Nilsson
Department of Electrical Measurements, Lund Institute of Technology, Lund University, P.O. Box 118, S-221 00 Lund, Sweden

P.L. Oroszlan
Zeptosens AG, Benkenstrasse 254, CH-4108 Witterswil, Switzerland

A. Palm
Department of Chemistry, Indiana University, Bloomington, IN 47405, USA

M. Palmblad
Division of Ion Physics, Uppsala University, Box 534, SE-751 21 Uppsala, Sweden

K. Sköld
Biological Mass Spectrometery Center and Department of Pharmaceutical Biosciences, Uppsala University, SE-75124 Uppsala, Sweden

P. Svenningsson
Department of Physiology and Pharmacology, Karolinska Institute, SE-17177 Stockholm, Sweden

M. Svensson
Biological Mass Spectrometery Center and Department of Pharmaceutical Biosciences, Uppsala University, SE-75124 Uppsala, Sweden

J. Zhang
Institute of Physics, NCCR Nanoscale Science, University of Basel, Klingelbergstrasse 82, CH-4056 Basel, Switzerland

Contents

G.A. Marko-Varga and P.L. Oroszlan (Editors)
Emerging Technologies in Protein and Genomic Material Analysis
Journal of Chromatography Library, Vol. 68

CHAPTER 1

Enabling Bio-analytical Technologies for Protein and Genomic Material Analysis and their Impact on Biology

GYÖRGY A. MARKO-VARGA[a] and PETER L. OROSZLAN[b]

[a]*Department of Analytical Chemistry, Lund University, Lund, Sweden*
[b]*Zeptosens AG, Witterswil, Switzerland*

1.1. BACKGROUND

Although the human genome was made public by the end of June 2000, it was not the complete human map; there were still parts that were not readily available. Currently, there is a plan that the National Human Genome Research Institute will publish the finished draft of the human-genome sequence in a scientific journal by 2003. During the period between the first launch of the human-genome sequences (June 2000) and today, the focus has turned towards the post-genomic area. High capacity and high performance technologies that are able to identify the expression of proteins are currently a research area attracting enormous attention. This is due to the fact that the DNA coding and 'the holy grail' of human life is no further than the nearest computer with an Internet connection. Almost the entire genome can be searched. These gene sequences hold only a limited percentage of the entire proteome that is present in cells. The reason is that, upon transcription and translation, proteins in many cases undergo a post-translational modifying step.

Proteomics has become the new 'hot' research where the protein expression area has received enormous attention scientifically as well as from investors in the stock market [1]. Several hundred million dollars have already been invested by the Pharma and Biotech companies and the business growth until 2005 is expected to be $5 billion [2].

Academia, the pharmaceutical industry and the biotech sector have moved their efforts and positions towards trying to complete the human proteome and what it holds. It seems that this task is not only a major effort to fulfil, but also difficult to define. What exactly the proteome as such holds in terms of gene products is one aspect of addressing the entire map of the human proteome. The other aspect is that the post-translational modifications that occur within the cell reflect a fundamental and very important process whereby additionally a large number of proteins is present. The post-translationally modified proteins may occur as many variants all sharing a larger degree of protein sequences, e.g. enzymes often occur as

References pp. 10

persist in the cell for hours or even days. However, ultimately, the major difference in approaching the protein expression profiling research area is that we lack the ability to produce a PCR technology for proteins whereby we are able to amplify any sequence we would like. Until we are able to make that ingenious discovery, which surely would be awarded an additional Nobel prize, we are left with two principle options: (1) develop novel technology that allows us to increase protein sequencing throughput, as well as make major improvements in analyte sensitivity detections, and (2) make use of larger amounts of biological starting materials that allows the sequence information to be determined. In many medical situations the latter possibility is ruled out since e.g. specific tissue material important for the disease is not available.

Today, there are two main strategic approaches that studies are performed according to: (i) global expression profiling, where a high number of proteins are readily sequenced, (here, tens of thousands of proteins can be identified), and (ii) a focused approach, where a given organelle in the cell is of particular interest because of a biological function occurring in a biological event of particular importance, or targeting a specific protein that is known to have a central role to play in a specific disease. In this respect, both the target protein itself and the neighboring protein alteration in the disease progression are of key importance.

Additionally, many critical functions of the cell are accomplished by a complex cascade of events, e.g. in signalling pathways, in which proteomics is an efficient and powerful technique for extracting specific details of signalling mechanisms. This has been explored and described in several interesting presentations by both Cellzome and MDS-Proteomics [6–8].

1.3.2. Multidimensional liquid chromatography

Traditional protein expression profiling using two-dimensional gel electrophoresis (2DGE) is still very useful in most proteomics laboratories all over the world. The combination of these two separation techniques is still the best method for analyzing intact proteins. 2DGE technology was originally developed early in the 1970s by O'Farell and Laemmli [9,10].

At present, there seems to be a movement towards 2D or multidimensional liquid chromatography replacing many experimental studies where 2D-gels are used currently. Still, 2D-gels will continue to have a place even in the future because of the inherent resolving power of displaying intact proteins at an impressive sensitivity level. A very interesting comparison of the above mentioned methods was a set of yeast studies carried out using both 2D-gels and 2D-LC. Gygi et al. [11] found that the capillary LC approach was superior to the gel technology, while Fey and Larsen, in a repeat of the experiment made at the Center for Proteome Analysis (CPA), Odense, Denmark, presented 2D-gel image mapping at an impressive level with even lower codon bias values, which exceeded the capacity of the 2D-capillary LC approach [12]. It is probable that further examples will occur in the future where 2D-capillary LC expression profiles show superior performances to that of 2D-gels. It is still early days regarding running proteomics with liquid phase separation. The quantitative aspect utilizing isotope labeling reagents such as ICAT and others should not be underestimated [12–15].

When it comes to integrated protein workstations, the most powerful separation technology interfaced to ESI-MS/MS is certainly multidimensional HPLC where research groups have been able to generate thousands of protein identities during a limited time range. An additional benefit is that all these steps are run in an automated mode that is operated with a high degree of reproducibility.

1.3.3. Fast and sensitive protein sequencing by mass spectrometry

For fast access to protein sequence information for verifying identities of proteins and protein complexes, the technology of choice for high sensitivity and high speed protein identifications is peptide mass sequencing by mass spectrometry with the various ionisation principles available [16]. In this volume, Grönborg and Noerregaard Jensen introduce new concepts whereby phosphoproteins can be analyzed. Post-translational modifications such as the phosphorylation stoichiometry using a combination of chromatographic enrichment strategies are outlined. Currently, there is an explosion of novel principles whereby protein sequencing by mass spectrometry is done. It is also anticipated that in the future more work will be focused on protein–protein interactions, since proteins mostly execute biological events as complexes.

This was highlighted when a recent milestone was achieved simultaneously by two groups using high-throughput protein–protein interaction sequencing. Both of the groups used yeast two-hybrid systems to map the interactions.

This experimental exercise can efficiently be utilized as a shortcut to human disease, where recent works by Gavin et al. [6] and Ho et al. [7], mapping protein–protein interactions in *Saccharomyces cerevisae*, resulted in the identification of more than 10 000 protein associations.

Protein–protein interactions are vital biological processes within the cell. No protein operates on its own; on the contrary, most seem to function within complicated cellular pathways, interacting with other proteins either in pairs or as components of larger complexes.

Both groups addressed the analysis of protein–protein interactions by mainly liquid phase separations, applied with liquid chromatographic approaches interfaced on-line to MS/MS. Gavin et al. [6] made an alteration using liquid chromatography in combination with 1D-gels, applying accelerated techniques that rapidly identify the protein function and characterize multidimensional protein complexes. Both groups made use of techniques that genetically pair engineered tags with sensitive automated mass spectrometry to probe multi-protein complexes in cells of yeast. Both single- and double-tagging approaches were applied to the yeast two-hybrid systems; however, neither the advantages nor the disadvantages of the two approaches are clear, even today.

Gavin et al. [6] used a gene-specific cassette containing a TAP tag at the $3'$-end of the genes. The TAP tag is used as a tandem-affinity approach in the purification of proteins. Using this approach, 1739 genes were tagged and 1167 genes were expressed in yeast.

References pp. 10

In the method of Ho et al. [7], recombinant-based cloning was utilized to add a flag epitope tag to 725 yeast genes. These clones were then transfected in yeast. Interactions were discovered for roughly 70% of the clones for a total of 1578 different interacting proteins, representing 25% of the yeast genome. However, it appears that the data generated in these studies also show that many examples of expected interactions are not present in the data sets, as well as showing defaults in the data from a yeast perspective [17]. The network of protein–protein interactions is made fully interactive through the database and the software allows highly complex hypothesis testing to be made [8]. It includes data from protein–protein interactions that were already known to exist in the literature as well as newly identified interactions performed in experimental studies.

As a new development for improving the speed, accuracy and sensitivity of protein sequencing, the development of the sequencing MALDI instrument, the ABI 4700 TOF/TOF workstation, is an example of a unique instrument for MS/MS sequence generation. The precursor ion is selected and fragmented in a high-energy collision-induced dissociation cell [18]. Ions will then be re-accelerated and analysed by the second TOF analyser. This ionisation principle offers coaxial MS/MS capability from a standard MALDI source. The pulsing laser fires 200 times per second (200 Hz) allowing simultaneous data acquisition to be performed. High sample throughputs can be generated with a sample speed of 400–1000 per hour.

At present, new Fourier transform ion cyclotron resonance mass spectrometry (FTICR) principles are being developed whereby peptide sequences can be obtained at low attomole levels [19].

Palmblad and Bergkvist provide developments in one chapter, made with FTICR, for peptide and protein determinations where examples are given of peptide sequencing in various human biofluids.

Today, researchers use mass spectrometry analysis to complement and enhance experimental results obtained by X-ray crystallography and NMR. Protein sequencing using mass spectrometry is a highly valuable technique in experimental protein structure studies. The analytical power of MS for studying proteins has improved significantly and is used routinely for both expression proteomics and functional proteomics studies.

In addition, the use of computational modeling techniques to generate 3D structures in-silico is also an important feature for understanding basic molecular interactions. In order to generate 3D models, these techniques rely on the existence of experimentally determined structures as well as primary amino acid sequences. It is also foreseen that computational approaches to protein structure determination will play a critical role in mining the vast library of gene sequence databases.

1.4. PROTEIN CHIP ARRAYS

With the capacity to generate a million or so antibodies that recognize every protein in the human proteome, new horizons can be envisioned. The next step would be to deposit and incubate a few microliters of biofluid from a patient or control onto the chip. Some of the sample's proteins will mate with antibodies on the chip and latch onto them. Because the proteins have been labeled with a fluorescent dye, they can be seen, with the help of a laser, as little glowing squares of red, green or yellow. The protein pattern that lights up will differ in a few critical ways from that of a normal person, identified by differential display analysis. By identifying those discrepancies and finding the proteins that cause them, it is now possible to rapidly identify the proteins that are present in normal patients and not in diseased samples. This identification may be the opening for a hypothesis where the link to the cause of the disease, either directly or indirectly, may be found by depositing a few drops of the sample fluid onto a chip and scanning for the protein changes that might indicate a change of differential protein expression.

This obviously will be a major challenge and undertaking since proteins are far more complex than strands of DNA; it is not just a simple transformation from DNA chips to protein chips.

Although the speed in research successes is impressive, it is anticipated, however, because of that the major breakthrough will not happen overnight e.g. the difficulties encountered in producing high quality specific antibodies to fight every variant of every protein in the human body.

Phage display technology is being used to make antibody libraries with the potential to make appropriate clone selection to select high affinity immunoreagents.

The big question is whether protein chip technologies will totally supplant the need for DNA chips, or rather be used as a strong complementary analytical tool.

Recently, a novel protein chip technology, which allows the high-throughput analysis of biochemical activities, has been used to analyse nearly all of the protein kinases from *Saccharomyces cerevisae*. Protein chips are disposable arrays of micro wells in silicone elastomer sheets placed on the top of microscope slides. The high density and small size of the wells allows for high-throughput batch processing and simultaneous analysis of many individual samples. Only small amounts of protein are required. Of a hundredfold known and predicted yeast protein kinases, somewhat more than a hundred were overexpressed and analyzed using 17 different substrates and protein chips. It was found that many novel activities and a large number of protein kinases are capable of phosphorylating tyrosine. Additionally, the tyrosine phosphorylating enzymes are found often to share common amino acid residues that lie near the catalytic region [19].

1.5. DNA/RNA ANALYSIS

Faced with the task of analyzing thousands of genes at once, researchers routinely turn to DNA chips to analyze the expression patterns in various biological materials of interest. The chips are one of the key innovations that have transformed genomics from a cottage industry into a large-scale automated business. Most genes serve as blueprints for the production of proteins, the workhorses of living cells. To understand diseases, it is compulsory to survey the activities and functions of these proteins.

Large-scale screening exercises where microorganisms or human clinical material are investigated is directed towards finding the genetic disorders where altering gene sequences in given chromosomes can be localized and linked to a specific given disease. In this way, several companies have made major efforts to use extremely large sets of, e.g. human biofluid banks, which hold the information from a large population. In many cases,

References pp. 10

these bio-banks have a track record of samples over time. This means that it is possible to follow a patient from an early age, when the disease was diagnosed, to a late state disease in an adult state. This makes these screening studies unique since it is possible to follow the change of disease over time, and thereby find the mechanisms that are involved in malfunctions.

1.5.1. Pharmacogenomics

Pharmacogenomics—the study of how the genetic code affects an individual response to drugs—holds the promise that drugs can be adapted to each person's own genetic makeup, as the key towards creating more powerful, personalized drugs with greater efficacy and safety. It requires the combination and understanding of traditional pharmaceutical sciences (e.g. biochemistry) with annotated knowledge of genes and proteins. Global and tailored gene expression analysis techniques such as genomic microarrays or gene chips have the potential to identify toxicology-related problems and simultaneously predict risks associated with the exposure of the drug candidate. Moreover, new technologies allowing simultaneous multiplexed analysis of biological markers (efficacy and surrogate markers), such as protein microarrays, are expected to deliver functional information on drug response and efficacy. This should allow short studies in the very early discovery phase to predict the long-term effect of the exposure and later phase clinical outcomes of a treatment. The integration and industrialization of these new techniques and disciplines represent an enormous potential leading to early selection of 'good' drug candidates by gaining information on the therapeutic index (i.e. the measure of the beneficial effect of a drug compared with its safety). Ideally, safety and efficacy should be studied simultaneously in animal models by global assessment of potential surrogate markers, resulting in the increase of efficiency with a scope of a significant return to the industry.

1.6. MINIATURIZED, MICRO-SCALE SYSTEMS

Large-scale shotgun sequencing using a high-throughput multi-capillary-based separation approach contributed enormously to the speeding up of sequencing efforts in the frame of the human genome project and continues to be the dominant technology in sequencing. As the next quantum leap in this direction, Guttman, in this book, proposes the dissemination and broad utilization of microfabricated, capillary-based integrated fluidic systems, to be used as stand-alone devices or integrated elements in hyphenated systems with relevance in genomics/proteomics. An important application practiced at the industrial scale and with potential for miniaturization is described by Irth et al.—the combination of homogeneous, continuous flow multi-protein biochemical assays with characterization of the interacting partners using mass spectrometry, thus allowing the simultaneous screening and affinity characterization of chemical compounds as well as the molecular mass determination of biologically active compounds in screening mode. Also in this volume, Palm presents an overview of 'Capillary Isoelectric Focusing Developments in Protein Analysis'.

New developments on the ability of different isoelectric focusing techniques to separate and determine proteins are presented with an overview of the field using both electrical and pressure driven capillary separation techniques. Additionally, silicon microtechnology developments for protein expression profiling—proteomics—is presented by Laurell, Nilsson and Marko-Varga where signal amplification is achieved. This is obtained by utilizing the ability to deposit multilayers of the protein samples onto the chip-target surface, whereby protein density is increased. MALDI-TOF MS is used for peptide mass fingerprinting, whereby the proteins are identified.

In another chapter of this issue, downscaling of liquid chromatography column dimensions to coupled column nano-flow LC is presented by Abian and Carascal, where the interfacing of the separation step to the mass spectrometer is presented for peptide analysis. Comparisons are made where the impact of chromatographic column dimensions are presented with applications made on endogenous peptides.

1.6.1. Nano-sciences and nano-scale analytical systems

Nanotechnology, and nano-scale science in general, is one of the most striking new directions and fields of expansion in science. The understanding of fundamental events in life is the driving force of efforts to investigate biomolecular interactions on the single molecule level. Hegner et al., in this volume, give an overview of recent advancements in single molecule analysis, discussing theory, development status and applications. They cover state-of-the-art force sensitive methods, optical tweezers and scanning force microscopy towards biosensing using micro-mechanical cantilever arrays, a promising new approach towards high-throughput multiplexed devices for analyzing biospecific interactions on the molecular level, the target for ultimate sensitivity in analytical disciplines. It is understood that investigation of single molecule interactions has the potential to provide information on the property or activity only of a certain individual subpopulation of the 'mass'. Nevertheless, that knowledge may lead to new insights into delicate biological systems and processes, such as fundamental molecular events leading to cancer. Besides potential analytical applications in life science research and diagnostics, the translation of biochemical recognition into, e.g. nano-mechanical motion, might have other highly speculative applications in the future, such as nanorobotics or biocomputation.

1.7. CONCLUDING REMARKS

Currently, there are a number of trends and analytical approaches available, but probably the toughest challenge that the analytical field has been called to solve is the one addressing powerful high resolution protein separation and sequencing of proteomes.

Considering the perspective of having potentially a million proteins to unravel, even more powerful technologies are needed for proteomics in the future, which will keep the bioanalytical research field highly active and attractive for several more decades.

References pp. 10

1.8. REFERENCES

1. International Economy 2001, April 24, p. 16
2. Study Foresees Proteomics Market Growing to $5.6B by 2006. GenomeWeb2001. Ref Type: Electronic Citation; http://www.genomeweb.com/articles/view.asp?Article = 20018156158
3. N.G. Anderson, A. Matheson and N.L. Anderson, Proteomics, 1 (2001) 3–12
4. HUPO; http://www.hupo.org/
5. SwissProt, Human Protein database; http://us.expasy.org/sprot
6. A.-C. Gavin, M. Bösche, R. Krause, P. Grandi, M. Marzioch, A. Bauer, J. Schultz, J.M. Rick, A.-M. Michon, C.-M. Cruciat, M. Remor, C. Höfert, M. Schelder, M. Brajenovic, H. Ruffner, A. Merino, K. Klein, M. Hudak, D. Dickson, T. Rudi, V. Gnau, A. Bauch, S. Bastuck, B. Huhse, C. Leutwein, M.A. Heurtier, R.R. Copley, A. Edelmann, E. Querfurth, V. Rybin, G. Drewes, M. Raida, T. Bouwmeester, P. Bork, B. Seraphin, B. Kuster, G. Neubauer and G. Superti-Furga, Nature, 415 (2002) 141–147
7. Y. Ho, A. Gruhler, A. Heilbut, G.D. Bader, L. Moore, S.-L. Adams, A. Millar, P. Taylor, K. Bennett, K. Boutilier, L. Yang, C. Wolting, I. Donaldson, S. Schandorff, J. Shewnarane, M. Vo, J. Taggart, M. Goudreault, B. Muskat, A.C. Dewar, L.Z. Michalickova, A.R. Willems, H. Sassi, P.A. Nielsen, K.J. Rasmussen, J.R. Andersen, L.E. Johansen, L.H. Hansen, H. Jespersen, A. Podtelejnikov, E. Nielsen, J. Crawford, V. Poulsen, B.D. Sùrensen, J. Matthiesen, R.C. Hendrickson, F. Gleeson, T. Pawson, M.F. Moran, D. Durocher, M.W.V. Mann, C.W.V. Hogue, D. Figeys and M. Tyers, Nature, 415 (2002) 180–183
8. D. Figeys, Curr. Opin. Mol. Ther., 4 (2002) 210–215
9. P.H. O'Farell, J. Biol. Chem., 250 (1975) 4007–4021
10. U.K. Laemmli, Nature, 227 (1970) 680–685
11. S.P. Gygi, G.L. Corthals, Y. Zhang, Y. Rochon and R. Aebersold, PNAS, 97 (2000) 9390–9395
12. S.J. Fey and P.M. Larsen, Curr. Opin. Chem. Biol., 5 (2001) 26–33
13. R. Aebersold and D.R. Goodlett, Chem. Rev., 101 (2001) 269–295
14. H. Zhou, J.A. Ranish, J.D. Watts and R. Aebersold, Nat. Biotech., 20 (2002) 512–515
15. M. Geng, J. Ji and F.E.J. Regnier, Chromatogr. A, 870 (2000) 295–313
16. M. Mann, R.C. Hendrickson and A. Pandey, Annu. Rev. Biochem., 70 (2001) 437–473
17. A. Kumar and M. Snyder, Nature, 415 (2002) 123–124
18. S. Martin, M. Vestal and K.R. Jonscher, Genom. Proteom., 2 (2001) 56–59
19. H. Zhu, J.F. Klemic, S. Chang, P. Bertone, A. Casamayor, K.G. Klemic, D. Smith, M. Gerstein, M.A. Reed and M. Snyder, Nat. Genet., 26 (2000) 283–289

G.A. Marko-Varga and P.L. Oroszlan (Editors)
Emerging Technologies in Protein and Genomic Material Analysis
Journal of Chromatography Library, Vol. 68

CHAPTER 2

DNA Sequencing: From Capillaries to Microchips

ANDRAS GUTTMAN

Torrey Mesa Research Institute, San Diego, CA 92121, USA

2.1. INTRODUCTION

The striking scientific discovery of the double helical structure of DNA by Watson and Crick in 1953 [1], and the following unveiling of the genetic code revealed the extremely important function of DNA in storing genetic information [2]. The sequence of more than three billion bases of the human genome represents the genetic blueprint of an individual, virtually holding all information required for the growth and development of that person [3]. Revealing the information encoded in the genome should give us a better understanding of genetically predetermined patterns, diagnostics of genetic diseases and human identification. DNA sequence information is considered to revolutionize our perception of how cells and organisms function.

Until the mid-1970s, DNA sequencing was a labor-intensive and time-consuming task. First Maxam and Gilbert [4] reported a method capable of sequencing up to 500 nucleotides by chemically cleaving end-labeled DNA restriction fragments at specific nucleotides. The resulting fragments were then separated by gel electrophoresis and detected by autoradiography. The sequence information was derived from the detected electrophoretic traces. A few years later, Sanger's group demonstrated the usefulness of the so-called dideoxy chain termination technique [5]. Due to the ease of use and the readiness for automation, this later method became the almost exclusive means of modern, large-scale DNA sequencing. The technique employs a so-called primer (a 15–25 bases long oligonucleotide), which anneals to the template DNA molecule initiating in vitro DNA synthesis. DNA polymerase enzyme is building up the chain, complementary to the template, using the four corresponding deoxynucleotides as building blocks. Addition of small amounts of dideoxy nucleotide triphosphates randomly terminates the polymerization reaction generating in this way an almost equal population of sequencing fragments. The resulting, so-called, sequencing ladders are then separated by means of polyacrylamide gel electrophoresis, which can resolve DNA molecules differing in length by just one nucleotide. Initially, ^{32}P-containing nucleotides were used for labeling; therefore, four lanes represented one sequence [6]. The fully developed slab gel was then subject to autoradiography. The sequence of the DNA template strand was read directly from the resulting autoradiogram (Fig. 2.1). Please note, at that time, autoradiograms were manually evaluated by tracing the individual bands one by one, and lane by lane.

2.2. THE EARLY DAYS OF AUTOMATED DNA SEQUENCING

Automated DNA sequencing was introduced in 1986 by Smith et al. [7] and resulted in significant improvements over radioactive labeling and manual data interpretation by using fluorophore labeled primers in the sequencing reaction and computerized data analysis. Fluorescence detection proved to be a very effective alternative to autoradiography. In this case the four primers, corresponding to the four termination reactions, were labeled by four different fluorophores, respectively, prior to the sequencing reaction. Upon completion of chain elongation, the four reaction products were combined together and size separated by polyacrylamide gel electrophoresis. The separated bands were then detected by their fluorescence and the sequence information was derived from the characteristic emission wavelengths. Fluorescence detection is appealing because it eliminates the use of radioactivity and offers easy automation of sequence reading. A scanning laser-induced fluorescence detection system interrogates the gel at a given distance from the injection point, in this way automating the extremely time-consuming manual gel reading procedure. The different emission spectra of the four fluorophores made automated reading possible. Later Prober and co-workers [8] introduced fluorescently labeled dideoxy nucleotides as chain terminators for the sequencing reaction, employing non-labeled primers. The resulting sequencing ladder fragments were then

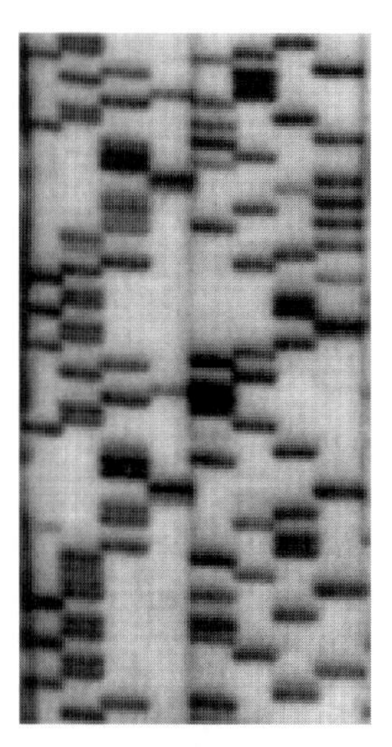

Fig. 2.1. Autoradiogram of a cross-linked polyacrylamide DNA sequencing gel. The Sanger sequencing approach was used to generate four sequencing ladders, which were analyzed in individual lanes. With permission from Ref. [46].

separated by high-resolution gel electrophoresis in single lanes, after pooling the four sequencing reaction mixtures.

DNA sequencing by separating and detecting individual DNA fragments differing by only one nucleotide is primarily an analytical challenge. High-resolution separations of DNA sequencing fragments were regularly accomplished by means of cross-linked polyacrylamide slab gels [9]. These gels needed to be quite large to accommodate the challenging separation task, involving labor-intensive manual processes. Loading the sequencing reaction samples onto those gels was also tedious and ergonomically challenging. The recently developed membrane-mediated loading technique successfully addressed this issue [10]. Problems with gel uniformity and proper dissipation of the so-called Joule heat being developed during the electrophoresis process also represented major problems. However, electric field-mediated separation technology has significantly advanced during the last two decades, primarily due to the better understanding of the electrophoretic separation process [11] and the rapid development of new microchannel-based separation technologies like capillary gel electrophoresis [12] and microchip electrophoresis [13].

2.3. THE ADVENT OF CAPILLARY GEL ELECTROPHORESIS

Capillary electrophoresis was introduced in the early 1980s as an instrumental approach to electrophoresis, using capillary tubes with an inner diameter of typically 20–100 μm, thus significantly increasing the separation performance [14]. The pioneering work of Cohen and Karger [15] revealed the possibility of filling narrow bore capillaries with cross-linked polyacrylamide gels. The first report on single nucleotide resolution separation of the pdA_{40-60} DNA ladder (Fig. 2.2) [16] opened up new horizons in modern capillary gel

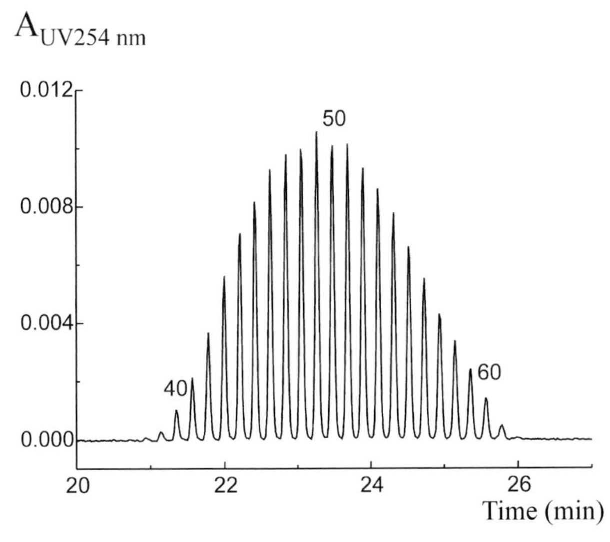

Fig. 2.2. The first capillary gel electrophoresis-based single stranded DNA separation (pdA_{40-60}) that opened up the possibility of large-scale DNA sequencing using microbore structures. With permission from Ref. [16].

electrophoresis-based DNA sequencing analysis and actually made possible the faster than anticipated completion of the sequencing of the humane genome [17,18]. This novel approach revealed the enormous separation power of capillary gel electrophoresis by resolving DNA molecules differing by only one nucleotide, exactly what was required for DNA sequencing [19]. In continuation of the successful introduction of gel filled capillary columns for high-resolution separation of DNA molecules, in 1990, Swerdlow and Gestaland reported the development of the first capillary gel electrophoresis-based DNA sequencer [20].

The lack of adequate stability of cross-linked polyacrylamide gels within microbore columns initiated a rapid development for novel, more capillary friendly sieving matrices. The first breakthrough during this race was reported by the Karger group demonstrating the usefulness of non-cross-linked polymeric solutions for rapid separation of single stranded DNA molecules, which also enabled the application of high temperatures during electrophoresis [21]. Based on their encouraging results, other groups started investigating the applicability of various linear polymers, such as polyethylene oxides [22], derivatized acrylamide [23] and cellulose molecules [24] as well as polyvinyl pyrolidones [25] for DNA sequencing applications. Some of these polymers also possessed self-coating characteristics that alleviated the otherwise necessary hydrophilic polymeric coating of the inner walls of the capillary columns. The introduction of narrow bore columns significantly reduced Joule heat generation enabling the application of orders of magnitude higher electric field strengths, resulting in rapid separation times. The only limit for the separation speed was the phenomena referred to as reptation, which reduced the resolution between longer sequencing fragments due to their chain length independent migration [11]. The use of capillary columns also enabled easier automation of the otherwise tedious sample injection and gel replacement procedures.

2.4. TOWARDS HIGHER THROUGHPUT: CAPILLARY ARRAY ELECTROPHORESIS

The next significant improvement in microbore channel-based DNA sequencing was the introduction of capillary arrays, thus, multiplexing the gel filled capillary columns [26]. This allowed the implementation of simultaneous analysis of up to 96 samples in an automated fashion [27]. Similar to the earlier slab gel-based sequencing systems, in this case a detector scans across an array of capillary columns acquiring the emission signal from each individual capillary (Fig. 2.3). Later a CCD based detection device was developed that overcame the limited duty cycle problem of the scanning setup. The sensitivity of the detection setup was further increased by the introduction of the sheath flow cuvette approach [28].

Choice of the appropriate parameters, such as temperature, for the electrophoresis of DNA sequencing fragments also played an important role in the development of high speed, high read-length separations. Temperatures as high as 80 °C were successfully applied to attain a sequencing speed of more than 1000 bases per hour [29]. Another issue is the high salt content of the DNA sequencing reaction, thus, sample preparation and cleanup turned out to be a significant issue in capillary gel electrophoresis-based DNA

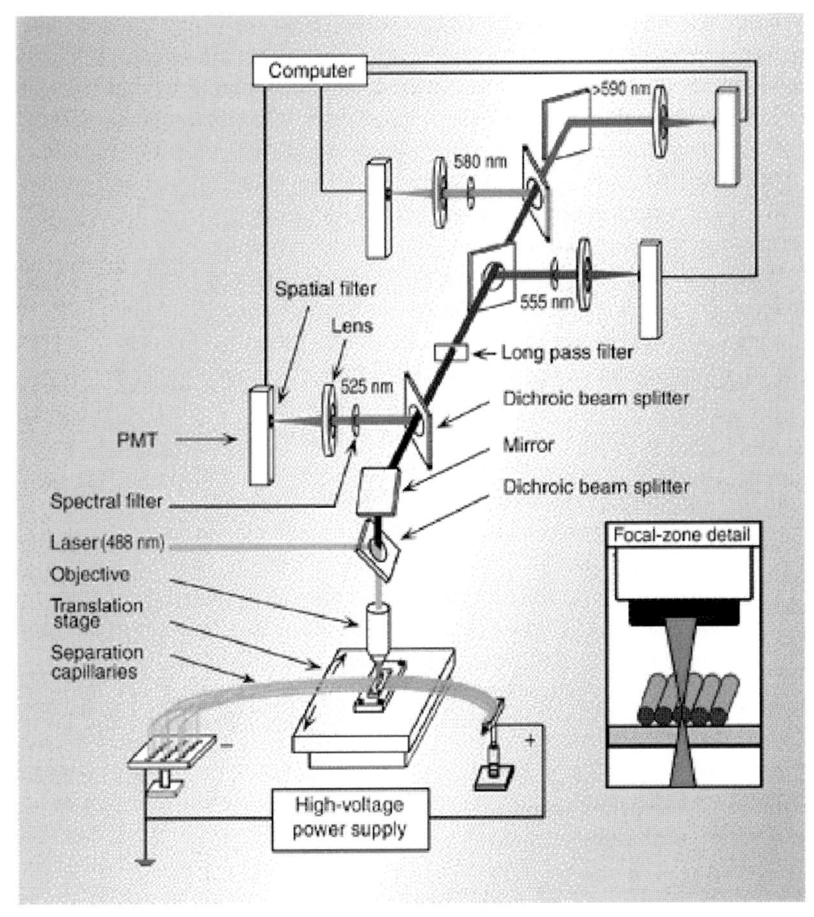

Fig. 2.3. Schematic representation of a four color confocal scanner for capillary array electrophoresis-based DNA sequencing. With permission from Ref. [30].

sequencing using an electrokinetic injection procedure. Conventionally, DNA sequencing samples are processed by ethanol precipitation, which is rather a labor-intensive step and also unreliable with regard to the remaining salt concentration in the samples. The use of ultrafiltration membranes and spin columns resulted in significant and consequent decreases in salt concentration. Other methods, such as liquid chromatography and size exclusion based techniques have also been attempted [30].

2.5. THE PROMISE OF MINIATURIZATION: MICROCHIPS

Microfabricated bioanalytical devices are very effective platforms for the simultaneous analysis of a large number of biologically important molecules, possessing great potential for large-scale DNA sequencing. In the last decade, microfabrication technologies, originally developed for the production of silicon-based chips for the microelectronics industry, have spread out in a variety of applications as biochemical analysis tools. Microfluidic devices are routinely used for transporting and manipulating minute amounts of fluids and/or biological entities through microchannel manifolds allowing integration of various chemical and biochemical processes into fast and automated monolithic microflow systems (Fig. 2.4) [31]. The various types of microfabricated chemical and biochemical separation devices are often referred to as lab-on-a-chip systems, which include miniaturized separation systems, microreactors, microarrays, combinations of the above, etc. [32–34]. Materials and microfabrication technologies, currently used to make microfluidic devices, as well as the main principles and formats of typical electrokinetic manipulations, separations and detection methods, have been described in detail in several reviews [13,35,36]. The progress in this area has been focused on the rapidly developing fields of genomics, albeit, also addressed proteomics and high throughput screening applications. The main advantages of miniaturizing bioanalytical separation tools are improved performance, speed and throughput, reduced costs and reagents consumption, and the possibility of parallel and integrated operations [37–40].

Integrated, microfluidic chip-based platforms hold a real promise to solve the sample preparation and injection problems [13]. In research-based, high throughput genomic facilities, such as production sequencing, the need to carry out a large number of complex experiments poses a real challenge in terms of efficiency, data quality and cost of analysis. Therefore, high throughput systems that perform large-scale experimentation and analysis, using minor amounts of reagents, are in great demand. The real bottleneck in microscale DNA sequencing is to find the proper interface between the chip and the external world.

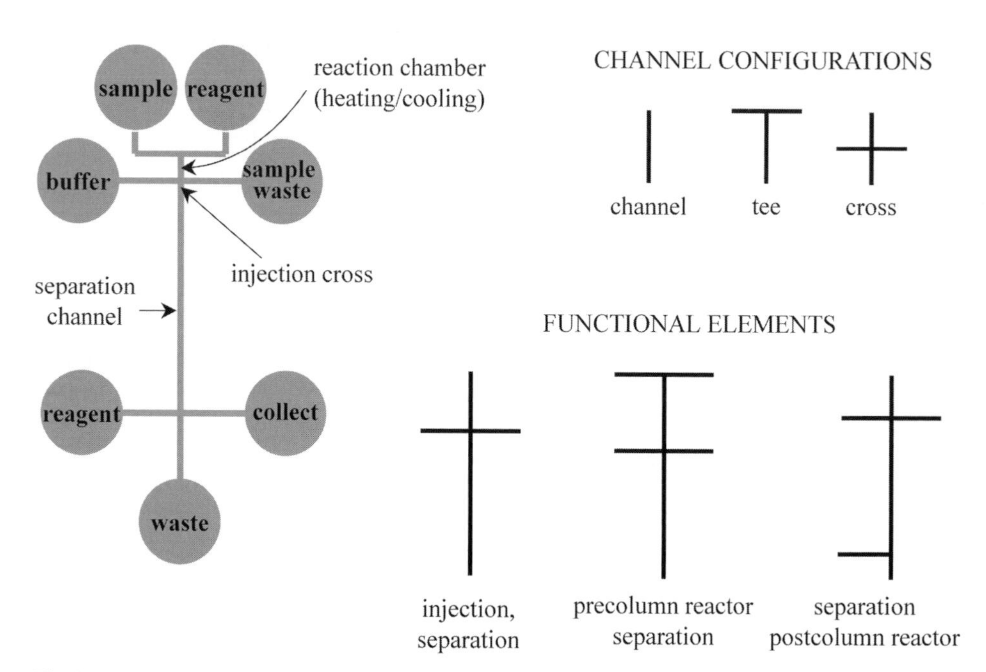

Fig. 2.4. Integrated electrophoresis microchip device and components. With permission from Ref. [31].

This is often required to automate sample purification/introduction/separation processes for high throughput applications and to provide continuous monitoring of the analyte molecules. Mathies and co-workers successfully addressed this issue by introducing a pressurized capillary array system to simultaneously load 96 samples into 96 sample wells of a radial microchannel array electrophoresis microplate for high-throughput DNA sizing (Fig. 2.5) [41].

The growing number of publications in this field proves the feasibility of DNA sequencing using microfluidic chips, one of the most challenging separation tasks in DNA electrophoresis, since very high resolving power is required for long reads and accurate base calling. High speed, four-color sequencing analysis has been demonstrated on-chip in 20 min with a read-length of over 500 bases [43]. Microfabricated devices were recently introduced to perform high throughput DNA sequencing using a 96-channel array format [41]. Based on the best results reported so far, roughly 1200 bases per channel per hour is possible, which would give a yield of 2.7 Mbase per day in a 24 h cycle [43], which corresponds to ~1 Gbase at 24/7 per year, using a single 96 channel chip system.

2.6. DEALING WITH THE DATA: BIOINFORMATICS

Precise base calling of the capillary array or multichannel microchip generated data still represented a real challenge, especially for high speed separation of the longer fragments that are usually not that well resolved [44]. Full automation of capillary gel electrophoresis-based DNA sequencing techniques in conjunction with highly computerized data analysis and storage (bioinformatics) have already facilitated rapid deciphering of the human genetic code [45]. Bioinformatics is a rapidly developing area dedicated to collecting, organizing and analyzing biological information, such as DNA sequences. Sequences are stored in several principal data banks at the GenBank system at Los Alamos National Laboratory in the USA and the EMBL Data Library at the European Moleular Biology Laboratory in Heidelberg. These databases make sequences available to geneticists, molecular biologists and bioinformaticians throughout the world using a computer network allowing thorough data mining. DNA sequence data deposition rate has recently been exponentially increasing [45].

2.7. CONCLUSIONS

The proposal of Venter and co-workers for large-scale shotgun sequencing employing the latest generation of highly automated multi-capillary based DNA sequencers and advanced signal processing algorithms further increased the speed of the sequencing of human and other genomes. By the year 2000, the last year of the 20th century, the sequence of the human genome was proudly announced by the groups of Venter and Collins [17,18]. Sequencing of other organisms, such as the recently announced rice genome [46] is a very important milestone in biotechnology, making such things possible as diet driven healthcare. The new millennium holds a great promise for the development and utilization of microfabricated chemical and biochemical analytical systems. Some of the main

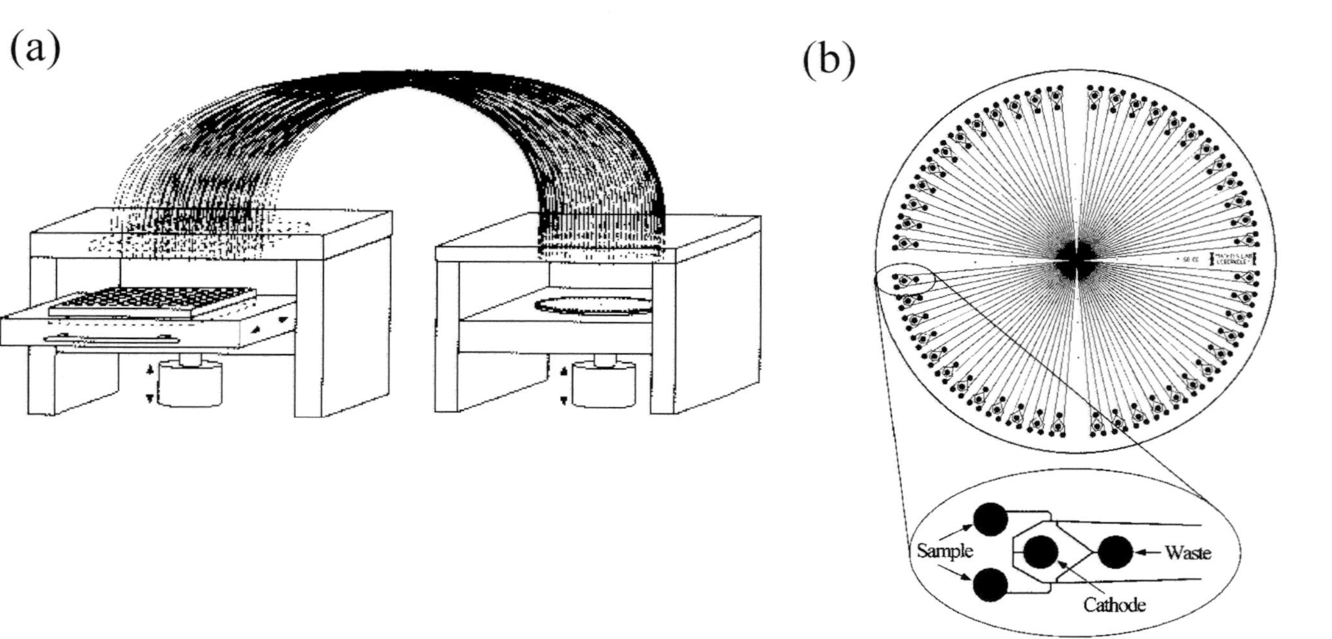

Fig. 2.5. Capillary array loader for a 96 lane electrophoresis microchip. Pressurization of the microtiter dish chamber (21 kPa, ~3 psi) is used to transfer 96 samples (transfer volume ~1 μl) to the sample reservoirs of the radial microplate (A). Mask pattern for the 96-channel radial capillary array electrophoresis microplate. The substrate is 10 cm in diameter. Channels are 110 μm wide and 50 μm deep. The effective separation distance from the injector (double-T) to the detector is 3.3 cm (B). With permission from Ref. [41].

advantages of miniaturization are reduced reagent consumption, improved performance, multifunctional possibilities with interconnected networks of channels, inherent mechanical stability of monolithic systems, possibility of parallelization and inexpensive mass production. System integration plays one of the most significant roles in miniaturization and will lead to greater process intensification. As miniaturization greatly improves heat and mass transfer, it opens new horizons and leads to revolutionary changes in bioanalytical sciences and separation technology. The driving force behind the development of microfabricated devices is certainly the commercialization of microfluidic technology with its tremendous potential of large-scale applications in genomics, proteomics, drug discovery and medical diagnostics.

The challenge of the 21st century will be working out the relevant analytical systems for the next quest of proteomics. Similar to DNA sequencing, this will also require a multidisciplinary effort of separation scientists, engineers and biologists using integrated microfabricated separation devices.

2.8. REFERENCES

1. J.D. Watson and F.H.C. Crick, Nature, 171 (1953) 964
2. M.W. Nirenberg and P. Leder, Science, 145 (1964) 1399
3. H. Lodish, D. Baltimore, A. Berk, S.L. Zipursky, P. Matsudaira and J. Darnell, Molecular Cell Biology, Scientific American Books, Inc, New York, 1995
4. A.M. Maxam and W. Gilbert, Proc. Natl Acad. Sci. USA, 74 (1997) 560
5. F. Sanger, S. Nicklen and A.R. Coulson, Proc. Natl Acad. Sci. USA, 74 (1997) 5463
6. L. Stryer, Biochemistry, W.H. Freeman and Co, New York, 1995
7. L.M. Smith, J.Z. Sanders, R.J. Kaiser, P. Hughes, C. Dodd, C.R. Connell, C. Heiner, S.B. Kent and L.E. Hood, Nature, 321 (1986) 674
8. J.M. Prober, G.L. Trainor, R.J. Dam, F.W. Hobbs, C.W. Robertson, R.J. Zagursky, A.J. Cocuzza, M.A. Jensen and K. Baumeister, Science, 238 (1987) 336
9. A.T. Andrews, Electrophoresis, Claredon Press, Oxford, 1986
10. A. Gerstner, M. Sasvari-Szekely, H. Kalasz and A. Guttman, BioTechniques, 28 (2000) 628
11. J. Noolandi, Annu. Rev. Phys. Chem., 43 (1992) 237
12. B.L. Karger, A.S. Cohen and A. Guttman, J. Chromatogr., 492 (1989) 585
13. J. Khandurina and A. Guttman, J. Chromatogr. A, 943 (2002) 159
14. J.W. Jorgenson and K.D. Lukacs, Science, 53 (1981) 266
15. A.S. Cohen and B.L. Karger, J. Chromatogr., 397 (1987) 409
16. A. Guttman, A. Paulus, A.S. Cohen, B.L. Karger, H. Rodriguez and W.S. Hancock, in: C. Schafer-Nielsen (Ed.), Electrophoresis'88, VCH, Weinheim, Germany, 1988, p. 151
17. J.C. Venter et al. Science, 291 (2001) 1304
18. E.S. Lander et al. Nature, 409 (2001) 860
19. A. Cohen, D.R. Najarian, A. Paulus, A. Guttman, J.A. Smith and B.L. Karger, Proc. Natl Acad. Sci. USA, 85 (1988) 9660
20. H. Swerdlow and R. Gesteland, Nucleic Acids Res., 18 (1990) 1415
21. M.C. Ruiz-Martinez, J. Berka, A. Belenkii, F. Foret, A.W. Miller and B.L. Karger, Anal. Chem., 65 (1993) 2851
22. E.N. Fung and E.S. Yeung, Anal. Chem., 67 (1995) 1913
23. R.S. Madabhushi, Electrophoresis, 19 (1998) 224
24. J. Bashkin, M. Marsh, D. Baker and R. Johnston, Appl. Theor. Electrophor., 6 (1996) 23
25. Q. Gao and E.S. Yeung, Anal. Chem., 70 (1998) 1382

26. R.J. Zagursky and R.M. McCormick, BioTechniques, 9 (1990) 74
27. X.C. Huang, M.A. Quesada and R.A. Mathies, Anal. Chem., 64 (1992) 2149
28. S. Takahashi, K. Murakami, T. Anazawa and H. Kambara, Anal. Chem., 66 (1994) 1021
29. O. Salas-Solano, E. Carrilho, L. Kotler, A.W. Miller, W. Goetzinger, Z. Sosic and B.L. Karger, Anal. Chem., 70 (1998) 3996
30. I. Kheterpal and R.A. Mathies, Anal. Chem., 72 (1999) A31
31. A. Guttman, in: C.W. Gehrke, R.L. Wixom and E. Bayer (Eds.), Integrated Microfabricated Device Technologies Chromatography—A Century of Discovery 1900–2000. The bridge to the Sciences/ Technology, Elsevier Science, Amsterdam, 2001, p. 200
32. M. Koch, A. Evans and A. Brunnschweiler, Microfluidic Technology and Applications, Research Studies Press Ltd, Badlock, Hertfordshire, England, 2000
33. A. Manz and H. Becker, Microsystem Technology in Chemistry and Life Sciences, Springer, Berlin, 1999
34. M. Heller and A. Guttman, Integrated Microfabricated Biodevices, Marcel Dekker, New York, 2001
35. C.S. Effenhauser, G.J.M. Bruin and A. Paulus, Electrophoresis, 18 (1997) 2203
36. V. Dolnik, S. Liu and S. Jovanovich, Electrophoresis, 21 (2000) 41
37. G.J.M. Bruin, Electrophoresis, 21 (2000) 3931
38. T. Chovan and A. Guttman, Trends Biotechnol., 200 (2002) 116
39. J.P. Kutter, Trends Anal. Chem., 19 (2000) 352
40. G.H.W. Sanders and A. Manz, Trends Anal. Chem., 19 (2000) 364
41. Y. Shi, P.C. Simpson, J.R. Schere, D. Wexler, C. Skibola, M.T. Smith and R.A. Mathies, Anal. Chem., 71 (1999) 5354
42. S.R. Liu, Y.N. Shi, W.W. Ja and R.A. Mathies, Anal. Chem., 71 (1999) 566
43. E. Carrilho, Electrophoresis, 21 (2000) 55
44. D. Brady, M. Kocic, A.W. Miller and B.L. Karger, Trans. Biomed. Eng., 47 (2000) 1271
45. N.J. Dovichi and J. Zhang, Angew. Chem. Int. Ed., 39 (2000) 4463
46. S.A. Goff et al. Science, 296 (2002) 92

G.A. Marko-Varga and P.L. Oroszlan (Editors)
Emerging Technologies in Protein and Genomic Material Analysis
Journal of Chromatography Library, Vol. 68

CHAPTER 3

Phosphoprotein and Phosphoproteome Analysis by Mass Spectrometry

MADS GRØNBORG and OLE NØRREGAARD JENSEN*

Protein Research Group, Department of Biochemistry and Molecular Biology, University of Southern Denmark, Odense, Denmark

3.1. INTRODUCTION

Completed and on-going genome sequencing projects provide a strong foundation for current biological and biomedical research. The focus has now shifted towards functional genomics and proteomics—the characterization of gene expression at the protein level [1]. Functional genomics and proteomics research is aimed at delineating the molecular details of intricate cellular processes mediated by proteins (gene products) in eukaryote as well as in prokaryote organisms. Since the mid-1990s, proteomics efforts were concentrated on developing effective and rapid methods for protein identification by mass spectrometry (MS) [2–4]. It is now evident that the cellular proteome is highly complex and dynamic due to sequence variations and (reversible) post-translational modifications (PTMs) of most proteins. Thus, the human genome may contain on the order of 35 000 genes but this set of genes may, in turn, be transcribed and translated into more than half a million different gene products due to alternative splicing and PTMs at the transcript and protein levels, respectively. One of the most important PTMs implicated in the modulation of protein activity and propagation of signals within cellular pathways and networks is by phosphorylation [5].

Phosphorylation by protein kinases, triggered in response to extracellular signals, provide a mechanism for the cell to switch on or off many diverse processes. These processes include metabolic pathways, kinase cascade activation, organelle and biomolecule transport and gene transcription [6–8]. In order to maintain a dynamic state within cells, dephosphorylation is catalyzed by protein phosphatases that are controlled by response to different stimuli so that phosphorylation and dephosphorylation are separately controlled events.

Protein phosphorylation is a regulatory mechanism in both eukaryotes and prokaryotes. In prokaryotes, phosphorylation mainly occurs on histidine, glutamic acid, and aspartic

*Corresponding author. *E-mail address:* jenseno@bmb.sdu.dk; http://www.protein.sdu.dk.

acid residues, whereas in eukaryotes it is usually restricted to serine, threonine and tyrosine residues [9]. Phosphorylation on serines and threonines is more abundant compared with tyrosine phosphorylation, which is quite rare. Indeed the ratio of phosphorylation on pSer:pThr:pTyr is 1800:200:1 in vertebrates, which emphasizes that tyrosine phosphorylation is tightly regulated [10].

The human genome contains 575 protein kinase domains, representing 2% of the total genome and making it the third most populous domain [11]. Perturbation of phosphorylation/dephosphorylation mechanisms in signal cascades and regulatory networks leads to a number of serious human diseases, including certain cancers and neurological and metabolic disorders.

In the present chapter, we will describe a series of MS based methods and strategies for phosphoprotein and phosphoproteome analysis (Fig. 3.1, Table 3.1). We also refer the interested reader to several recent reviews on this topic [12–16].

3.2. PHOSPHOPROTEIN ANALYSIS: A CHALLENGE FOR MASS SPECTROMETRY

PTMs lead to a mass increment (e.g. phosphorylation and glycosylation) or a mass deficit (e.g. truncation, processing) of proteins and peptides relative to their theoretical molecular mass determined from the full length amino acid sequence. MS is now the preferred method for direct characterization of PTMs in proteins, particularly phosphorylation [4,16]. This is due to the specificity of accurate mass determination of proteins and peptides by MS. However, MS is rarely used as a stand-alone technique. It is often advantageous to combine it with complementary analytical or biochemical methods, such as chromatography, enzyme-assays or immunological techniques.

High mass accuracy and the ability to analyze complex peptide mixtures are the foremost advantages of MS. The sensitivity of modern MS-based strategies using MALDI [17] and electrospray ionization (ESI) [18] to generate biomolecular ions allows protein identification at the low femtomole (fmol) level (corresponding to an abundance of 100–1000 copies/cell in a culture of 10^6–10^7 cells) or, as in the case of neuropeptides, even at the zeptomole (10^{-21} mol) level [19–22]. Determination of PTMs requires, however, significantly more protein starting material than for the identification of a protein. Comprehensive characterization of proteins by MS currently requires on the order of 1–20 pmol per protein species.

MS was initially applied for assignment of phosphorylation sites using fast atom bombardment ionization and sector mass analyzers [23,24]. Although several different MS-based strategies for characterization of phosphorylation sites have been devised during the last decade, all methods generally involve proteolytic digestion of the protein followed by MALDI-MS and/or ESI-MS analysis of the resulting peptides to identify the specific phosphorylation site(s) in the protein. Tandem mass spectrometry (MS/MS) is now commonly used for amino acid sequencing of phosphopeptides, as discussed below.

Several analytical challenges are encountered in the analysis of phosphoproteins by MS: (i) The stoichiometry of phosphorylation is generally low, i.e. only a small fraction of the available intracellular pool of a protein is phosphorylated at any given time in response

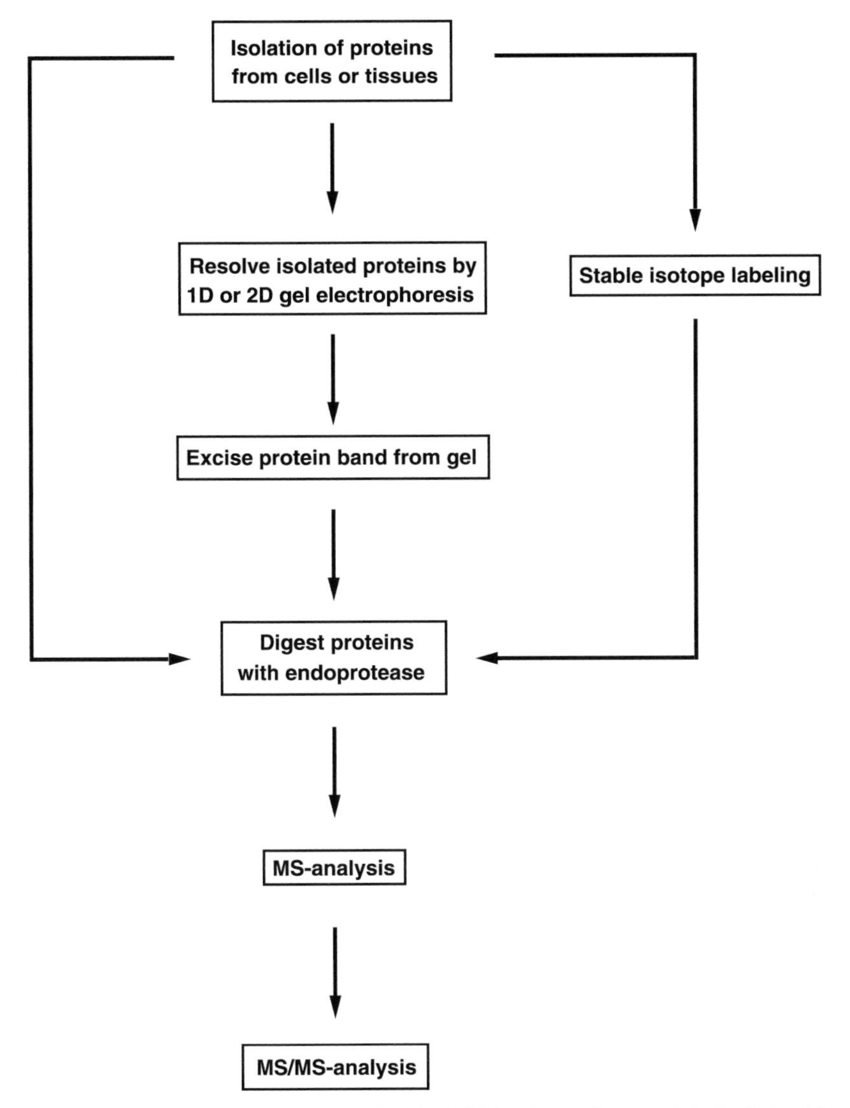

Fig. 3.1. Strategies for phosphoprotein analysis by MS. Phosphoproteins are enriched or isolated from either cells (culture) or tissues, for example by immunoprecipitation with phosphoamino acid specific antibodies. Isolated proteins are separated by one- or two-dimensional gel electrophoresis and protein band(s) of interest are subsequently excised from the gel and digested by a sequence specific endoprotease (e.g. trypsin, Lys-C, Asp-N). The generated peptides are analyzed by MS and MS/MS to identify potential phosphopeptides and specific phosphorylation sites. When using LC-MS/MS, the gel electrophoresis may be circumvented. Relative quantitation of phosphoproteins isolated from two different cellular stages (e.g. stimulated and non-stimulated cells) can be achieved by stable isotope labeling (metabolically or chemically) followed by MS and MS/MS analysis of differentially labeled peptides. MS analysis can be conducted for characterization of intact phosphoproteins in combination with enzymatic treatment (*in vitro* alkaline phosphatase or kinase treatment). See Table 3.1 and main text for further details.

TABLE 3.1

OVERVIEW OF MASS SPECTROMETRY BASED METHODS FOR PHOSPHOPEPTIDE AND PHOSPHOPROTEIN ANALYSIS. OUR JUDGMENT OF THE APPLICABILITY AND EFFICIENCY OF VARIOUS METHODS FOR CHARACTERIZATION OF PHOSPHOSERINE (pSer), PHOSPHOTHREONINE (pThr) AND PHOSPHOTYROSINE (pTyr) ARE LISTED. THE METHODS ARE GRADED FROM (–) (NOT APPLICABLE) TO (+++) (EXCELLENT)

	Sample preparation				Mass spectrometry				
	Enrichment		Derivatization		MS		MS/MS		
	Ab (proteins)	IMAC (peptides)	β-Elimination	p-Amidate	Alkaline phosphatase	Metastable decay	Product ion scan	Precursor ion scan	Neutral loss scan
pSer	+	++	++	++	+++	++	+++	+++ (−79[a])	++ (98 Da)
pThr	+	++	+	++	+++	++	+++	+++ (−79[a])	++ (98 Da)
pTyr	+++	++	–	++	+++	+	+++	+++ (216,043)	+ (80 Da)

[a]Detection of PO_3^- (79 Da) is conducted in negative ion mode.

to a stimulus. Analysis of low abundance phosphoproteins implicated in signal transduction is highly challenging. (ii) Proteins can exist in several different phosphorylated forms, implying that any given phosphoprotein is heterogeneous. (iii) Phosphatases could dephosphorylate residues unless precautions are taken to inhibit their activity during sample preparation. In addition to these 'biological difficulties', phosphoprotein analysis by MS is further complicated in several ways due to the properties of the phosphate moiety itself. The ionization efficiency of phosphorylated proteins and peptides is significantly reduced as compared with the non-phosphorylated species when analyzed by MALDI and ESI. The signal to noise ratio for a tyrosine phosphorylated peptide was shown by Chen et al. to be 10 times lower compared with the non-phosphorylated species in MALDI-MS [25]. Similar studies with ESI were done by Miliotis et al. who showed a reduction in the signal to noise ratio of a tyrosine phosphorylated peptide compared with the non-phosphorylated peptide [26]. The reduction of ionization efficiency in MALDI- and ESI-MS is thought to be the result of increased hydrophilicity and acidity due with the negative charges on the phosphate group. MALDI-MS or ESI-MS in the negative ion mode sometimes produce improved ion signals as compared with the positive ion mode [27]. Addition of ion-pairing agents, such as ammonium acetate/citrate may also aid in the detection of phosphopeptides by MS [28].

Additional problems arise by the use of sequence specific endoproteases for the generation of peptides from phosphoproteins. Phosphorylated amino acid residue(s) interfere with the cleavage specificity of the protease (e.g. trypsin, Lys-C, Asp-N, Glu-C), which leads to the production of larger peptides containing 'missed cleavage sites', which are more complicated to analyze and sequence by MS/MS because of their size. The use of more promiscous proteases, such as elastase in combination with MS/MS is a viable method for the detection and sequencing of phosphopeptides [29].

3.3. MASS SPECTROMETRY BASED PHOSPHATASE/KINASE ASSAYS

Phosphate groups are attached to proteins and peptides by protein kinases. The addition of one phosphate group leads to a mass increase of 79.97986 a.m.u. of the polypeptide. Similarly, removal of phosphate groups by the action of phosphatases results in a mass decrease of 79.97986 or multiples thereof. MS is a convenient technique to monitor such reactions because mass determination provides direct evidence of the enzyme catalyzed addition or removal of modifying groups [30,31].

An MS-based phosphatase or kinase assay using MALDI-MS or ESI-MS analysis of intact phosphoproteins is a simple and straightforward method for monitoring the state of PTM, particularly for abundant (recombinant) proteins [32,33]. Similarly, an MS-based phosphatase assay is a useful tool to identify phosphopeptides in crude peptide mixtures generated by enzymatic proteolysis of phosphoproteins [13,34,35]. The removal of the negatively charged phosphate groups from phosphopeptides results in an improved response from the cognant peptides in MALDI and ESI-MS in the positive ion mode. It is, however, an indirect method for the detection of phosphopeptides and a more detailed analysis of the native phosphopeptides by MS/MS is required to localize the exact phosphorylation sites.

MS also enables the detection of phosphopeptides and phosphoamino acid residues by their 'diagnostic' metastable decomposition via gas phase β-elimination of phosphoric acid during time-of-flight mass analysis [35–37]. Phosphoserine and phosphothreonine residues undergo partial gas phase β-elimination of phosphoric acid (H_3PO_4, 98 Da), in MALDI-TOF-MS and ESI-MS resulting in conversion into dehydroalanine residue (69 Da) and dehydroaminobutyric acid residue (83 Da), respectively. A partial loss of phosphate (HPO_3, 80 Da) is seen in MALDI reflector TOF MS in the case of tyrosine phosphorylated peptides [30,36]. The degree of metastable decay in MALDI is matrix dependent, and is mostly observed when a 'hot' matrix, such as α-cyano-4-hydroxycinnamic acid is utilized.

3.4. PHOSPHOPROTEIN AND PHOSPHOPEPTIDE ENRICHMENT STRATEGIES

It is often advantageous to use affinity-based methods for enrichment of post-translationally modified proteins and peptides from crude mixtures prior to their analysis by MS. PTM of proteins and peptides such as phosphorylation and glycosylation can allow selective affinity purification of these functionally modified peptides/proteins. This will circumvent some of the analytical problems associated with the suppression of modified species in MS and increase the sensitivity and dynamic range of the analysis. By reducing the excess of unmodified proteins/peptides present in a sample, phosphoprotein/peptide ions are generated and detected more efficiently. Several strategies have successfully been used for enrichment of phosphoproteins or phosphopeptides. Phosphoproteins can be enriched by immunoprecipitation (IP) with antibodies that recognize different phosphospecific consensus sequences in the protein. Several studies have been published where high-affinity anti-phosphotyrosine antibodies have been used for IP prior to SDS-PAGE and MS-analysis leading to the discovery of numerous novel phosphoproteins involved in signal transduction [38–40]. Although these antibodies have been effective at enriching and identifying tyrosine phosphorylated proteins, they are generally not very good at enriching phosphopeptides [41].

A more general method applies immobilized metal affinity chromatography (IMAC) (also called metal-chelate affinity chromatography) for phosphoprotein purification, as first described in 1986 by Andersson and Porath [42]. Phosphoprotein and phosphopeptide enrichment by IMAC exploits the high-affinity of phosphate groups towards the metal-chelated stationary phase, e.g. Fe(III)-NTA or Fe(III)-IDA [43–45] or Ga(III)-IDA [46]. This strategy has been demonstrated for the identification and mapping of phosphorylation site(s) using protein samples from gels combined with MS and MS/MS experiments [35].

Phosphopeptide analysis by IMAC proves ineffective in some cases due to the loss of small or very hydrophilic peptides as well as by interference from unspecific binding of non-phosphorylated peptides [47]. Elimination of 'false positives' in the IMAC eluate (i.e. non-phosphorylated peptides) can be easily accomplished by enzymatic dephosphorylation by alkaline phosphatase after initial MALDI-TOF-MS analysis or by MS/MS based sequencing [35]. The extent of non-specific binding to the IMAC resin may be minimized by O-methylesterification of peptide carboxyl groups [48].

3.5. CHEMICAL DERIVATIZATION OF PHOSPHOPEPTIDES AND PHOSPHOPROTEINS

Several chemical derivatization methods may be useful for phosphopeptide analysis by MS. By removing and replacing the phosphate group by a more stable and less acidic group, the MS analysis of phosphopeptides and interpretation of the results becomes more straightforward [47,49,50]. Phosphopeptides or phosphoproteins undergo β-elimination under strong alkaline conditions (e.g. NaOH) resulting in conversion of phosphoserine or phosphothreonine residues into the reactive dehydroalanine or dehydroamino-2-butyric acid, respectively [51]. The α,β-unsaturated residue that is generated can readily react with a variety of nucleophiles, for example, ethandithiol or other alkylation reagents [52].

One weakness of this strategy is the inability to derivatize phosphotyrosine-containing peptides due to the high stability of phosphotyrosine because of the presence of an aromatic ring, which makes derivatization impractical [49]. A second complication arises from the potential reactivity of cysteine residues in the peptides. *S*-alkylated cysteine residues may also be affected by β-elimination. This problem can be minimized by conversion of cysteine residues to cysteic acid residues by oxidation with performic acid. It is also important to keep in mind that *O*-glycosylated Ser and Thr undergo β-elimination and the end products are the same as for the phosphorylated Ser and Thr residues.

3.6. SELECTIVE DETECTION AND SEQUENCING OF PHOSPHOPEPTIDES BY TANDEM MASS SPECTROMETRY

Tandem mass spectrometry is one of the most powerful strategies to identify and sequence phosphopeptides for localization of phosphorylation sites. Three scan modes can be applied in tandem mass spectrometry experiments, including (i) product ion scan, (ii) precursor ion scan and (iii) neutral loss scan. Product ion scan can be conducted on any kind of tandem mass spectrometer, whereas precursor ion scan and neutral loss scan can only be carried out on mass spectrometers with two (scanning) mass analyzers, for example, a triple quadrupole instrument. Recent advances in software and hardware enable all of these three scan modes on hybrid quadrupole/TOF instruments.

The most frequently used strategy to obtain amino acid sequence information by MS/MS is by product ion scanning (Fig. 3.2). Fragment ions are generated by collision-induced dissociation (CID) of an isolated phosphopeptide precursor ion using an inert gas. The resulting fragment ions are then mass analyzed to obtain sequence information for identification of phosphorylation site(s) in the peptide. Phosphoserine and phosphothreonine often undergo gas phase β-elimination during CID leading to the loss of phosphoric acid (H_3PO_4, 98 Da) (Fig. 3.2) and subsequent production of the 'diagnostic species' dehydroalanine and dehydroaminobutyric acid, respectively. These residues are used to pinpoint the exact location of the phosphorylation site in the peptide. As mentioned previously, phosphotyrosine does not readily undergo gas phase β-elimination of phosphoric acid and is usually observed in the MS/MS spectrum as the intact residue of 243 Da. In some cases, a loss of phosphate (HPO_3, 80 Da) and water is seen in the case of phosphotyrosine.

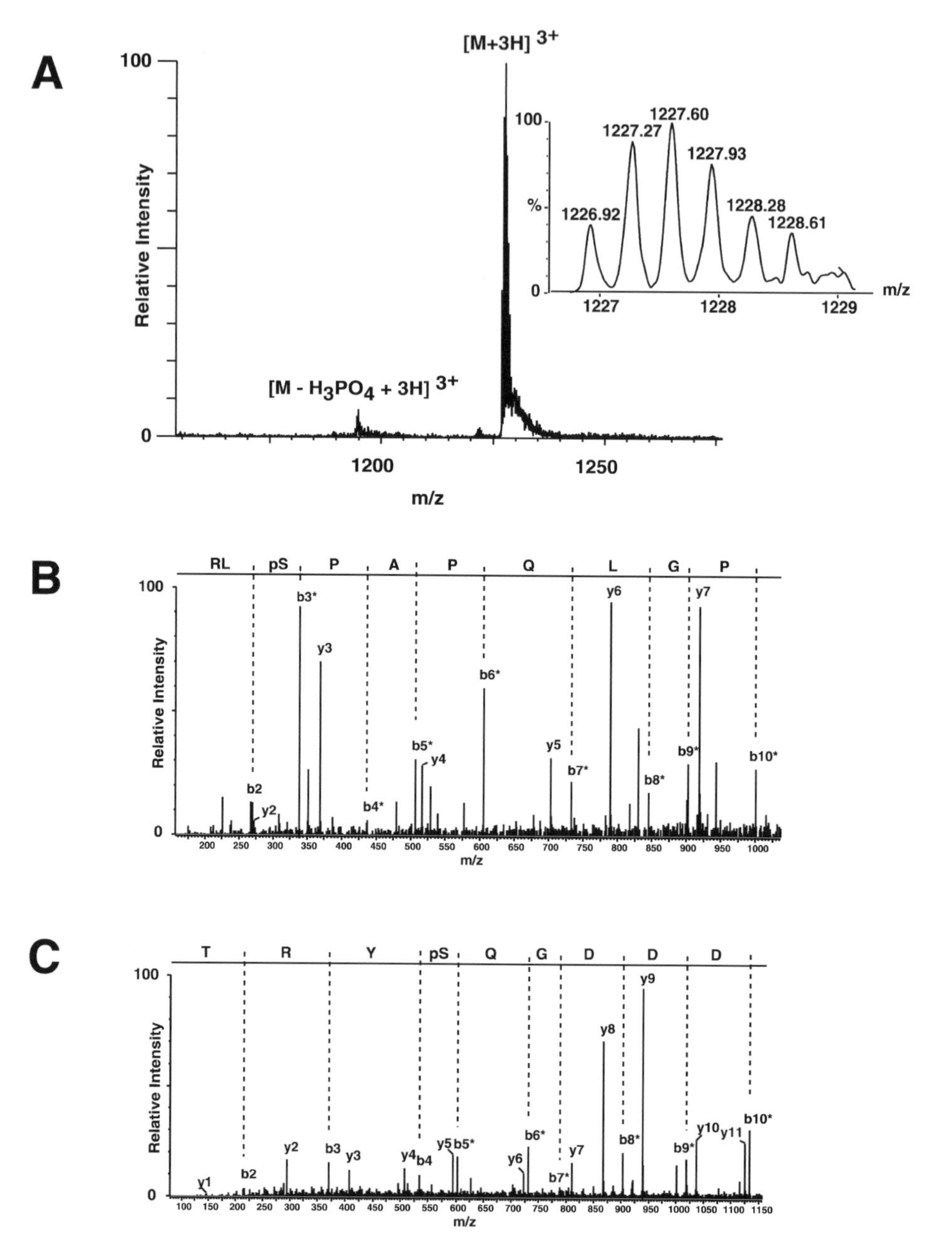

Fig. 3.2. Phosphopeptide analysis and sequencing by ESI-MS/MS. Localization of phosphorylation sites in a novel human phosphoprotein, Frigg (KIAA 0668), isolated in a phosphoproteomics experiment using an anti-phosphoserine antibody [66]. (A) Part of a MS/MS spectrum of a triply charged peptide ion at *m/z* 1226.93 ([52]RLpSPAPQLGPSSDAHTSYYSESLVHESWFPPR[82]; Mr 3677.76). Note the gas phase β-elimination of phosphoric acid (H_3PO_4) from the precursor ion. (B) Part of an MS/MS spectrum of a quadruply charged phosphoserine-containing peptide RLpSPAPQLGPSSDAHTSYYSESLVHESWFPPR (*m/z* 920.45; Mr 3677.76). The sequence derived from N-terminal (b) and C-terminal (y) peptide fragment ions are indicated [85]. Ions labeled

Selective and specific detection of phosphopeptides in crude mixtures can be accomplished by either MS/MS precursor ion scanning or MS/MS neutral loss scanning. Both of these scanning modes take advantage of the gas phase properties of phosphopeptide ions, i.e. the neutral loss of phosphoric acid H_3PO_4 or the presence of the phosphate ion PO_3^- (negative ion mode) as a unique 'fingerprint' to identify phosphopeptides. In precursor ion scanning, the first mass analyzer scans the entire *m/z*-range of interest, to transmit all ion species, in turn, into the collision cell where fragmentation (CID) takes place. The second mass analyzer is used as a mass filter, which allows only a particular fragment ion (reporter ion) to reach the detector. Precursor ion scanning can be conducted in positive and negative ion modes. In the negative ion mode one can scanning for different phosphate-specific 'reporter ions', including $H_2PO_4^-$ (97 Da), PO_3^- (79 Da) and PO_2^- (63 Da) [53–55]. One drawback of precursor ion scanning in negative ion mode is that effective sequencing of peptides is not possible. The instrument has to be switched to positive ion mode and the pH of the sample has to be adjusted prior to MS/MS analysis.

Precursor ion scanning in the positive ion mode for detection of phosphotyrosine-containing phosphopeptides on a Q-TOF hybrid tandem mass spectrometer was recently demonstrated [56]. This method is based on accurate mass determination of immonium ions (216.043 Da) that are unique for phosphotyrosine residues.

Neutral loss scanning is used to detect peptide modifications, which give rise to the loss of neutral fragments. Two mass analyzers are scanned simultaneously but with a specific *m/z*-offset corresponding to the diagnostic 'neutral fragment' generated by CID in the collision cell. Identification of phosphorylated peptides by neutral loss scanning involves detection of H_3PO_4 (98 Da) and HPO_4 (80 Da) [29]. However, monitoring of the neutral loss involves scanning for different *m/z* values for different charge states of the precursor (e.g. 98 for singly charged, 49 for doubly charged, 32.7 for triply charged precursor ions). Neutral loss scanning can be coupled on-line to LC-MS/MS, where the MS/MS data are screened for mass differences of 98 and/or 80 Da. One limitation of this method is the discrimination of phosphotyrosine detection due to the high stability of the arylphosphates with respect to the loss of phosphoric acid (− 98 Da) as compared with alkylphosphoesters (phosphoserine and phosphothreonine).

3.7. LIQUID CHROMATOGRAPHY—TANDEM MASS SPECTROMETRY (LC-MS/MS)

Capillary liquid chromatography (capLC) coupled with ESI or MALDI tandem mass spectrometry (MS/MS) is a powerful tool for the analysis of complex peptide mixtures and

---

with an asterisk (*) indicate fragment ion species where phosphoserine was converted to dehydroalanine by gas phase β-elimination of phosphoric acid. Identification of dehydroalanine in the MS/MS spectrum is used to 'pin-point' the exact location of the phosphorylation site in the peptide/protein sequence. The phosphorylation site was localized to the serine in position 3 in the peptide (serine 54 in Frigg). (C) Part of an MS/MS spectrum of a triply charged phosphoserine-containing peptide LTRYpSQGDDDGSSSSGGSSVAGSQSTLFK (*m/z* 987.76; Mr 2960.27). The sequence derived from N-terminal (b) and C-terminal (y) peptide fragment ions are indicated. The phosphorylation site was localized to the serine in position 5 of the peptide (serine 12 in Frigg).

for phosphopeptide analysis. LC-MS/MS encompass peptide separation followed by (on-line) mass determination and automatic MS/MS sequencing of the eluting peptides. The problem associated with detection of phosphopeptides, because of the naturally low stoichiometry of phosphorylation can, to a certain degree, be overcome by LC-MS/MS because of separation and concentration of peptides prior to MS and MS/MS analysis. By loading the peptides on a reverse phase nanocolumn (usually C_{18} material), and subsequent gradient elution, the complexity of the peptide sample delivered to the ion source is reduced considerably. Peptides are typically eluted from the column at a flow rate of 100–200 nl per minute and the chromatographic peak width is on the order of 15–30 s. Automated data-dependent MS to MS/MS switching enables highly efficient and fast sequencing of peptides. The strategy allows MS/MS analysis and identification of up to hundreds of peptides during one HPLC run [57]. In phosphoprotein analysis, it is sometimes useful to analyze the neutral loss of phosphoric acid by post-processing and reconstruction of LC-MS/MS data obtained from ion trap or Q-TOF instruments. The combination of multidimensional chromatography with MS/MS provides a powerful tool for analysis of very complex peptide mixtures. Yates and co-workers recently introduced a pre-fractionation of complex peptide mixtures, using strong cation exchange capillary liquid chromatography and subsequent separation by reverse phase capillary LC coupled on-line to a mass spectrometer for peptide sequencing of a yeast cell proteolysate [3,58]. The use of IMAC in combination with RP-HPLC-MS/MS is a promising technology for phosphoproteome analysis [48] (see below).

3.8. PHOSPHOPROTEOME ANALYSIS BY MASS SPECTROMETRY

Two-dimensional gel electrophoresis (2DE) is a high-resolution technique for proteome analysis [59]. 2DE is unique because it enables the separation of various post-translationally modified forms of proteins. Thus, the combination of 2DE and MS for phosphoproteome analysis is appealing because the former can resolve various phosphorylated forms of a protein. 2DE combined with immunoblotting or 32-P labeling has been used to separate and detect phosphoproteins in cell lysates, followed by protein identification by MS and MS/MS [12,14,59]. In a few cases, phosphorylation sites were identified by MS and MS/MS of peptides derived from individual protein spots in 2DE gels [35,60].

Several MS driven phosphoproteomics strategies that circumvent the need for 2DE gels have been reported during recent years. The method reported by Oda et al. [61] involves the replacement of Ser/Thr phosphate groups by a biotinylated affinity tag via β-elimination and Michael addition. The derivatized phosphoproteins were enriched by affinity chromatography using Avidin beads prior to their analysis by MS and MS/MS. The efficiency of the method towards analysis of complex mixtures of proteins still needs to be investigated. An alternative derivatization approach that, in principle, enables the detection of phosphotyrosine-containing peptides was suggested by Zhou et al. [62]. This strategy involves selective isolation of phosphopeptides via a sequence of chemical reactions to generate a phosphoamidate species, followed by solid phase capture, elution and MS/MS analysis. The strategy was applied on whole lysate of yeast

(*Saccharomyces cerevisiae*) to test its performance in highly complex peptide mixtures. Affinity purification of the tryptic peptides from yeast and subsequent analysis by automated LC-MS/MS led as the identification of 24 phosphopeptides originating from abundant phosphoproteins. None of the identified phosphopeptides were phosphorylated on tyrosine residues, most likely due to the natural low abundance of this species in yeast. The phosphopeptide recovery efficiency after affinity purification was determined as approximately 20%, which excludes the use of this strategy as a global approach for phosphopeptide enrichment. Importantly, the strategies proposed by Oda et al. [61] and Zhou et al. [62] can both be used to quantify the relative phosphorylation state of proteins from two different sources by differential labeling of the phosphoprotein populations with stable isotopes (see below).

An IMAC and MS/MS based phosphoproteomics methodology was recently reported by Ficarro and co-workers [48]. Protein mixtures were digested by trypsin and the resulting peptides were then converted to methyl esters [63] to minimize the problem of interference from non-phosphorylated peptides during IMAC-purification of phosphopeptides. Thus, this derivatization should improve the selectivity of IMAC for phosphopeptides and eliminate confounding binding through free carboxylate groups. The strategy was tested on whole cell lysate from *S. cerevisiae* where more than 1000 peptides were detected, out of which 216 peptide sequences defining 383 phosphorylation sites were identified, making it the most successful phosphoproteome analysis to date, although no quantitative information on phosphorylation site occupancy was obtained.

A targeted phosphoproteomics approach for studying cell signaling events was demonstrated by Pandey and co-workers [64]. Using a combination of anti-phosphotyrosine antibodies and MS, a majority of the known components of the EGF signaling pathway was determined in a single experiment. In addition, several novel signaling proteins were discovered. In a follow-up study, several more phosphoproteins were identified and characterized by MS, including the determination of phosphorylation sites [65]. A proteomics application of phosphoserine and phosphothreonine antibodies for enrichment of phosphoproteins was recently reported [66] and lead to the identification by MS of several known human structural phosphoproteins and a novel human phosphoprotein, Frigg, of unknown function. In a similar experiment, a 'differential display' proteomics study demonstrated that proteins could be enriched by an antibody directed against the epitope of protein kinase B (PKB/Akt, a serine/threonine kinase) leading to the discovery of a novel substrate for PKB [67].

In all of the above strategies aimed at comprehensive phosphoprotein analysis, it remains to be established whether the analytical techniques enable the detection and relative quantitation of low abundance phosphoprotein species. Further refinement of these new approaches is clearly needed.

3.9. EMERGING MS METHODS FOR PHOSPHOPROTEIN ANALYSIS

Electron capture dissociation (ECD) combined with Fourier transform ion cyclotron resonance (FTICR) mass spectrometry [68] is a new method for sequencing of small intact proteins and peptides as well as for the study of PTMs in proteins. ECD induces more

References pp. 34–38

extensive polypeptide backbone fragmentation compared with CID, consequently providing greater sequence coverage. At the same time ECD is a gentle fragmentation method that leaves PTMs intact. In the case of phosphopeptides, no loss of phosphoric acid, phosphate or water is seen from the parent ion when ECD-based sequencing is utilized [69,70]. The superior resolution seen in FTICR-MS makes it possible to study large post-translationally modified peptides and proteins, which are not amenable to conventional MS analysis.

MALDI tandem mass spectrometry (MALDI-MS/MS) is another rapidly evolving technology. MALDI ion sources have been combined with Q-TOF [71], ion trap [72] FT-ICR [73] and TOF-TOF tandem analyzers [74] and their utility for analysis of post-translationally modified peptides, including phosphopeptide analysis has been demonstrated [72,75–77].

3.10. RELATIVE QUANTITATION OF PHOSPHORYLATION SITE OCCUPANCY BY STABLE ISOTOPE LABELING OF PROTEINS

Stable isotope labeling of proteins and peptides facilitates the use of MS to read out and quantify relative protein concentrations in a mixture of differentially labeled protein populations. Incorporation of a stable isotope of C, N, H or O (^{13}C, ^{15}N, D or ^{18}O, respectively) in one of the protein populations leads to a mass increment of peptides from that pool (Fig. 3.3). Mixing of the normal and isotopically labeled proteins or peptides followed by MS analysis will give rise to two ion signals from each peptide species: One signal from the normal peptide and an adjacent signal originating from the isotopically labeled but otherwise identical peptide. Relative protein quantitation is accomplished by determining the peak intensity ratio for each peptide pair (Fig. 3.3).

The stable isotope labeling technique facilitates a ‘differential experiment’ where two or more sets of proteins are compared, e.g. a ‘perturbed’ system versus a ‘control’ system [78,79]. Several strategies are based on that approach and include Isotope-Coded Affinity Tagging (ICAT, only enrichment of cysteine containing peptides) [78], Phosphoprotein Isotope-Coded Affinity Tags (PhIAT, only enrichment of serine/threonine phosphorylated peptides) [80], N- or C-terminal labeling [48,81,82] and metabolic labeling [79,83,84].

Since phosphorylation sites may occur in any given peptide that contains a substrate recognition motif for a kinase, phosphoprotein quantitation relies on isotope labeling of either all peptides derived from a phosphoprotein or of each individual phosphate group present in a protein. The former task can be accomplished by metabolic labeling using stable isotope labeled amino acids or growth media or by N- or C-terminal derivatization techniques. The latter task requires efficient chemical or enzymatic conversion of the phosphate group into a labeled compound, preferably with an affinity tag to facilitate purification. Thus, a generally applicable and robust method for MS-based relative quantitation of phosphorylation sites has still to be established.

3.11. CONCLUSION

MS analysis of low picomole amounts of purified phosphoproteins is becoming almost routine in many laboratories due to the development of a number of complementary

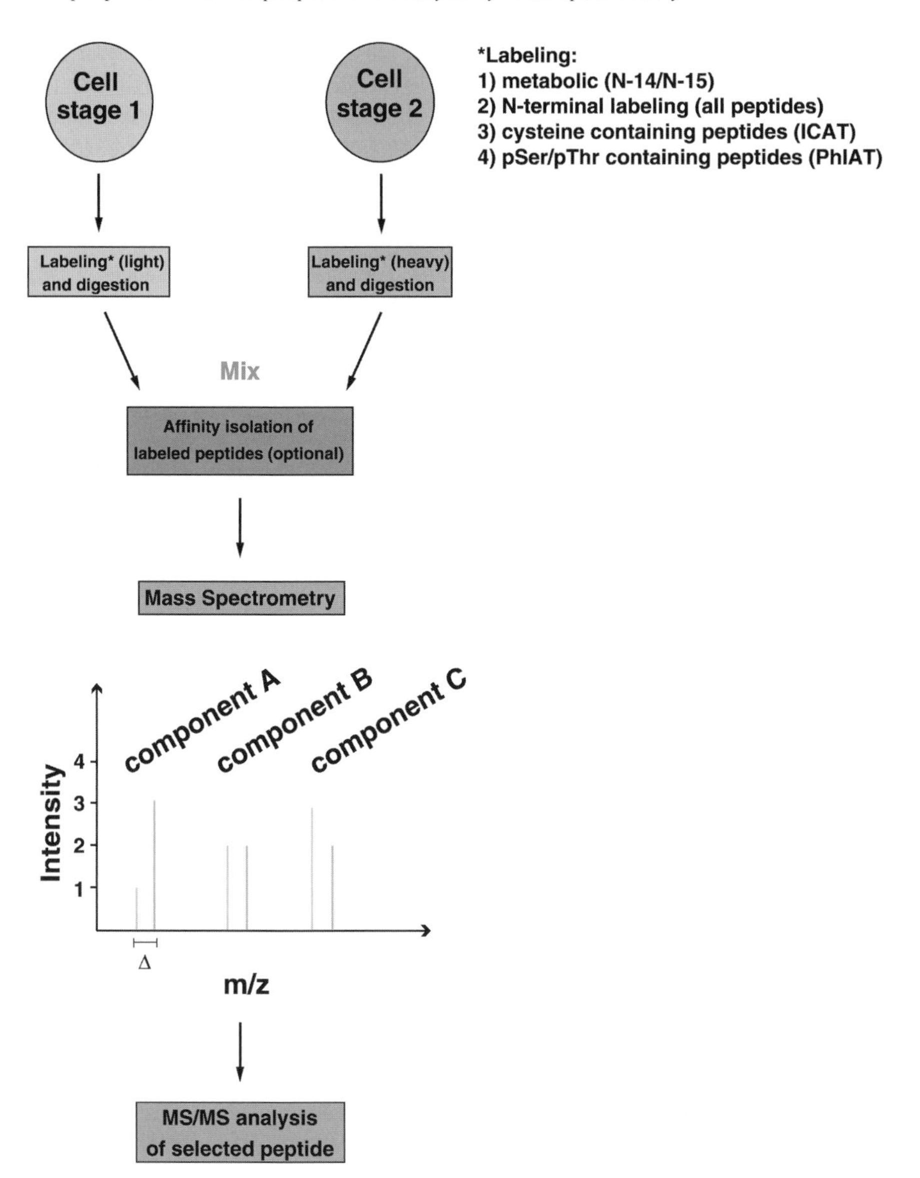

Fig. 3.3. MS based quantitation of peptides by differential labeling via incorporation of stable isotopes in peptides/proteins. Relative quantitation of proteins from two different cellular stages can be accomplished by the use of stable isotope incorporation in peptides or proteins in combination with MS. Proteins from one cellular stage (cell stage 1) are labeled with 'light' compound, whereas the other cellular stage (cell stage 2) is labeled with a 'heavy' compound. The light and heavy compounds differ by mass and can be distinguished by the mass spectrometer. After labeling, the two sets of protein populations are mixed and digested with a sequence-specific endoprotease (e.g. trypsin). Optionally, some chemically derivatized peptides can be affinity purified prior to MS-analysis (see text). After digestion and enrichment, the labeled peptides are analyzed by MS. The relative expression level in the two cellular stages is determined by the ratio of the ion abundances of light and heavy peptide species.

methods for phosphopeptide enrichment, detection, mass determination and sequencing. Systematic, quantitative phosphoprotein analysis remains a major challenge for proteomics research. MS is a key technology for the identification, characterization and quantitation of phosphorylation events but much work lies ahead in order to enable future time-resolved, global analysis of *in vivo* phosphorylation events at the whole cell level. The low abundance of most phosphoproteins and the inherent dynamic nature of reversible protein phosphorylation governing basal cellular processes such as cell cycle and metabolism will probably require completely new approaches. Rapid advances in MS hardware and software technology and improved, miniaturized sample preparation methods for MS analysis of proteins are promising developments that may pave the way for global phosphoprotein profiling sooner than most researchers anticipate.

3.12. ACKNOWLEDGEMENTS

We thank all members of the Protein Research Group at the Department of Biochemistry and Molecular Biology at University of Southern Denmark for their contributions to the development of methods for protein and proteome analysis by mass spectrometry.

3.13. REFERENCES

1. A. Pandey and M. Mann, Proteomics to study genes and genomes. Nature, 405 (2000) 837–846
2. A. Shevchenko, O.N. Jensen, A.V. Podtelejnikov, F. Sagliocco, M. Wilm, O. Vorm, P. Mortensen, H. Boucherie and M. Mann, Linking genome and proteome by mass spectrometry: large-scale identification of yeast proteins from two dimensional gels. Proc. Natl. Acad. Sci. USA, 93 (1996) 14440–14445
3. M.P. Washburn, D. Wolters and J.R. Yates III, Large-scale analysis of the yeast proteome by multidimensional protein identification technology. Nat. Biotechnol., 19 (2001) 242–247
4. R. Aebersold and D.R. Goodlett, Mass spectrometry in proteomics. Chem. Rev., 101 (2001) 269–295
5. P. Cohen, The regulation of protein function by multisite phosphorylation—a 25 year update. Trends Biochem. Sci., 25 (2000) 596–601
6. C.A. Koch, D. Anderson, M.F. Moran, C. Ellis and T. Pawson, SH2 and SH3 domains: elements that control interactions of cytoplasmic signaling proteins. Science, 252 (1991) 668–674
7. T. Hunter, 1001 protein kinases redux—towards 2000. Semin. Cell Biol., 5 (1994) 367–376
8. J.D. Graves and E.G. Krebs, Protein phosphorylation and signal transduction. Pharmacol. Ther., 82 (1999) 111–121
9. J.X. Yan, N.H. Packer, A.A. Gooley and K.L. Williams, Protein phosphorylation: technologies for the identification of phosphoamino acids. J. Chromatogr. A, 808 (1998) 23–41
10. T. Hunter, The Croonian Lecture 1997. The phosphorylation of proteins on tyrosine: its role in cell growth and disease. Philos. Trans. R. Soc. Lond. B: Biol. Sci., 353 (1998) 583–605
11. J.C. Venter, M.D. Adams, E.W. Myers, P.W. Li, R.J. Mural, G.G. Sutton, H.O. Smith, M. Yandell, C.A. Evans, R.A. Holt, J.D. Gocayne, P. Amanatides, R.M. Ballew, D.H. Huson, J.R. Wortman, Q. Zhang, C.D. Kodira, X.H. Zheng, L. Chen, M. Skupski, G. Subramanian, P.D. Thomas, J. Zhang, G.L. Gabor Miklos, C. Nelson, S. Broder, A.G. Clark, J. Nadeau, V.A. McKusick, N. Zinder, A.J. Levine, R.J. Roberts, M. Simon, C. Slayman, M. Hunkapiller, R. Bolanos, A. Delcher, I. Dew, D. Fasulo, M. Flanigan, L. Florea, A. Halpern, S. Hannenhalli, S. Kravitz, S. Levy, C. Mobarry, K. Reinert, K. Remington, J. Abu-Threideh, E. Beasley, K. Biddick, V. Bonazzi, R. Brandon, M. Cargill, I. Chandramouliswaran, R. Charlab, K. Chaturvedi, Z. Deng, V. Di Francesco, P. Dunn, K. Eilbeck, C. Evangelista, A.E. Gabrielian, W. Gan, W. Ge, F. Gong, Z. Gu, P. Guan, T.J. Heiman, M.E. Higgins, R.R. Ji, Z. Ke, K.A. Ketchum, Z. Lai, Y. Lei,

Z. Li, J. Li, Y. Liang, X. Lin, F. Lu, G.V. Merkulov, N. Milshina, H.M. Moore, A.K. Naik, V.A. Narayan, B. Neelam, D. Nusskern, D.B. Rusch, S. Salzberg, W. Shao, B. Shue, J. Sun, Z. Wang, A. Wang, X. Wang, J. Wang, M. Wei, R. Wides, C. Xiao, C. Yan, A. Yao, J. Ye, M. Zhan, W. Zhang, H. Zhang, Q. Zhao, L. Zheng, F. Zhong, W. Zhong, S. Zhu, S. Zhao, D. Gilbert, S. Baumhueter, G. Spier, C. Carter, A. Cravchik, T. Woodage, F. Ali, H. An, A. Awe, D. Baldwin, H. Baden, M. Barnstead, I. Barrow, K. Beeson, D. Busam, A. Carver, A. Center, ML. Cheng, L. Curry, S. Danaher, L. Davenport, R. Desilets, S. Dietz, K. Dodson, L. Doup, S. Ferriera, N. Garg, A. Gluecksmann, B. Hart, J. Haynes, C. Haynes, C. Heiner, S. Hladun, D. Hostin, J. Houck, T. Howland, C. Ibegwam, J. Johnson, F. Kalush, L. Kline, S. Koduru, A. Love, F. Mann, D. May, S. McCawley, T. McIntosh, I. McMullen, M. Moy, L. Moy, B. Murphy, K. Nelson, C. Pfannkoch, E. Pratts, V. Puri, H. Qureshi, M. Reardon, R. Rodriguez, Y.H. Rogers, D. Romblad, B. Ruhfel, R. Scott, C. Sitter, M. Smallwood, E. Stewart, R. Strong, E. Suh, R. Thomas, N.N. Tint, S. Tse, C. Vech, G. Wang, J. Wetter, S. Williams, M. Williams, S. Windsor, E. Winn-Deen, K. Wolfe, J. Zaveri, K. Zaveri, J.F. Abril, R. Guigo, M.J. Campbell, K.V. Sjolander, B. Karlak, A. Kejariwal, H. Mi, B. Lazareva, T. Hatton, A. Narechania, K. Diemer, A. Muruganujan, N. Guo, S. Sato, V. Bafna, S. Istrail, R. Lippert, R. Schwartz, B. Walenz, S. Yooseph, D. Allen, A. Basu, J. Baxendale, L. Blick, M. Caminha, J. Carnes-Stine, P. Caulk, Y.H. Chiang, M. Coyne, C. Dahlke, A. Mays, M. Dombroski, M. Donnelly, D. Ely, S. Esparham, C. Fosler, H. Gire, S. Glanowski, K. Glasser, A. Glodek, M. Gorokhov, K. Graham, B. Gropman, M. Harris, J. Heil, S. Henderson, J. Hoover, D. Jennings, C. Jordan, J. Jordan, J. Kasha, L. Kagan, C. Kraft, A. Levitsky, M. Lewis, X. Liu, J. Lopez, D. Ma, W. Majoros, J. McDaniel, S. Murphy, M. Newman, T. Nguyen, N. Nguyen, M. Nodell, S. Pan, J. Peck, M. Peterson, W. Rowe, R. Sanders, J. Scott, M. Simpson, T. Smith, A. Sprague, T. Stockwell, R. Turner, E. Venter, M. Wang, M. Wen, D. Wu, M. Wu, A. Xia, A. Zandieh, X. Zhu. The sequence of the human genome. Science, 291 (2001) 1304–1351

12. O.N. Jensen, Modification-specific proteomics: strategies for systematic studies of post-translationally modified proteins. In: W. Blackstock and M. Mann (Eds.), Proteomics: A Trends Guide, Elsevier Science, London, 2000, p. 36–42
13. M.R. Larsen and P. Roepstorff, Mass spectrometric identification of proteins and characterization of their post-translational modifications in proteome analysis. Fresenius J. Anal. Chem., 366 (2000) 677–690
14. J. Godovac-Zimmermann and L.R. Brown, Perspectives for mass spectrometry and functional proteomics. Mass Spectrom. Rev., 20 (2001) 1–57
15. A. Sickmann and H.E. Meyer, Phosphoamino acid analysis. Proteomics, 1 (2001) 200–206
16. M. Mann, S.E. Ong, M. Gronborg, H. Steen, O.N. Jensen and A. Pandey, Analysis of protein phosphorylation using mass spectrometry: deciphering the phosphoproteome. Trends Biotechnol., 20 (2002) 261–268
17. M. Karas and F. Hillenkamp, Correspondence: laser desorption ionization of proteins with molecular masses exceeding 10 000 Daltons. Anal. Chem., 60 (1988) 2299–2301
18. J.B. Fenn, M. Mann, C.K. Meng, S.F. Wong and C.M. Whitehouse, Electrospray ionization for mass spectrometry of large biomolecules. Science, (1989) 246
19. P.E. Andrén, M.R. Emmett and R.M. Caprioli, Micro-electrospray zeptomole/attomole per microliter sensitivity for peptides (short communication). J. Am. Soc. Mass Spectrom., 5 (1994) 867–870
20. A. Shevchenko, M. Wilm, O. Vorm and M. Mann, Mass spectrometric sequencing of proteins silver-stained polyacrylamide gels. Anal. Chem., 68 (1996) 850–858
21. M. Wilm, A. Shevchenko, T. Houthaeve, S. Breit, L. Schweigerer, T. Fotsis and M. Mann, Femtomole sequencing of proteins from polyacrylamide gels by nano-electrospray mass spectrometry. Nature, 379 (1996) 466–469
22. K.A. Resing and N.G. Ahn, Applications of mass spectrometry to signal transduction. Prog. Biophys. Mol. Biol., 71 (1999) 501–523
23. P. Petrilli, P. Pucci, H.R. Morris and F. Addeo, Assignment of phosphorylation sites in buffalo beta-casein by fast atom bombardment mass spectrometry. Biochem. Biophys. Res. Commun., 140 (1986) 28–37
24. P. Cohen and D.G. Hardie, The actions of cyclic AMP on biosynthetic processes are mediated indirectly by cyclic AMP-dependent protein kinase. Biochim. Biophys. Acta, 1094 (1991) 292–299
25. J. Chen, Y. Qi, R. Zhao, G.W. Zhou and Z.J. Zhao, Assay of protein tyrosine phosphatases by using matrix-assisted laser desorption ionization time-of-flight mass spectrometry. Anal. Biochem., 292 (2001) 51–58
26. T. Miliotis, P.O. Ericsson, G. Marko-Varga, R. Svensson, J. Nilsson, T. Laurell and R. Bischoff, Analysis of regulatory phosphorylation sites in ZAP-70 by capillary high-performance liquid chromatography coupled

proteins by enrichment with phospho-specific antibodies: identification of a novel protein, Frigg, as a protein kinase A substrate. Mol. Cell Proteomics, 1(7) (2002) 517–527.
67. S. Kane, H. Sano, S.C. Liu, J.M. Asara, W.S. Lane, C.C. Garner and G.E. Lienhard, A method to identify serine kinase substrates. Akt phosphorylates a novel adipocyte protein with a Rab GTPase-activating protein (GAP) domain. J Biol. Chem., 277 (2002) 22115–22118
68. F.W. McLafferty, D.M. Horn, K. Breuker, Y. Ge, M.A. Lewis, B. Cerda, R.A. Zubarev and B.K. Carpenter, Electron capture dissociation of gaseous multiply charged ions by Fourier-transform ion cyclotron resonance. J. Am. Soc. Mass Spectrom., 12 (2001) 245–249
69. A. Stensballe, O.N. Jensen, J.V. Olsen, K.F. Haselmann and R.A. Zubarev, Electron capture dissociation of singly and multiply phosphorylated peptides. Rapid Commun. Mass Spectrom., 14 (2000) 1793–1800
70. S.D. Shi, M.E. Hemling, S.A. Carr, D.M. Horn, I. Lindh and F.W. McLafferty, Phosphopeptide/phosphoprotein mapping by electron capture dissociation mass spectrometry. Anal. Chem., 73 (2001) 19–22
71. A. Shevchenko, A. Loboda, W. Ens and K.G. Standing, MALDI quadrupole time-of-flight mass spectrometry: a powerful tool for proteomic research. Anal. Chem., 72 (2000) 2132–2141
72. J. Qin and B.T. Chait, Identification and characterization of posttranslational modifications of proteins by MALDI ion trap mass spectrometry. Anal. Chem., 69 (1997) 4002–4009
73. T. Solouki, J.A. Marto, F.M. White, S. Guan and A.G. Marshall, Attomole biomolecule mass analysis by matrix-assisted laser desorption/ionization Fourier transform ion cyclotron resonance. Anal. Chem., 67 (1995) 4139–4144
74. K.F. Medzihradszky, J.M. Campbell, M.A. Baldwin, A.M. Falick, P. Juhasz, M.L. Vestal and A.L. Burlingame, The characteristics of peptide collision-induced dissociation using a high-performance MALDI-TOF/TOF tandem mass spectrometer. Anal. Chem., 72 (2000) 552–558
75. C.H. Lee, M.E. McComb, M. Bromirski, A. Jilkine, W. Ens, K.G. Standing and H. Perreault, On-membrane digestion of beta-casein for determination of phosphorylation sites by matrix-assisted laser desorption/ionization quadrupole/time-of-flight mass spectrometry. Rapid Commun. Mass Spectrom., 15 (2001) 191–202
76. K.L. Bennett, A. Stensballe, A.V. Podtelejnikov, M. Moniatte and O.N. Jensen, Phosphopeptide detection and sequencing by matrix-assisted laser desorption/ionization quadrupole time-of-flight tandem mass spectrometry. J. Mass Spectrom., 37 (2002) 179–190
77. M.A. Baldwin, K.F. Medzihradszky, C.M. Lock, B. Fisher, T.A. Settineri and A.L. Burlingame, Matrix-assisted laser desorption/ionization coupled with quadrupole/orthogonal acceleration time-of-flight mass spectrometry for protein discovery, identification, and structural analysis. Anal. Chem., 73 (2001) 1707–1720
78. S.P. Gygi, B. Rist, S.A. Gerber, F. Turecek, M.H. Gelb and R. Aebersold, Quantitative analysis of complex protein mixtures using isotope-coded affinity tags. Nat. Biotechnol., 17 (1999) 994–999
79. Y. Oda, K. Huang, F.R. Cross, D. Cowburn and B.T. Chait, Accurate quantitation of protein expression and site-specific phosphorylation. Proc. Natl. Acad. Sci. USA, 96 (1999) 6591–6596
80. M.B. Goshe, T.P. Conrads, E.A. Panisko, N.H. Angell, T.D. Veenstra and R.D. Smith, Phosphoprotein isotope-coded affinity tag approach for isolating and quantitating phosphopeptides in proteome-wide analyses. Anal. Chem., 73 (2001) 2578–2586
81. M. Munchbach, M. Quadroni, G. Miotto and P. James, Quantitation and facilitated de novo sequencing of proteins by isotopic N-terminal labeling of peptides with a fragmentation-directing moiety. Anal. Chem., 72 (2000) 4047–4057
82. O.A. Mirgorodskaya, Y.P. Kozmin, M.I. Titov, R. Korner, C.P. Sonksen and P. Roepstorff, Quantitation of peptides and proteins by matrix-assisted laser desorption/ionization mass spectrometry using (18)O-labeled internal standards. Rapid Commun. Mass Spectrom., 14 (2000) 1226–1232
83. R.D. Smith, L. Pasa-Tolic, M.S. Lipton, P.K. Jensen, G.A. Anderson, Y. Shen, T.P. Conrads, H.R. Udseth, R. Harkewicz, M.E. Belov, C. Masselon and T.D. Veenstra, Rapid quantitative measurements of proteomes by Fourier transform ion cyclotron resonance mass spectrometry. Electrophoresis, 22 (2001) 1652–1668
84. S. Ong, B. Blagoev, I. Kratchmarova, D.B. Kristensen, H. Steen, A. Pandey and M. Mann, Stable isotope labeling by amino acids in cell culture, SILAC, as a simple and accurate approach to expression proteomics. Mol. Cell Proteomics, 1(5) (2002) 376–386

G.A. Marko-Varga and P.L. Oroszlan (Editors)
Emerging Technologies in Protein and Genomic Material Analysis
Journal of Chromatography Library, Vol. 68

CHAPTER 4

Quantitative Peptide Determination Using Column-Switching Capillary Chromatography Interfaced with Mass Spectrometry

J. ABIAN* and M. CARRASCAL

Structural and Biological Mass Spectrometry Unit, Department of Medical Bioanalysis, Instituto de Investigaciones Biomédicas de Barcelona-Consejo Superior de Investigaciones Científicas (IIBB-CSIC), Institut d'Investigacions Biomèdiques August Pi i Sunyer (IDIBAPS), Rosselló 161, 7ª Planta, 08036 Barcelona, Spain

4.1. INTRODUCTION

The mass spectrometry characterization of peptides and proteins has become an essential tool in proteomics and other areas of biomedicine. The minute amounts of these compounds often available for analysis, especially when endogenous molecules are handled, require efficient sample preparation and isolation procedures along with highly sensitive detection methods.

Until now, immunological assays were the only assays able to quantify with sufficient selectivity peptides and proteins at the attomole/femtomole level in complex biological matrices [1]. One of the main disadvantages of immunoassays is cross-reactivity, i.e. the reaction of the antibody with one or more compounds, such as isotypes or degradation products, leading to erroneous results. One way to circumvent these problems is to use liquid chromatographic (LC) fractionation before immunological analysis [2–7]. Antigen–antibody interactions can also be used as a highly specific sample purification step before peptide quantitation by other methods. For example, immunoaffinity columns in combination with matrix-assisted laser desorption ionization time-of-flight (MALDI-TOF) mass spectrometry (MS) have been used to analyze peptides at low femtomole levels [8]. A simpler system based on immunoprecipitation and MALDI-TOF MS has been used to analyze soluble amyloid beta peptides in cell culture supernatants at picomolar concentration [9].

Several other non-immunological techniques for quantifying peptides in biologically relevant matrices have been developed. For example, liquid chromatography is used on-line with fluorescence detection to achieve detection limits of 1 ng/ml beta-endorphin fragments in plasma [10]. In another application, methionine enkephalin is quantified in pituitary

*Corresponding author. Tel.: +34-93-3638312; fax: +34-93-3638301; *e-mail address:* jambam@iibb.csic.es.

References pp. 70–73

tissue at the same levels using fast atom bombardment (FAB) MS for the off-line analysis of chromatographic fractions [11]. A more complex procedure makes use of LC separations in combination with enzyme cleavage and GC-MS for the analysis of peptides in biological matrices with 50 ng/ml detection limits [12]. LC separations have also been used off-line for the quantification of the peptide DALDA by continuous flow-FAB [13]. In addition, cyclosporin is directly quantified in blood after liquid extraction by secondary ion TOF-MS (SIMS-TOF) and MALDI-TOF MS with limits of detections (LODs) of 7 and 23 ng/ml, respectively [14].

The use of liquid chromatography in combination with on-line atmospheric pressure ionization (API) interfaces for MS has proved to be a powerful tool for peptide analysis [15–32]. Several authors have shown detection limits below the ng/ml level in biological fluids such as blood and urine [19,24,27]. Electrospray ionization (ESI)-based API sources have become the method of choice, together with MALDI-TOF MS, for the analysis of peptides and proteins. ESI-MS analysis gives accurate masses of peptides and proteins. Furthermore, the CID spectra of peptides facilitate the identification of their amino acid sequence. One of the advantages of ESI is that it can easily be interfaced with separation techniques such as LC and capillary electrophoresis (CE). Therefore, although some off-line ESI-MS methods have been used for the direct quantification of peptides [33,34], ESI-MS is mainly used coupled with separation techniques.

By miniaturizing ESI sources to micro and/or nanodimensions, detection limits down to the attomole range can be reached [35–39]. These sources have, therefore, great potential in bioanalysis because they can reach quantitation limits similar to those obtained in immunoassays. Nanospray (nESI) sources are popular in those areas where only a limited amount of sample is available, for instance in the characterization of proteins and peptides from polyacrylamide 2D gels [40–43]. Micro- and nanoelectrospray methods can cope with flow rates from about 20 to 2000 nl/min [36,38,39,44–49] and on-line coupling of these interfaces to capillary LC (capLC) creates highly sensitive LC-MS methods for peptide analysis [38,50,51].

The practical use of capLC-MS for peptide analysis in biological samples has been partially restricted by its limited sample volume loadability. Injection of sample volumes in the submicroliter range requires injection of relatively high concentrations to achieve adequate detector response. In addition, nanoliter sample handling has proven difficult. The use of gradient elution or the injection of the sample in a train of low eluotropic strength solvent increases the amount of sample injected into microscale separation methods. A more effective procedure involves the use of column-switching techniques, by which several microliters can be injected into a pre-concentration column (PC) and desorbed to the analytical capLC column. Concentration detection limits in the low pmol/l range have been achieved [49,52–56].

This chapter addresses certain aspects of the PC-capLC-MS system and its applications to the quantitative analysis of peptides in biological samples.

4.1.1. Capillary liquid chromatography

Capillary liquid chromatography utilizes columns with inner diameters below 0.5 mm and flow rates from some few μl/min down to the nl/min range (Table 4.1). Miniaturization

TABLE 4.1

CLASSIFICATION OF LC COLUMNS BY COLUMN INTERNAL DIAMETER. MEDIUM-LARGE SCALE PREPARATIVE COLUMNS (>10 MM I.D.) ARE NOT INCLUDED

	Column diameter	Flow-rate
Conventional		
Wide-bore (preparative)	>4.6 mm	>3 ml/min
Normal-bore (analytical)	3–4.6 mm	0.5–3 ml/min
Narrow-bore	1–2 mm	0.02–0.3 ml/min
Capillary		
Microbore	0.15–0.8 mm	2–20 μl/min
Nanobore	20–100 μm	0.1–1 μl/min

of chromatographic procedures using capLC offers substantial advantages over conventional LC separation methods. Capillary columns show increased separation efficiency, minimal solvent consumption, higher peak concentration at the detector and higher peptide recovery. Several reviews on the development and applications of capLC can be found in the literature [32,57–59].

Nowadays, capLC is a widespread technique in laboratories involved in the analysis of peptides and proteins, especially in the proteomics area. CapLC can be coupled to a wide range of detectors including UV, fluorescence and electrochemical detectors. The low flow rates employed in capLC allow both the use of expensive solvents (i.e. deuterated) and coupling with detectors that show their best performance at the nanoliter per minute range.

CapLC is probably to LC the same as, some decades ago, open tubular capillary chromatography was to gas chromatography. Horvart first reported capLC in 1967 [60]. The preparation, analytical characteristics and some applications of capLC columns were described in the late 1970s by Scott and Kucera [61], Hirata and Novotny [62] and Ishii and co-workers [63,64]. Two main classes of capLC columns were developed: wall-coated open tubular (OTLC) columns and packed columns. Several authors [65] have shown the applicability of OTLC columns, which have been coupled to ESI-MS [66] and APCI [67]. However, open tubular columns have considerable practical drawbacks in extended use. Their lower loading capacity and capacity factors, the nanoliter injection volume requirements and the complexity of column preparation have relegated OTLC to a more restricted use than packed columns.

Packed capLC columns consist of metal or fused-silica capillaries filled with chromatographic supports with particle sizes ranging from 1.0 to 50 μm. CapLC can be divided into micro (0.8–0.15 mm i.d.) and nano (20–100 μm i.d.) LC depending on the inner diameter of the column (Table 4.1), although other classifications are mentioned in the literature [32,58,68].

Capillary columns usually show overall performance and efficiency comparable with or better than conventional columns. Jorgenson and colleagues reported 40 000 theoretical plates for resorcinol ($k' = 2.7$) on a 25 μm i.d. column, 33 cm in length, filled with 5 μm particles. These authors observed that column efficiency improved as the column diameters fell from 50 to 12 μm [69,70]. This improvement was explained as the result of the decrease of flow dispersion and resistance to mass transfer with smaller column diameters. Wall effects make a higher contribution in capLC than in conventional LC columns. This leads to relatively higher band-broadening at high flow rates. Band-broadening can be diminished using electro-driven flows, as performed in capillary electrochromatography (CEC), instead of pressure forces. CEC columns can reach efficiencies of over 300 000 theoretical plates/m [71], far more than the equivalent pressure-driven capLC. Since electro-driven flows are less affected by particle size than pressure-driven flows, smaller particles and longer columns can be used; and the efficiency obtained, increased. The separation and detection of peptide mixtures at the attomole level by CEC were described recently [72]. Lubman reported the analysis of protein digests by pseudo-electrochromatography, a pressurized CEC mode. This technique is intermediate between capLC and CEC. The preparation of CEC columns and the applications of this technique were reviewed recently [59,73–75].

4.1.2. Interfacing LC with electrospray MS

The on-line combination of liquid chromatographs with mass spectrometers has been a long-term goal with origins that can be traced back to the early 1960s [76]. First attempts were based on mechanical transport of the evaporated eluent into the ion source, such as in the moving belt interface [77,78], or alternatively on the use of very low flows compatible with the MS vacuum, as in the direct liquid introduction (DLI) [79,80] and continuous-flow fast atom bombardment (CF-FAB) interfaces [81].

Due to these flow-rate requirements, capillary columns were already used in the early LC-MS experiments with the moving belt [82], DLI [83,84] or CF-FAB [85,86] interfaces. At that time, however, very low flows were seen as unsuitable by chromatographers. The ability of LC-MS interfaces to cope with conventional chromatographic flow rates was, therefore, a major pre-condition for general acceptance. Developments in vacuum technology and the advent of interfaces such as thermospray (TSP) [87] and particle beam (PB) [88] greatly enhanced the compatibility of LC with MS requirements. These interfaces, with optimum flow rates in the range of 0.5–2 ml/min, are still used in some laboratories. In fact, TSP has been for about a decade the dominant ionization technique for effective on-line LC-MS. Unfortunately, TSP rarely showed the necessary efficiency for peptide analysis [89].

The most important revolution in the field of LC-MS has probably been the development of modern API sources [90,91], based on ESI ionization [92,93] and discharge-induced chemical ionization (APCI) [94–96]. The first ESI sources were reported independently by the Fenn group [92,97,98] and by Alexandrov et al. [93] in the mid 1980s. ESI proved to be a much more efficient ionization technique for high-mass, highly polar compounds than TSP and was compatible with common reversed-phase LC solvent mixtures. The best performance of early ESI interfaces was found at flow rates of around 0.5–5 μl/min. In terms of its coupling to separation techniques, this flow-rate range was not suitable for most people at that time for the reasons indicated above. Coupling with chromatography required the use of post-column flow splitters or fused-silica capillary columns. Additionally, the physical requirements for electrospraying made it difficult to work with certain common chromatographic eluents with high conductivity and/or high surface tension.

Practical solutions for LC coupling rapidly came in the form of assisted electrospray interfaces. Pneumatically assisted electrospray (ionspray, ISP) [99] and ultrasonically assisted electrospray [100] raised flow-rate limits to 1 ml/min. At the same time, CE was coupled to MS by using either a coaxially added sheath liquid [101] or the liquid junction device [102]. This device allowed the electric circuit of the electrophoretic system to be closed and served as a post-column T-piece for solvent addition in order to reach the optimum flow-rate for the electrospray process.

Later ESI developments brought about the microelectrospray [38] (μESI) and nESI devices [35,36]. The first μESI interface developed by Emmett and Caprioli used 50 μm i.d. needles and worked in the 0.3–0.8 μl/min range. This design enabled 50 fmol of enkephalin (full scan) to be detected by use of an integrated concentration-desalting device in the form of a small length of capillary packed with C18 solid phase. The success of this configuration brought back the early ESI probe configuration in the form of several optimized designs, sometimes also called μESI interfaces. These coped with flow rates in the range of ca. 200 nl/min to 4 μl/min, with optima depending on the particular design.

The nESI ion source employs capillaries with spraying orifices of 1–2 μm inner diameter and produces a stable electrospray signal with flow rates in the low nanoliter per minute range. As will be shown later, nESI emitters afford enhanced mass and concentration sensitivity over other ESI interfaces. Nanospray interfaces have found their most popular applications in the peptide analysis field [36,40,41,103,104]. A few microliters of sample can be electrosprayed for more than an hour. During this time, full-scan spectra and MS/MS spectra can be obtained with a minimum consumption of sample.

The need for effective coupling of LC with μESI and nESI ion sources has probably been the driving force behind capLC development during the last decade. In the early 1990s, Hunt and colleagues sequenced subpicomole amounts of antigenic peptides bound to the major histocompatibility complex (MHC) molecules using capLC columns and commercially available ESI interfaces [105,106]. Further downscaling of the LC-MS method by coupling nanoLC to dedicated μESI interfaces resulted in detection limits down to the attomole range [38,49,54,56] and quantitative determinations of nanomolar amounts of peptides were possible using internal standards [56].

Nowadays, capLC-ESI MS has become a very powerful analytical technique in proteomic studies. Enzymatic digests of proteins can be separated on-line and sequenced by capLC-ESI MS with automatic scanning procedures in triple quadrupole, ion trap or hybrid quadrupole/TOF mass spectrometers [43,107,108] (Fig. 4.1). Protein sequence coverage higher than 50% can be routinely obtained from well-defined Coomassie stained spots (about 1–5 pmol total protein). Gatlin et al. demonstrated 99% coverage on the characterization of mutant globins (15 kDa) when analyzing peptide mixtures obtained by

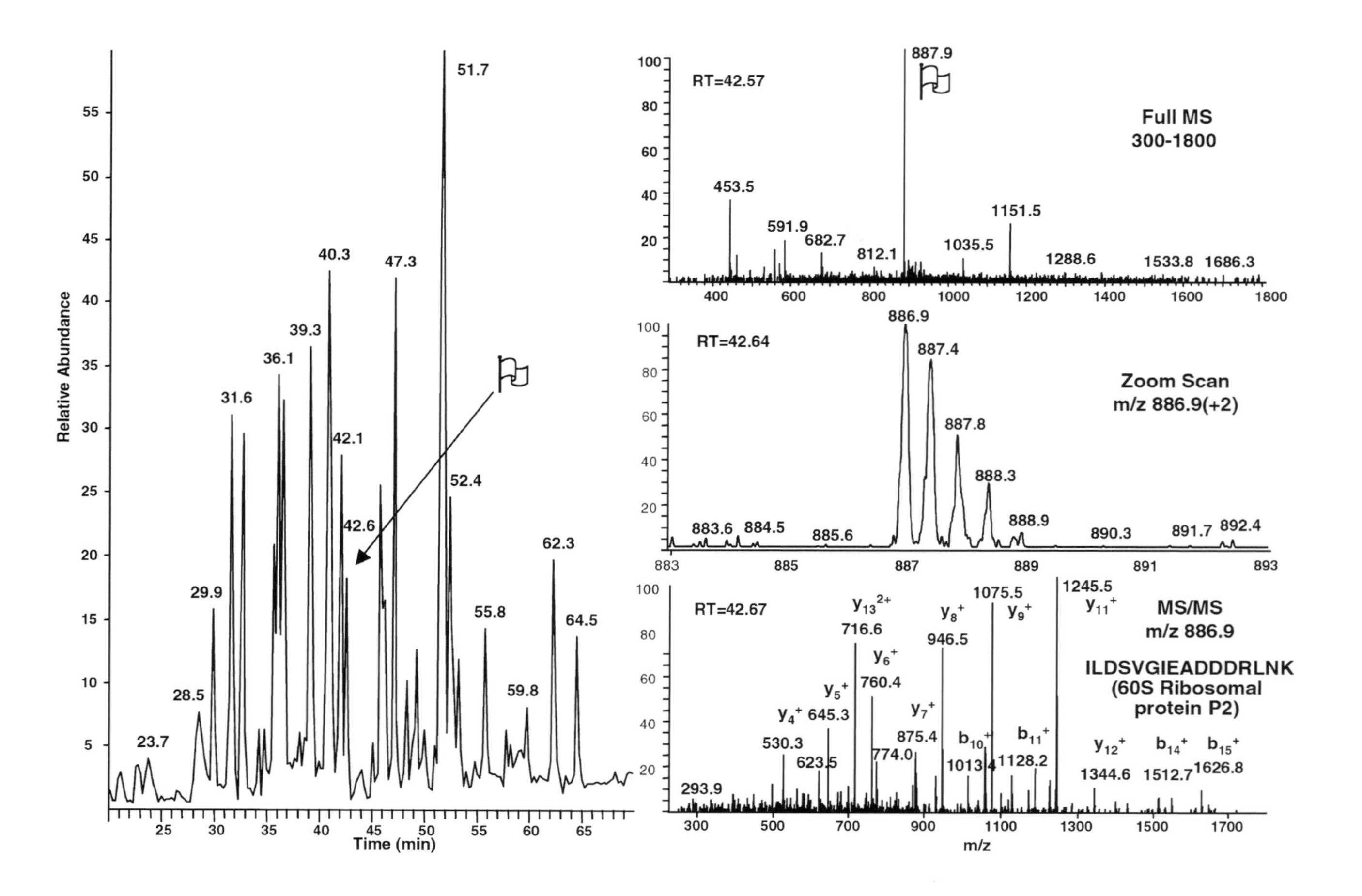
Relative Abundance
Time (min)
RT=42.57
Full MS
300-1800
RT=42.64
Zoom Scan
m/z 886.9(+2)
RT=42.67
MS/MS
m/z 886.9
ILDSVGIEADDDRLNK
(60S Ribosomal
protein P2)
m/z

mixing several different enzymatic digestions of the protein [43]. The same group has recently reported the large-scale analysis of the yeast proteome [109]. In this study 5540 peptides from a digested cell protein extract were detected and sequenced automatically by multidimensional capLC-MS.

4.1.3. Advantages of miniaturization in LC-MS

Commercially available HPLC columns with a variety of internal diameters that allow a wide range of working flow rates can be purchased. Columns can be classified in three main classes: conventional, narrow-bore and capillary (Table 4.1).

Each column class demands very different technical specifications for the chromatographic system (liquid chromatograph and detector). It is possible to build or to buy flow splitting devices that allow the use of conventional HPLC pumps and gradient systems with narrow-bore or capillary columns [54]. However, dedicated detection systems are needed to fit the corresponding flow-rate range.

Typical operational characteristics of analytical (3–4.6 mm) and narrow-bore (1–2 mm) columns are summarized in Table 4.2. These columns are the ones used most often for LC-MS coupling. The reason for their popularity can mainly be found in the ease of use of larger-bore columns. These do not require any major modification of standard equipment and facilitate large volume injections of diluted samples in cases where the concentration of the analyte is low and sample availability is not limited. Conventional LC can be coupled with ESI interfaces by either post-column splitting or optimized ESI interfaces capable of attaining flows of up to 2 ml/min [110]. However, the first choice results in significant sample losses and the latter does not give noticeably higher

TABLE 4.2

HPLC COLUMN CHARACTERISTICS. ADAPTED FROM TOMER ET AL. [57]. ALL COLUMNS ARE 25 CM LONG

Column ID	Volume	Flow-rate (per min)	Injection volume	Rel. conc. at detector	Relative loading capacity
4.6 mm	4.1 ml	1 ml	100 μl	1	8469
2.0 mm	783 μl	0.2 ml	19 μl	5.3	1598
1.0 mm	196 μl	47 μl	4.7 μl	21.2	400
320 μm	20 μl	4.9 μl	485 nl	206	41
50 μm	490 nl	120 nl	12 nl	8459	1

Fig. 4.1. CapLC-μESI MS analysis of a peptide mixture from an in-gel tryptic digest. The ThermoQuest LCQ ion trap was programmed for automatic data-dependent acquisition. A full spectrum (top right), zoom (center right) and product ion scans (bottom right) centered on the most abundant peptide signal (indicated with a flag in the full spectrum) were used consecutively to obtain sequence information over the entire chromatographic elution. The spectral data shown in this figure was acquired in the elution zone indicated by the asterisk in the chromatogram (left) (from Ref. [132]).

sensitivity. Narrow-bore columns are best suited for this purpose. They have optimum flow rates that fit the ISP source requirements and can be coupled to ESI sources, using reasonable flow-split ratios.

The characteristics of capLC columns are well documented and their advantages recognized by chromatographers [57,65,111–114]. Despite this, until recently their practical use was restricted to a few laboratories. The need for specialist instruments has also acted as a barrier to the laboratory implementation of these kinds of miniaturized applications. In addition, the commercial availability of these columns was very limited in the past. For this reason, then and even now, capillary columns were often home-built using available stationary phases and slurry-packing procedures [57]. These considerations explain in part the relatively fewer applications of capLC than of conventional or narrow-bore LC-MS, especially in the case of nanobore capillaries.

However, capillary separations can be performed with normal LC equipment after some painless adaptations. These entail the use of flow splitters, miniaturized injectors and miniaturized detection devices. Kits to upgrade normal-bore HPLC equipment to micro- and/or nanoscale separation are commercially available. An extensive review of the specific needs and possible solutions for capillary separations has been published [58].

4.1.3.1. Chromatographic considerations

It has been repeatedly mentioned in the literature that miniaturization of chromatographic column parameters brings advantages in terms of solvent consumption, improved detection sensitivities and hence reduced sample consumption [54,58].

A reduction in column diameter causes higher sample peak concentration in the detector. The maximum peak concentration of the sample in the column eluate (C_{max}) depends on the absolute amount of sample loaded on the column, and the column efficiency. C_{max} is inversely proportional to the column dead volume (V_0) and the retention factor [115]. Since V_0 is a function of the column internal diameter (i.d.), it can be assumed that the C_{max} ratio for two different i.d. columns is equal to the inverse ratio of the squares of their respective i.d. values. For example, when the same mass of sample is injected, miniaturization from a 4.6 mm analytical column to a 320 μm capillary column results in a theoretical concentration gain at the detector of more than two orders of magnitude, simply by diminishing column diameter by about one tenth (Table 4.2).

4.1.3.2. Mass spectrometric considerations

The above gain in peak concentration results is an improvement in response when using detectors such as the UV or refractive index detectors, with which the signal obtained within the working range is proportional to the concentration of the sample.

However, mass spectrometers are mass flow-dependent detectors and they behave like the flame ionization detector or the electrochemical detector for gas and liquid chromatography, respectively. These detectors have a response proportional to the total number of molecules being detected per unit of time. Accordingly, the equation for a mass flow-sensitive detector includes sample concentration (C_{max}) and the flow actually entering the mass spectrometer. This can be expressed in terms of total flow-rate (F) and a split ratio variable (S):

$$R \propto C_{\max} F S$$

Thus, if we continuously introduce ions into a MS, the signal will decrease with lower flow rates. As the gain in analyte concentration obtained by using small diameter columns occurs at the expense of a lower flow-rate entering the detector, the higher peak concentration does not necessarily result in a greater response in the mass spectrometer.

In practice, the effect of the flow-rate on the mass spectrometer response is highly dependent on interface characteristics. For example, our homemade μESI sources produce a response that increases with flow-rate within a given flow range (Fig. 4.2) [56]. This effect has also been described for a megaflow ESI interface on the high flow-rate range [110]. In both cases, a plateau is reached where the response is not mass-flow dependent. In other cases, reverse mass-flow dependency behavior, such as that reported by Hopfgarner for an ESI interface, is observed. In this example, the signal diminished when the flow-rate rose to the range of 1–100 μl/min [110]. Similar behavior was reported by Banks et al. who showed how the ESI ion signal obtained across a wide range of LC-MS flow rates actually decreased by about 30% when the flow-rate increased from 1 μl/min to 1 ml/min, even though in the latter case 1000 times more sample was consumed [116].

The explanation for these disparate behaviors resides in the different efficiency of the diverse ESI source designs for ion sampling and ion production [117]. In electrospray, ions are produced at atmospheric pressure and have to be introduced into the mass spectrometer. The quality of ion sampling from the atmospheric pressure plume strongly affects the total number of ions reaching the MS detector. The spray tip of the interface becomes a splitter since only a portion of the spray plume has access to the MS inlet capillary or skimmer. It has been estimated that the overall efficiency for ion sampling and detection using a triple quadrupole TAGA instrument with an ESI source working at 3–6 μl/min is about one ion in 100 000 [118]. In terms of ion transfer to the quadrupole

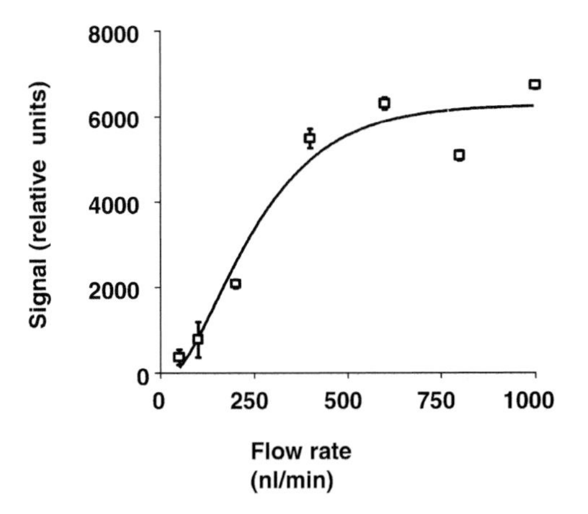

Fig. 4.2. Influence of the flow-rate on the response of the mass spectrometer, using continuous infusion of endothelin-1. Experimental conditions: distance, 3 mm; electrical voltage, 1.6 kV.

analyzer, only one ion in 10 000 was introduced for MS analysis and detection. Droplet space charge repulsion puts a limit on the maximum droplet density in the spray, resulting in a wide spray plume. Consequently, only a small part is sampled by the MS. Production of microsprays with small plumes enhances sampling efficiency [35,90]. As low-flow ESI capillaries can be placed very close to the inlet of the MS, so a larger part of the ESI plume leaving the ESI needle is focused into the analyzer.

The size and positioning of the spray cone determine the number of ions actually entering the mass spectrometer, whereas several parameters of the ESI emitter, for instance the inner and outer diameters of the spray tip, can be optimized to improve ionization efficiency. Flow-rate and tip diameters determine the formation and size of the droplets from which ions will be desorbed [119]. Droplet formation at the Taylor cone is dependent on the flow-rate [35]. By decreasing the flow-rate, smaller droplets are emitted from the ESI source. Small droplets possess a higher surface-to-volume ratio, enhancing the desorption of ions into the gas phase. To stabilize the Taylor cone, the capillary diameter has to be reduced in parallel with the flow-rate reduction. Early conventional Analytica of Branford ESI interfaces operating at 1–5 μl/min were made of stainless-steel capillaries of about 500 μm i.d. MicroESI needles, as developed by Caprioli, used fused-silica capillaries with 20–50 μm i.d. and show optimum flow rates in the range of 200–800 nl/min. Finally, fused-silica or glass capillaries used for nanospray have diameters of only 1–2 μm at the tip exit and produce stable sprays in the range of 20–200 nl/min [35–37]. Conventional 500 μm i.d. interfaces produce initial droplets with average sizes in the μm range [119]. Nanoelectrospray tips produce much smaller droplets with a radius of approximately 200 nm. Mann et al. estimated that, at a concentration of 1 pmol/μl, such small droplets would each contain on average only one analyte molecule.

The combination of improved ion sampling efficiency and spray performance has a dramatic effect on the performance of the ESI source. Emmett et al. placed a 20 μm ESI capillary (flow-rate 200 nl/min) a few mm from the MS inlet to obtain attomole detection limits [38,120]. Mann et al. estimated the efficiency of an ESI source fitted with a 1–2 μm i.d. nESI tip operated at ca. 20 nl/min and placed 1–2 mm in front of the MS entrance [37]. In a conventional interface, only one in 200 800 analyte molecules in the solvent were detected by the MS detector, similar to the 1/100 000 value estimated by Smith et al. (see above). The nESI tip improved this ratio to one in 390 molecules detected by the mass spectrometer, which was 510-fold better. With these miniaturized emitters, mass sensitivity is greatly enhanced: spectra have been obtained with low femtomole amounts of sample, and attomole and zeptomole detection limits have been reached [39]. Despite the much lower flow rates, concentration sensitivity is also enhanced: 2–3 times higher ion currents are reported for a given analyte concentration than for a conventional ESI source.

Ionization and ion sampling efficiency increase inversely to flow-rate. The smaller the plume of microdroplets and the closer this is to the inlet orifice, the higher the sampling ratio. Likewise, the smaller the microdroplets sprayed, the higher the ion production rate. Nowadays, it is generally recognized that miniaturized systems designed for working at reduced flow-rate, which use small diameter columns and an optimized interface for this flow, give enhanced mass and concentration sensitivities.

Nevertheless, the combination of LC-ESI MS has not yet given detection limits at the sub-attomolar level, like those obtained in off-line nanoelectrospray analysis (see above).

One of the reasons is that there is no routine LC methods with flow rates down to the low nl/min. Thus LC separation methods are operated under a relatively high flow-rate: 100–500 nl/min versus the 5–20 nl/min for off-line infusion nanoelectrospray. Accordingly, on-line nanoseparation mass spectrometry techniques appear less sensitive than continuous infusion nanoelectrospray tips when absolute detection limits are compared.

In comparison with conventional LC-ESI MS, the practical benefits of low-flow LC-ESI MS systems lie on the lower limits of detection that can be obtained (needed both for the analysis of samples available in limited amounts or in very low concentrations). On the other hand, their perceived drawbacks in terms of commercial availability are not that significant in view of the relative simplicity of production of custom-made capillary columns and low-flow spray devices. Although reproducibility and ruggedness are still to be proven, it can be predicted that the use of miniaturized LC-MS systems will steadily increase in the near future.

4.1.4. Drawbacks of capillary chromatography

Concentration detection limits attainable by diminishing column diameter are limited by the optimum injection volume and loading capacity of the specific column. In principle, the parameters are reduced proportionally to the square of the columns inner diameter. For column diameters of less than 1 mm, injection volumes lie in the submicroliter range. For the same sample concentration, this proportionally reduces the total amount of sample that can be injected (Table 4.2).

Although it is possible to concentrate the sample by solvent evaporation, there are practical difficulties in dealing efficiently with low injection volumes. In most cases, the initial sample volume is also small, so that volume reduction and sample handling become unmanageable. For these reasons, with narrow-bore and especially with capillary columns, some kind of on-line sample pre-concentration becomes necessary.

Simple solutions are either to inject the sample into a solvent with lower eluotropic strength, to achieve on-column analyte focusing, or to use a solvent gradient to pre-concentrate the sample at the head of the column before elution. In this way, it is possible to inject a relatively large volume of sample into a capillary column without deleterious chromatographic effects, provided that sample components are strongly retained under the initial conditions.

It is also likely that, when biological extracts are injected onto a miniaturized LC system, the background matrix exceeds the capacity of the capillary column. Therefore, before injection, an extraction procedure to isolate the target analytes has to be employed. Typical solutions are liquid–liquid extraction [52], immunoaffinity isolation [53] or micro-dialysis [50].

Another important issue in capLC is the increased dead times appearing as a consequence of the low flow rates used. Solvent gradients take longer to reach the column due to the greater effect of the dead volumes of the chromatographic system. Sample loading is also slower than in conventional chromatography: the injection of a 10 μl sample on the head of a 75 μm column working at 500 nl/min would take 20 min.

References pp. 70–73

4.1.5. The use of pre-columns and column-switching in CapLC

On-line pre-concentration columns can be used to circumvent the injection volume, loading capacity and dead time limitations of micro and nanoscale separation techniques. A given sample can be concentrated and washed on a narrow-bore or capillary pre-column before the retained material is desorbed in small volume to the capillary column. The use of on-line pre-concentration microcolumns improves the concentration detection limits and simplifies overall sample handling. Pre-columns can be directly coupled to the analytical column, although in most cases the use of column-switching procedures affords a more effective and flexible set-up (Fig. 4.3). When a column-switching system is used, the pre-column exit can be disconnected from the analytical column so that sample loading and washing steps can be performed at flow rates much higher than those optimal for the analytical column, thus reducing the time needed for the procedure. In addition, analytical column elution and washing can be performed independently of sample loading, so that one sample can be loaded while the previous sample is still being separated on the analytical column. This also contributes to increased sample throughput.

The pre-column and analytical column can contain similar reversed-phase sorbents. Optimal conditions require that the peptides be equally or more retained in the analytical column than in the concentration column. Otherwise, peptides desorbed from the pre-column would pass the analytical column with no chromatographic effect. This can be accomplished by selecting the appropriate reversed-phase sorbent for each column. In most cases, however, it would be preferable to use a different kind of chromatographic interaction in the pre-column than in the analytical column. This combination deals with complex biological samples very efficiently [56,121,122]. For example, on-line size exclusion chromatography (SEC) simplifies the peptide mixture before analysis. However, as SEC does not pre-concentrate the peptide sample, it has to be coupled to another pre-concentration column [121]. Ion exchange pre-columns can be used for the efficient

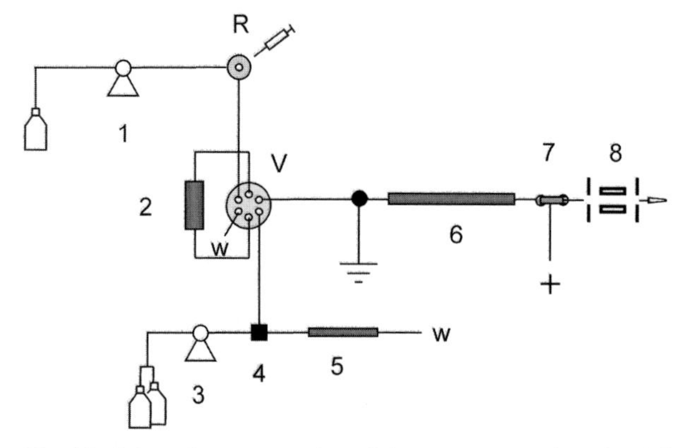

Fig. 4.3. Schematic representation of the pre-concentration-microLC-μESI MS set-up: 1. pre-concentration pump; 2. pre-concentration column; 3. gradient pump for analytical column mobile phase; 4. flow splitter; 5. pressure balance column; 6. microLC column; 7. μESI interface; 8. mass spectrometer. R, injection valve; B, switching valve; W, waste.

on-line removal of SDS from peptide mixtures [122]. This kind of sorbent also selectively pre-concentrates and desorbs peptide mixtures before reversed-phase capLC [56]. A similar, more sophisticated approach is used in comprehensive multidimensional chromatography [109,123]. Jorgerson et al. demonstrated zeptomole detection limits for the analysis of peptide digest from porcine thyroglobulin using a two-dimensional anion exchange-reversed-phase capLC system and fluorescence detection. The peak capacity of this 2D method was estimated to be over 2000 components per analysis [123].

4.1.5.1. Practical advantages of PC with miniaturized LC-MS systems

Table 4.3 illustrates, in a comparative manner, the step-by-step increase in sensitivity that can be achieved by stepwise reduction in column internal diameter and sample pre-concentration techniques. For this purpose, we took our own data generated in the course of a study on the LC-ESI MS detection of low endogenous levels of endothelins [56]. By using 1 mm i.d. columns operating at a flow-rate of 100 μl/min, we achieved 50 fmol detection limits for 100 μl injection of standard solutions. This corresponds to a concentration sensitivity of 0.5 nmol/l. However, when working with real samples, the detection limit was up to 500 fmol and for this we had to process 15 ml samples of HUVEC supernatants with a resulting concentration sensitivity of 5 nmol/l [165].

The application of the concepts considered above led us to try the same approach, but instead used 250 μm i.d. capillary columns at a flow-rate of 5 μl/min. The LOD for standards was reduced to 2 fmol but any further gains were limited by the 10 μl maximum injection permitted on these columns (data not shown).

In a further scale-down step to nanoLC, we used 75 μm i.d. columns operated at a flow-rate of 0.5 μl/min and found a higher mass sensitivity with a LOD of only 0.5 fmol but low concentration sensitivity (0.2 nmol/l) due to the restricted 2.5 μl sample injection. This, however, could be readily increased to 200 μl injected by applying on-line sample pre-concentration. Although the mass sensitivity was similar to that obtained with capillary columns (1.5 fmol), the concentration sensitivity improved significantly to 30 pmol/l for real samples. In the latter case, only 500 μl are processed, as opposed to the 15 ml needed initially with the microbore system [56].

TABLE 4.3

DETECTION SENSITIVITY AS A FUNCTION OF COLUMN INTERNAL DIAMETER AND SAMPLE PRE-CONCENTRATION PROCEDURES. STD: USE OF STANDARDS (NO MATRIX)

System	Column i.d. (μm)	Flow (μl/min)	Sample	Vol. Inj. (μl)	LOD (fmol)	Conc. LOD (nmol/l)
Narrow-bore LC-MS	1000	100	Std	100	50	0.5
	1000	100	15 ml	100	500	5
Narrow-bore LCMS/MS	1000	100	Std	100	100	–
μLC-MS/MS	250	5	Std	10	2	0.2
μPC microLC-MS/MS	250	5	Std	100	2	0.02
nanoLC-MS/MS	75	0.5	Std	2.5	0.5	0.2
μPC nanoLC-MS/MS	75	0.5	500 μl	200	1.5	0.03

4.1.6. Quantitative analysis

Quantitative peptide analyses by LC-MS have mainly used narrow-bore LC columns [15–27,124,125]. Typically, 2.1 mm i.d. columns working at flow rates around 75–300 μl/min are used for this purpose. Quantification is generally carried out by using the corresponding deuterated peptides as internal standards. In other cases, peptide analogs derived from the substitution of one or more amino acids in the sequence are employed [21,27]. Murphey et al. make use of a ^{13}C, ^{15}N doubly labeled standard for the analysis of bradykinin 1–5 in human blood by LC-ESI MS/MS. With selected reaction monitoring (SRM), they obtained detection limits of 4 fmol/ml blood. Darby et al. used ovine insulin for the quantitation of human and porcine insulin in plasma at a low ng/ml level.

Peptides are commonly analyzed by ESI in the positive ion mode. Fierens et al. reported, however, a quantitative method for C-peptide (Mr 3017) in urine using negative ion ESI ionization and SRM.

Reports on the use of capLC-MS for peptide quantitation are scarce [56,124,126]. Capillary column-switching techniques have also been employed, but mainly for qualitative analyses often related to proteomics studies. Despite this, capLC-ESI-MS in combination with on-line pre-concentration has already shown its value in quantitative bioanalysis [49,52,53,56,126]. For example, Cai and Henion combined immunoextraction with a PC-microLC-MS method [53]. An immunoaffinity column was used to pre-concentrate LSD and its metabolites in diluted urine samples from volumes as large as 50 ml. The bound compounds were then released into a reversed-phase trapping column with dimensions of $0.5 \times 15\ mm^2$. From this column, the compounds were eluted in the back-flush mode towards the analytical column (300 μm × 150 mm) and analyzed using a conventional ESI interface. Thus, LSD and its analog could be detected at the 2.5 pg/ml level.

Zell et al. described a validated method for the analysis of pharmaceutical compounds in blood [52]. Here, 1 ml of blood was extracted with dichloromethane. After evaporation to dryness, the sample was reconstituted in 70 μl of 20% methanol. Then, the sample was injected into an analytical system consisting of a 0.8 mm i.d. × 5 mm reversed-phase trapping column, a 300 μm i.d. × 150 mm analytical column and an ESI interface with a spray needle of small internal diameter. With the use of such a PC-microLC-MS method, detection limits of 1 pg/ml in 1 ml blood were obtained (LOD 1 pg).

Oosterkamp et al. reported a PC-nanoLC-MS method for the detection of the peptide endothelin in cell culture supernatants [56]. To this end, an analytical method using custom-made nanoLC columns (75 μm i.d.) and a microelectrospray interface was developed. Thus, to analyze large sample volumes in order to cope with low biological analyte concentrations, the capLC-ESI MS method was coupled to a strong anion exchange PC column (0.8 mm × 5 mm). The cell supernatant was deproteinated with acetonitrile, and the supernatant was evaporated to dryness. In this way up to 200 μl of the reconstituted sample (equivalent to 50 μl of supernatant) could be injected into the analytical system. A similar method has been reported by the same group for the analysis of bradykinin and some of its metabolites in plasma. In this case, the PC was performed with a reversed-phase sorbent. Details on these methods will be given in Section 4.3.

4.2. INSTRUMENTATION

4.2.1. Commercial and custom-made capLC instrumentation

The optimal flow-rate for a 150 μm column is around 1 μl/min, which requires special qualities of the pumping system, especially when gradient elution is needed. Syringe pumps obtain reproducible flow rates in the microliter per minute range. Lee and colleagues describe a brilliant procedure to prepare and store a solvent gradient before the analytical step using syringe pumps [127]. They also apply a 'peak parking' procedure by controlling the pressure of the syringe pump. This enables an efficient characterization of peptide mixtures by tandem mass spectrometry [128,129]. Simpler set-ups are based on conventional HPLC gradient pumps and flow splitting. Flow splitters can be custom-made using a T-piece and some kind of restrictor. Alternatively, calibrated flow splitters giving reproducible flows are also available from commercial sources.

Currently, several companies distribute miniaturized chromatographic systems, including pumps, automatic microinjectors and microcolumns as well as suitable cells or interfaces for photometric or MS detection. Most ESI systems can be upgraded for capLC-MS work. Low-volume UV cells for micro (35 nl cell, LC Packings) and nano (3 nl cell) LC are also available for most commercial detectors. These systems can also be built using conventional laboratory equipment with minimal effort and at low cost [130–132].

4.2.2. Capillary column preparation

The first attempts to develop LC capillary columns occurred at the beginning of the 1980s [65,69,133–136]. Nowadays, capillary columns can be bought from various manufacturers. As a result of their high market price and the ease of preparing high-quality columns in the laboratory, many groups are still packing their own miniaturized columns.

Before Takeuchi and Ishii demonstrated the quality of fused-silica capillaries [137], early attempts to prepare packed capLC columns used metal or PTFE tubing [60,63]. Although Teflon [138] or PEEK lines [139] are still used to prepare capillary columns, fused silica is the material that most authors choose. Fused-silica capillary columns with internal diameters from 500 to about 50 μm can be easily prepared using slurry-packing methods [32,47,57,74,130,140]. Recently, Colon reviewed several other methods for column packing [73].

There are several protocols for the packing of micro- and nanobore columns [57,69,130, 141–144]. Usually, the body of the capillary column is made of commercially available untreated fused silica.

The first step in column packing is to build a frit at the exit end of the fused-silica capillary column. The capillary column is then coupled to a reservoir filled with a slurry of the packing material. This is pushed under pressure into the column inlet by a flowing stream. The frit allows the solvent to pass through the column and the solid phase fills the column in a process similar to filtration. A practical guide for this packing procedure is described by Tomer et al. [57].

The frit in fused-silica capillaries is made by tapping the analytical column into a pile of silica particles, filling approximately 1 mm of the outlet of the column [69]. Then, the end is heated by a torch that sinters the silica particles and forms a frit, which is stable at relatively high pressures [57]. The sintered glass method is probably the most commonly employed. In our experience, sintered glass frits are stable when capillaries with outer diameters greater than 300 μm are used. Columns with smaller outer diameters (180 μm) are more vulnerable to the heating process that makes the capillary too fragile to handle.

Alternatively, the frit can be made by pushing the capillary column through a porous membrane sheet two or three times, resulting in a stable plug at the outlet of the column. Membrane frits using Teflon [47], PVDF [130] or glass-fiber sheets [32] are also possible. Alternatively, Lubman used a Valco microbore column-fitting with a very small amount of glass wool as a frit. It was 0.1 mm thick and proved to be more stable than glass frits [145]. A zero dead volume Valco union can also be used to hold a stainless mesh frit [74].

We routinely use two glass fiber filters sandwiched between two capillaries (the column and the outlet capillary) inside a holder capillary [56]. This frit was shown to be stable at the pressure necessary for analysis (Fig. 4.4).

After the frit is ready, the capillary column is coupled to a reservoir filled with the slurry of the packing material. The slurry is pushed under pressure into the column inlet by a flowing stream. The frit allows the solvent through the column and the solid phase fills it in a process similar to filtration. Liquid solvents [56,57,70,74,146] and supercritical fluids [140,147,148] are used for slurry-packing. Likewise, gas streams are used for dry-packing procedures [144,149].

Columns with diameters as small as 20 μm have been packed successfully by using these procedures [69]. However, as smaller sizes are difficult to fill, the smallest common diameters found in the literature are around 50–75 μm. Despite this, Hsi and Jorgensson reported the preparation by slurry-packing methods of 12 μm i.d. capLC columns filled with 5 μm particles. The particles had to be filtered through an 8 μm filter to prevent the column from plugging with the larger particles in the commercial material [70].

Five micron particles are the most commonly used for column filling. Recently, MacNair et al. introduced ultrahigh pressure capLC (UHPL), in which non-porous particles of 1.0–1.5 μm are used to fill 29–100 μm i.d. capLC columns [150]. More than 200 000 theoretical plates ($k' = 1$) can be generated using 25–50 cm long columns.

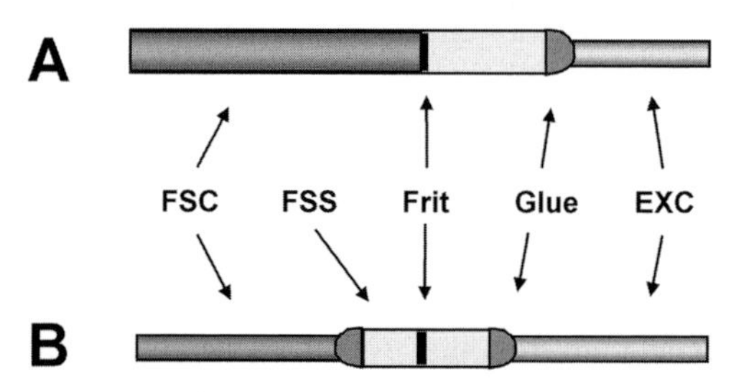

Fig. 4.4. General scheme for the preparation of fused-silica capLC columns with internal diameter (A) above and (B) below 180 μm. FSC, fused-silica capillary column; FSS, fused-silica sleeve; EXC, fused-silica exit capillary.

UHPLC can also produce high-speed separations (less than 100 s) with column efficiencies over 20 000 plates [151]. MacNair demonstrated the efficacy of this technique in analyzing peptides and proteins with an ion trap MS detector [152]. To take full advantage of the high speed available, Lee and colleagues coupled UHPLC to a fast scanning time-of-flight MS detector [151].

Recently, other special classes of packed columns with immobilized stationary phases [153] or a continuous porous bed (monoliths) [154,155] have been developed. One major advantage of these columns is that they do not need the end frit, which conventional packed capLC columns need in order to retain the stationary phase.

4.2.3. Pre-columns and column-switching

Pre-concentration columns can be prepared in the same way as analytical columns from 250 to 530 i.d. FSC tubes. Higher column diameters can be used when high volumes have to be pre-concentrated in a reasonable time or when complex samples could produce column overloading. Capiello et al. [139] described how to produce suitable columns with diameters up to 0.76 mm from PEEK tubing. Conventional ZDV peak or stainless-steel connectors are used as column ends with 1/16 nuts and ferrules. The membrane frit at the outlet end is introduced into the ZDV union and sealed in place by pressing the column end against it and then tightening the nut. In the examples given later in this study we used a commercial kit consisting of a stainless-steel holder and 800 μm i.d. cartridges. The same holder could load cartridges with i.d. from 200 to 800 μm. Cartridges available include C18 and ion exchange materials.

The basic structure of a pre-concentration system is depicted in Fig. 4.3. An auxiliary HPLC pump loads the sample into the pre-column and washes out non-retained compounds. The auxiliary pump is permanently working at 50–200 μl/min depending on the pre-column, so that samples are loaded very fast even with high volume samples. Sample loops from 20 to 200 μl are usually employed.

The pre-column is connected to a motor-driven Valco switching valve. When the valve is in the 'analysis' position, the pre-column is connected in series with the gradient HPLC system and the analytical column. Valve switching is manual or automatic through the relay contacts on the HPLC system.

On-line desorption from the PC column can be performed under backward or forward-flush conditions. Backward-flush desorption is very vulnerable to plugging and pressure problems. Small particles, which are pre-concentrated on the pre-column, can be desorbed onto the analytical column, even with in-line filters. Under forward-flush conditions, though, the analytes pass through the entire PC column before reaching the analytical column with the consequent band-broadening. This effect is diminished when a solvent gradient is used for chromatography, in which case, forward-flush conditions are preferable.

4.2.4. NanoESI and microESI sprayers and interfaces

Unprocessed, commercially available stainless-steel and fused-silica capillaries of adequate internal diameters have been used for ionspray, and for conventional and microESI. To optimize spray formation at the lower flow rates used in microelectrospray interfaces, some changes are usually made to the capillary tip. Fused-silica capillaries with small i.d. are commonly etched with HF [38] or the tip is rubbed with sandpaper, resulting in very sharp ESI tips. Electropolishing has also been used with metal capillaries to produce similarly sharp tips [156]. ESI tips have been constructed with i.d. values down to 10 μm. By using a micropipette puller, spray tips as small as a few μm can be prepared from glass or fused-silica capillaries [35,39]. We adopt the same approach in our laboratory to obtaining microspray capillaries by using a torch and letting the capillary stretch under gravity. Although not used for LC-MS, the approach of Valaskovic et al. [39] should be mentioned. They obtain reproducible tip diameters down to 2 μm by first pulling capillaries and later etching them with HF. These tips can be operated at flow rates from 0.1 to 20 nl/min, depending on the diameter.

4.2.4.1. Voltage application

In metallic needles, electrical voltage can be applied directly to the spraying needle. When using non-conductive glass or fused-silica capillaries, the voltage at the spray tip is applied in two different ways, either by liquid contact with an electrode at some point before the needle tip [39,56,157], or by coating of the capillary with gold [35,36,158] or silver [159]. In tips employing a liquid contact, the voltage is conducted through the mobile phase to the ESI tip. For example, as shown in Fig. 4.5, an operational simple microESI interface can be put together with a 50 μm i.d. inlet capillary connected to a 25 μm i.d. outlet spray tip. The electrical contact is made at the stainless-steel holder and the frit helps to produce an effective liquid contact in the gap. This custom-made emitter assembly is mounted on a commercially available micropositioner, as shown in Fig. 4.6. Microwires have also been used to produce the electrical contact, but only for CE-ESI MS coupling [160]. Gold-coated ESI tips have been made by a gold sputtering procedure. Subsequently, a clamp attaches an electrical voltage to the spray. The latter method is less sensitive to different mobile phases, as recently shown by Vanhoutte et al. [55]. The connection of nanoscale separations to MS has been accomplished with various interfaces.

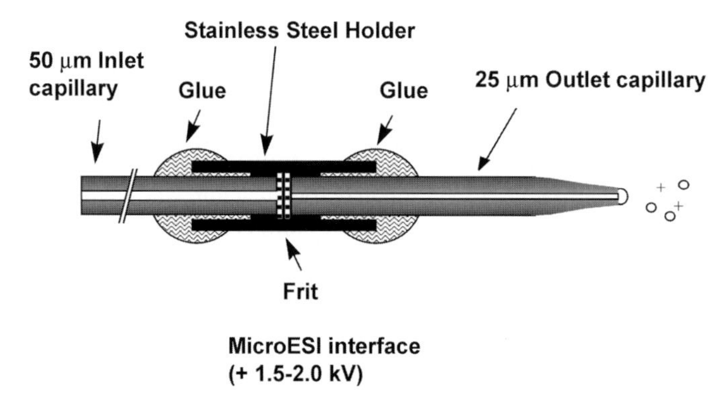

Fig. 4.5. Diagram of a micro-electrospray interface (μESI). Inlet capillary: 50 μm i.d., 190 μm o.d.; stainless-steel capillary: 250 μm i.d., 350 μm o.d.; glass fiber frits; outlet capillary: 25 μm i.d., 190 μm o.d.

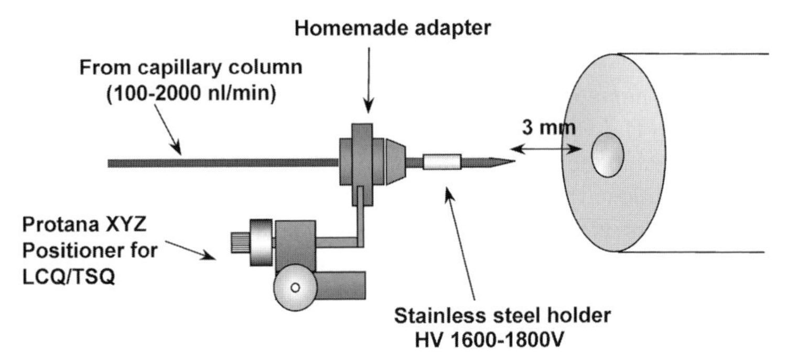

Fig. 4.6. Practical set-up of the μESI interface.

Hunt et al. used nanoLC columns in combination with a normal ESI interface for structural elucidation of subpicomole amounts of peptides bound to the MHC molecules [105,106]. On-line CEC-MS has also been accomplished with normal-scale ESI interfaces [161–163]. By modifying the electrospray interface, an improvement in detection limits down to 2 fmol protein digest can be obtained [127]. Further downscaling of the LC-MS method by coupling of nanoscale LC to dedicated microESI interfaces has resulted in detection limits down to the attomole range [49,54,56].

4.2.4.2. Practical set-up

Figure 4.5 shows the scheme of a sheathless micro-electrospray ionization (μESI) interface based on a liquid junction [48] commonly used in our laboratory. In comparison with other set-ups where the spray needle is built as part of the capillary column, this interface is independent of the capLC system. This allows for column-switching and replacement without changing source conditions. Furthermore, it is possible to monitor the column eluate on-line by UV detection before MS analysis.

This interface uses a stainless-steel capillary as electrical contact. The electrical contact with the eluent takes place in the gap produced between two fused-silica capillaries inside the metal sleeve. An on-line filter inserted between the fused-silica capillaries also helps produce the electrical contact and at the same time reduces the dead volume of the connection. To build the on-line filter, the end of the 250 mm i.d. stainless-steel capillary was twisted against a Whatman GF/A glassfiber filter to cut out a piece of filter corresponding to the diameter of the capillary. The filter was sandwiched between two 190 μm o.d. capillaries of 50 and 25 μm i.d., which were glued into a stainless-steel capillary of 250 μm i.d. as inlet and outlet capillary, respectively.

The end of the outlet capillary has to be previously tapered to create the μESI needle tip. For this, about 10 cm of fused-silica capillary is cut. A light weight (10 g) is fixed to one end of the capillary. The other end is fixed to a support, leaving the capillary in a vertical position with the weight hanging free. The flame of a torch is applied to the center of the line and the capillary stretches under gravity. Two capillaries with sharp tips are obtained in this way. The needle tips are carefully cut to obtain an inner bore of about 5 μm.

The bore diameter can be monitored with a stage micrometer (Graticules Ltd, Tonbridge, Kent, UK) under a microscope.

The μESI interface was mounted into an *x–y–z*-manipulator from a Protana Nanoelectrospray Source Kit (The Protein Analysis Company, Odense, Denmark) and the power supply was connected to the stainless-steel capillary of the μESI tip (Fig. 4.6). The μESI-tip was carefully positioned at 3–5 mm from the inlet of the Finnigan LCQ mass spectrometer (ThermoQuest, Barcelona, Spain). For normal analyses, an electrospray voltage between 1.6 and 2.0 kV was applied. For continuous infusion experiments, the inlet capillary of the μESI was connected via an injector to a syringe pump.

The conventional electrospray interface of the Finnigan instruments can be used instead of this dedicated interface when capLC columns with inner diameter larger than 180 μm at flow rates higher than 3–5 μl/min are used. When working at flow rates close to or lower than 5 μl/min, sensitivity can be improved by reducing the default distance between the needle tip and the MS entrance (typically 3 cm) to 5 mm.

4.3. EXAMPLES OF QUANTITATIVE PEPTIDE ANALYSIS

Our laboratory is developing quantitative methods for the determination of endogenous peptides in matrixes such as cell culture supernatants, biological fluids and tissues. These methods use custom-made μESI interfaces and 75–250 μm i.d. capLC columns. Columns are operated under nanoLC (550 nl/min) or microLC (1–5 μl/min) conditions and are on-line coupled to ion exchange or reversed-phase PC columns. In pre-concentration procedures sample volumes up to 200 μl can be injected. These on-line PC-capLC-μESI MS systems analyze the peptide class of endothelins in supernatants of human umbilical vein endothelial cell (HUVEC) cultures as well as the peptide bradykinin and its metabolites in human and rat plasma.

Generally, SRM procedures were applied for maximum selectivity. The implementation of tandem MS methods also favors the simplification of sample handling, for instance utilizing solely protein precipitation by acidifying the sample or by adding an organic solvent such as acetonitrile. We used these methods to deproteinate cell culture extracts, plasma and total blood samples before PC-capLC-μESI MS analysis.

The three endogenous components of the endothelin family, ET-1, ET-2 and ET-3, are potent vasoconstrictors of 21 amino acids with molecular weights around 2.5 kDa [164]. Bradykinin is also a vasoactive non-apeptide produced from kininogen in plasma that is rapidly degraded to shorter, inactive metabolites [20,126].

Endothelins have been analyzed in HUVEC supernatants by narrow-bore LC-ESI MS in our laboratory and by others [165,166]. The analysis of bradykinin and its metabolites in plasma has also been reported recently [20,126].

4.3.1. Endothelin extraction from HUVEC

For a typical extraction, 10 μl of a 100 nmol/l alaET internal standard solution and 100 μl of diluted ET standards were added to 500 μl HUVEC in polypropylene tubes. The growth medium for HUVEC culture consists of Medium 199 supplemented with 20% fetal calf serum. HUVEC supernatants were treated with acetonitrile (1 ml) and the tubes were vortexed for 1 min.

The tubes were centrifuged at 4000*g* for 5 min in a Beckman TJ-6 Centrifuge (Palo Alto, CA), and the supernatants were pipetted into new tubes. The supernatants were evaporated close to dryness using a rotavap and were frozen at −20 °C until injection. Just before injection, 100 μl methanol:water (50:50, v/v) and 1.9 ml of the pre-concentration mobile phase was added. Of this mixture, 200 μl was injected into the analytical system.

4.3.2. Bradykinin extraction from plasma

Five microliters of a 10 nmol/l Leu-BK1-8 internal standard solution and 5 μl of diluted BK standard mixtures were added to 100 μl human or rat plasma. Subsequently, 200 μl acetonitrile were added and the tubes were vortexed for 1 min. The tubes were centrifuged at 12 000*g* for 5 min in an Eppendorf Centrifuge (Palo Alto, CA), and the supernatants were pipetted into new tubes. The supernatants were evaporated close to dryness using a rotavap and were frozen at −20 °C until injection. Solvent evaporation is mandatory in reversed-phase separation, since excess of organic modifier would destroy retention on the PC column.

Before injection, the samples were diluted with PC mobile phase and centrifuged for 1 min to eliminate any solid residue present in the sample.

4.3.3. CapLC columns

Capillary columns (10 cm length, 75 μm i.d./190 μm o.d. and 180 μm i.d./375 μm o.d.) were prepared from fused-silica capillaries, as described above. For the 75 μm column a frit was produced sandwiching two glass-fiber filters between the column capillary and the capillary outlet inside a capillary sleeve (Fig. 4.4). For this purpose, the fused-silica capillary to be used as column was glued into a holder capillary (fused silica, 250 μm i.d., 450 μm o.d., 20 mm long). As a frit, the end of the 250 mm i.d. capillary was twisted against a Whatman GF/A glass-fiber filter in order to cut out a piece of filter with the diameter of the holder capillary. This action was repeated to give a frit with a double layer of glass-fiber filters. Then, an outlet capillary (50 μm i.d., 190 μm o.d.) was glued into the other end of the 250 μm i.d. capillary, sandwiching the two filters between the two capillaries. In the case of the 180 μm column, the frit is produced at the end of the column capillary in the same way. The outlet capillary (50 × 150) is introduced into the FSC capillary to serve as the outlet line. The frit is pushed 5 mm into the FSC by pressing with the EXC and both tubes are held together with a drop of glue.

The frit was tested by applying a flow-rate of 0.1–0.2 ml/min and, subsequently, the column was slurry-packed at 15 MPa using 5 μm Kromasil C18 particles (Akzo-Nobel, Arnhem, The Netherlands). The columns thus developed were much more stable in use.

References pp. 70–73

For method set-up, the outlet of the column was connected to an ABI UV-detector equipped with a CZE cell holder for UV detection. The capillary flow cell was homemade by burning the polyimide coating of a 250 μm i.d. fused-silica capillary. The inlet and outlet capillaries, 75 μm i.d. and 180 μm o.d., were glued into the capillary, leaving a 5 mm space. The total volume of the flow cell was approximately 250 nl.

4.3.4. Micro-ESI interface

A micro-electrospray ionization (μESI) interface based on a liquid junction [47] was produced as described above (Figs. 4.5 and 4.6).

The performance of the μESI interface was tested under continuous infusion conditions. We investigated the relationship between the flow-rate and the peak height by continuous infusion of ET-1 (Fig. 4.2). Up to flow rates of 400 nl/min, the signal increased linearly in relation to the flow-rate, showing a mass dependent response of the signal. However, when the flow-rate rose from 400 to 1000 nl/min, no further increase of the signal was observed. At these flow rates, a liquid rain at the entrance of the MS was visible, which indicated a bad spray performance. The size and number of these microdroplets increased with flow. Thus, the lack of mass dependency at high flow rates could be attributed to the decreasing ionization efficiency of the bigger droplets [167].

To increase the sensitivity at these flow rates further, the outlet of the ESI capillary needs to be reduced by tapering the point in combination with adjustment of the needle distance [37,39,46]. We observed improved spray stability over a broader range of mobile phase compositions after tapering of the needle tip [56].

4.3.5. CapLC-μESI MS

A conventional HPLC system (Kontron 325 System, ABI 140A Solvent Delivery system or an Agilent HP1100) delivered a gradient to the analytical separation at a flow-rate of 0.25–0.3 ml/min. This flow was reduced to 550 nl/min (75 μm column) or 2 μl/min (180 μm column) using a flow splitter consisting of a T-piece connected to an analytical column (dimensions: 2 × 200 mm, 5 μm particles) as back-pressure regulator. These values resulted in linear flow rates of approximately 2 mm/s. According to the h/u curve [54], these columns were operated a little above their optimal flow rates.

For endothelin analyses, Solvent A consisted of acetonitrile:water:acetic acid:trifluoro-acetic acid (TFA) (10:90:1:0.05, v/v), and Solvent B of acetonitrile:methanol:water:acetic acid:TFA (45:45:10:1:0.05, v/v).

For bradykinin analyses, Solvent A consisted of water:acetonitrile:acetic acid (90:10:1, v/v). Solvent B consisted of water:methanol:acetonitrile:acetic acid (20:40:40:1, v/v). An elution gradient from 40 to 60% B was used, unless otherwise stated.

For injection, a 5 μl loop was used, which is considerably higher than the optimal injection volume of a 75 μm column (20 nl). Most peptides, however, are eluted in a gradient elution mode and so are pre-concentrated on the top of the column. In a typical analysis, the injection loop was put 5 min on-line.

The outlet of the capLC column was coupled to the μESI interface by a Teflon connector [54]. The electrospray positive voltage (1.6 kV) was put on the needle and the injection port was grounded. Under these conditions, an electric field exists along the chromatographic column in the same way as in pressurized-flow electrochromatography (pEC) [167]. The effects of the electric potential on separation and resolution cannot be neglected. Although lower potentials are used than in common electrochromatographic separations (approximately 20 kV), such low potentials influence separation efficiency [167]. Thus, their effect has to be kept in mind when, for example, optimizing chromatographic conditions off-line by using a UV detector instead of the ESI-MS. Here, the separation was optimized in a straightforward way, by maintaining the electrospray voltage constant for optimum ionization and modifying mobile phase composition to optimize separation.

The selectivity of the LC-MS method was enhanced by applying MS/MS conditions. Endothelins and bradykinins were analyzed with two SRM programs. In the case of endothelins, the selected reaction from the triple-charged molecular ion $[M + 3H]^{3+}$ to the double-charged fragment b_{20}^{2+} was followed. This is a highly selective cleavage resulting from the elimination of the terminal tryptophan amino acid. The following reactions were monitored: ET-1, 831.5/1144.5; ET-2, 849.9/1171; ET-3, 881.9/1220; alaET, 810.8/1113. In the case of BKs, the reactions monitored corresponded to the fragmentation of the $[M + 2H]^{2+}$ ion of this compound to the singly charged $y_8'^{+}$ (BK) and b_6^{+} (BK1-8, Leu-BK1-8, BK1-7) fragments. These reactions were: BK, 531/904; BK1-8, 453/642; Leu-BK1-8, 435.6/642; BK1-7, 379/642. To increase sensitivity, the isolation window of parent and daughter were set as high as 2.5 and 3 u, respectively. Optimal MS/MS collision energy was found at a setting of 20% energy in the LCQ.

4.3.6. On-line pre-concentration (PC)-capLC-μESI MS

For pre-concentration of relatively high biological sample volumes (>50 μl), PC columns with a sufficient capacity have to be used. Therefore, we chose pre-columns of 0.8 × 5 mm, also used by other authors [49,52,53]. These column can be operated at relatively high flow rates (up to 200 μl/min), so decreasing the pre-concentration time necessary to load high volume samples.

To perform effective pre-concentration and desorption, a strong anion exchange (SAX)-PC μ-pre-column (0.8 × 5 mm; LC Packings, The Netherlands) was selected. Pre-concentration was performed at neutral or slightly basic pH, where the carboxyl acid groups of the peptides are negatively charged. For desorption, a pH lower than three is chosen, at which the peptide carboxyl groups are protonated. Thus, the peptide loses its affinity for the SAX-PC sorbent and is eluted at the solvent front. Desorption can be performed with the acidic mobile phases, such as trifluoroacetic acid or acetic acid, normally employed in peptide analysis. This system allows efficient coupling between the PC solvents and the TFA or formic acid containing mobile phases commonly used for peptide analysis.

Ion exchange-PC columns have several advantages over reversed-phase (RP) PC columns. In the solvent gradient conditions commonly employed for the LC separation of petides, RP-PC columns have to be kept on-line during the entire gradient in order to desorb all the retained compounds to the analytical column. As desorption in ion exchange-PC columns can be achieved independently of the proportion of organic solvent in the LC eluent, the PC column can be switched off-line before the analytical gradient is started. In addition, any dilution due to band-broadening at the on-line PC column desorption step can be eliminated by pre-concentration of the eluting peptides on top of the analytical column using a solvent of low eluotropic strength. An important advantage of ion exchange PC is that it introduces different selectivity into the pre-concentration method.

Briefly, the method used a pre-concentration solvent that consisted of acetonitrile: 10 mmol/l ammonium acetate, pH 7.0 (10:90, v/v). A Waters M-6000 (Milford, MA) pump delivered the pre-concentration solvent at a flow-rate of 0.2 ml/min. An injection coil of 200 μl was used to inject standards and biological samples onto the PC column. The column was washed for 5 min with pre-concentration solvent. By switching the pre-column on-line with the analytical column, the compounds were desorbed onto the analytical column. During this step the flow-rate of the LC pump was increased to 0.7 μl/min to reduce the time needed for peptide desorption. This flow resulted in a breakthrough time of 10 min for endothelins. To ensure quantitative desorption, the pre-column was switched back after 15 min. Immediately, the gradient program was started. The chromatographic time events are shown in Table 4.4.

To reduce the overall analysis time further, the sample was pre-concentrated during the analysis of the previous sample.

Using standard injections, the recovery of the on-line PC-nanoLC-MS was between 75 and 90% for the three endothelins (50 fmol injected). The theoretical limits of detection were 0.5 fmol (2.5 pmol/l, 200 μl injection; signal-to-noise = 3).

Initially, a high memory effect was observed (>5%). This appeared because the analytes stuck to the connection capillaries of the pre-concentration part of the system. To eliminate these compounds, the injection port was cleaned with acetonitrile:water:acetic acid:TFA (50:50:1:0.5, v/v), which was also injected into the pre-concentration part of the system. This clean-up injection reduced the memory effect to below 0.1%.

TABLE 4.4

TIME EVENT OF THE SAX-PC-NANOLC-μESI MS METHOD. FLOW-RATE OF PC PUMP IS CONSTANT AT 200 μL/MIN

Time (min)	Concentration of Solvent B (%)	Flow-rate (ml/min) of LC pump	Column-switching valve
0	10	0.3	LC pump to capLC
5	10	0.4	LC pump to PC
15	10	0.4	
15.1	20	0.3	
20		0.3	LC pump to capLC
25	60	0.3	
25.1	80	0.3	
30	80	0.3	
30.1	10	0.3	

The SAX-based PC method showed, however, only minor retention of BKs (breakthrough volume <0.2 ml) when using a PC mobile phase of 5 mM ammonium acetate, pH 7.0. This lack of retention could be explained by the presence of only one negative group in the BK peptide, which is the terminal –COO– group, and of three positive groups: the amino terminal and two basic arginine residues. To circumvent these problems, a method using a C18 pre-column (Hypersil C18) was tested for these compounds.

BK and specially the more polar BK fragments showed a relatively weak interaction with the C18 material. For this reason, special care was taken in the selection of PC and elution solvents. The analytical column should retain the analytes stronger than the PC column. Thus, when desorbed from the PC column by the mobile phase, the target analytes will be concentrated on top of the analytical column, thereby reducing band-broadening during desorption. We tested a PC solvent that consisted of 5% acetonitrile and 0.05% TFA. For this mobile phase composition, the breakthrough volume for BK1-7 was 2 ml. BK, BK1-8 and Leu-BK1-8 were still retained after passing this volume of eluent. Back-flush isocratic desorption towards the capillary column with a high organic modifier content (60% acetonitrile) resulted in peak widths of less than 0.5 min. Elution with the analytical mobile phase used for BK separation showed similar peaks as for direct injection, indicating that desorption was effective and no excessive band-broadening in the connection between the PC and the capillary column occurred.

It was observed that the use of TFA as LC additive was responsible for important losses in ESI sensitivity due to the charge-neutralizing effect of ion pairing. For this reason, TFA was eliminated from the mobile phase of the analytical separation and replaced with acetic acid (1% v/v). This change afforded good sensitivity, although solvent containing TFA had better chromatographic performance and BK1-3 showed no retention under these conditions.

For PC, TFA was maintained but its concentration in the PC mobile phase dropped to 0.01%. TFA could not be totally eliminated from the PC mobile phase, since it was required for BK retention on the PC column. Acetic acid was used to acidify the PC mobile phase and to protect the compounds from binding to the connecting capillaries. In addition, a small percentage of methanol was used to eliminate polar matrix components. The optimized PC mobile phase consisted of 3.5% methanol, 0.5% acetic acid and 0.01% TFA. Under these PC conditions, BK1-7 started to break through at approximately 1.6 ml, whereas Leu-BK1-8 and BK1-8 had higher breakthrough volumes (>2.5 ml, $n = 3$). Unfortunately, no retention was observed for the metabolites BK1-3 and BK1-5 [16].

The PC was switched on-line to the analytical column on exceeding 0.8 ml PC mobile phase (4 min).

4.3.7. Matrix effects and loading capacity of the PC-capLC system

Sample treatment was reduced in these procedures to protein precipitation. Protein precipitation can be performed by acidifying the sample or by adding an organic solvent, e.g. acetonitrile. TFA is one of the acids that can be used for acidifying biological samples and is directly compatible with μPC-microLC-MS. We noticed, however, that TFA precipitation gave very 'dirty' samples, probably due to the ion-pair effect of TFA. We did not try other acids such as perchloric acid, since this acid is not compatible with MS. For these reasons, precipitation by acetonitrile addition was the method finally selected. One of the advantages of using acetonitrile is that the dilution caused by its addition to the sample can be reduced by evaporation.

The direct injection of a precipitated human plasma sample spiked with BKs on a PC-capLC-UV system caused peak heights that were approximately 3–5 times lower than those caused by a standard solution (no matrix). Variation of the injected sample volume revealed that, with up to 25 μl reconstituted plasma, PC recovery was around 70–80% for BK, Leu-BK1-8 and BK1-8 (Table 4.5). By increasing the sample volume to 200 μl, recovery dropped constantly towards 30% (Fig. 4.7). Repetitive sample injection rapidly reduced the BK signals to below 1% of the original signal. The recovery of acetonitrile extracts from whole blood was in the same range of plasma. Nor did the system allow repetitive injection of more than five samples. This pointed to a saturation of the PC and/or analytical column with highly apolar matrix components. The signal could be recovered by employment of a wash procedure after every plasma sample (200 μl acetonitrile were injected onto the PC column). These results indicate that saturation mainly occurred at the PC column and could be reversed easily. The sample amount injected onto the PC column was consequently reduced to 25 μl in an injection volume of 100 μl (100 μl of human or rat plasma was used for precipitation and 25 μl of the reconstituted plasma was diluted 4-fold to facilitate manual handling). To avoid gradual saturation of the analytical column, a gradient of acetonitrile/methanol in water after every run was used. The resulting analytical method was stable for at least 100 injections. The PC-microLC/UV system had detection limits of 20 nmol/l BKs and was linear from 50 to 2000 nmol/l BKs in plasma ($r^2 > 0.99$, $n = 3$).

Similar matrix effects were observed in the analysis of ET from HUVEC. When using concentrated extracts (re-solution of the extract in the same volume of initial HUVEC), lower peak heights than those expected theoretically were obtained. A subsequent four-fold dilution (the equivalent of 50 μl HUVEC supernatant in a 200 μl injection) resulted in

TABLE 4.5

RECOVERY AND REPEATABILITY DATA FOR ACETONITRILE PRECIPITATION AND PC-MICROLC-UV OF 500 NM BKS

Plasma by LC-UV	Recovery (in percentage, $n = 5$)			Repeatability (in percentage, $n = 6$)
	Acetonitrile precipitation[a]	Matrix effect on PC[b]	Total[c]	
Bradykinin	75	70	53	3.3
LeuBK1-8	86	75	65	5.0
BK1-8	86	87	75	3.6

[a]Comparison of peak areas of plasma samples that are spiked before and after acetonitrile precipitation.

[b]Comparison of peak areas of plasma samples spiked after acetonitrile precipitation and matrix-free injected standards.

[c]Comparison of peak areas of plasma samples spiked before acetonitrile precipitation and directly injected standards.

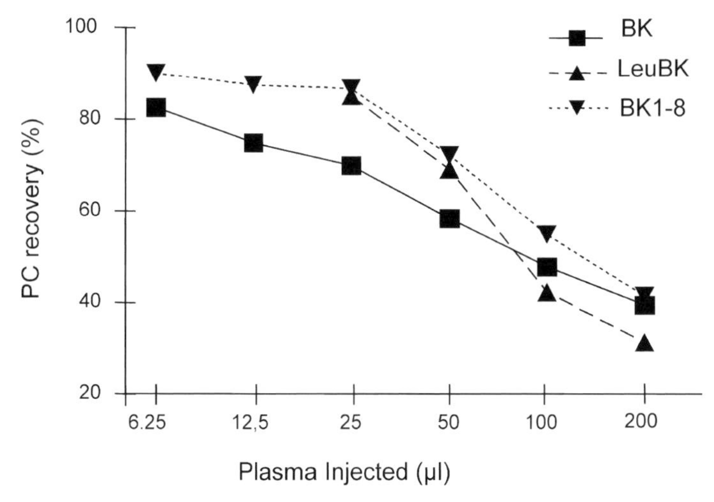

Fig. 4.7. Effect of the total volume of plasma injected on BK recovery by PC-microLC/UV. Recovery was calculated relative to the UV peak area obtained from the same amount of BK injected in PC mobile phase (no plasma present).

better recoveries. Additionally, an extensive wash procedure was employed after every sample. Methanol, acetonitrile:water:acetic acid (50:50:1, v/v) and acetonitrile were consecutively injected into the analytical system. To prevent a similar process on the analytical column, a gradient of acetonitrile/methanol in water was programmed. With these changes, the analytical method could be used for more than two weeks without noticeable loss in sensitivity.

Matrix interference also needs to be taken into account if the appropriate analytical column is to be selected. Maximum sensitivity for endothelin analysis in HUVEC supernatants was achieved by using 75 µm columns at 250 nl/min. Comparison showed there was no advantage in using 75 µm columns instead of 180 µm columns when analyzing bradykinins in plasma. It even appeared that in this configuration the 75 µm column showed increased band-broadening, probably due to column overloading of the smaller column with plasma extracts, more concentrated in matrix components than HUVEC supernatants.

It should be noted that the presence of some matrix in the sample could, in some cases, have beneficial effects on recovery. ET standard disappeared rapidly from its solutions when these compounds were at femtomole levels and in the absence of matrix or other additives. Peptide losses were evident as quickly as 10 s after preparation of standard solutions. These losses were probably due to non-specific binding of the peptide to the wall of the sample tube. This problem was solved after addition of 0.1–0.2 mg/ml bovine serum albumin to the solutions of ETs or BKs.

4.3.8. Roughness and stability of the µESI-MS interface

The injection of reconstituted plasma samples (1/4 dilution, 100 µl injection, see above) into the PC-capLC-ESI MS/MS method broke down BK signals fast. Spray

solutions of endothelins directly analyzed by capLC-MS/MS (no PC) was 0.5 fmol (signal-to-noise = 3 for a 2.5 μl injection volume).

The method was linear from 50 pmol/l to 2.0 nmol/l for ET-1 and ET-2 (200 μl injection, $r^2 = 0.99$). Accuracy and reproducibility data for both peptides are given in Table 4.6. The ET-1 concentration in HUVEC was determined as 0.2 ± 0.03 nmol/l. ET-3, however, showed less stability than the other compounds ($r^2 = 0.96$), which is probably due to the relative difference between the retention time of the peak and the IS. The method was stable for more than 100 injections.

Ashby et al. [166] obtained detection limits of 6 pmol/l, which is five times better than this method. However, absolute detection limits are 330-fold higher. Ashby et al. used 500 ml HUVEC supernatant, whereas the present method only needs 500 μl, of which 50 μl are in fact injected.

In the case of BKs in aqueous samples, the limits of detection were around 1 fmol, which corresponds to 5 pmol/l (200 μl injections, signal-to-noise = 3). Linearity for standards was from 5 to 2000 fmol (25–10 000 pmol/l, 200 μl injections).

In the analysis of plasma samples (Fig. 4.10), reproducible peaks were obtained for BK and BK1-8, with a variation in retention times of less than 5% ($n = 8$). However, BK1-7 had less stability, probably as a result of the initial breakthrough at the PC column. The determination limits in plasma were 50 pmol/l for BK and BK1-8. The method was linear from 100 to 2000 pmol/l ($r^2 > 0.99$, $n = 5$).

4.4. RESUME

With on-line pre-concentration for trace enrichment and an internal standard to correct for spray instability, peptides in biological samples can be quantitatively analyzed in the femtomole range. The use of different methods for pre-concentration (ion exchange) and analysis (reversed-phase) improves selectivity in the pre-concentration of peptides from complex matrices. This, in addition to the selectivity obtained from the high sensitivity full-scan MS/MS analysis available on the ion trap mass spectrometer, provides the analyst with an invaluable tool for peptide bioanalysis.

In the case of endothelins, the injection of 200 μl of diluted extracts allowed LODs down to 30 pmol/l in HUVEC cultures. These values were similar to those in immunoassays.

TABLE 4.6

PRECISION AND ACCURACY DATA OF ET-1 AND ET-2 ($N = 3$)

	Concentration (nmol/l)	Precision (%)	Accuracy (%)
ET-1	0.1	17	105.5
	0.5	13.4	89.2
	2	10	98.3
ET-2	0.1	13.2	96.4
	0.5	26	86.6
	2	6.5	97.5

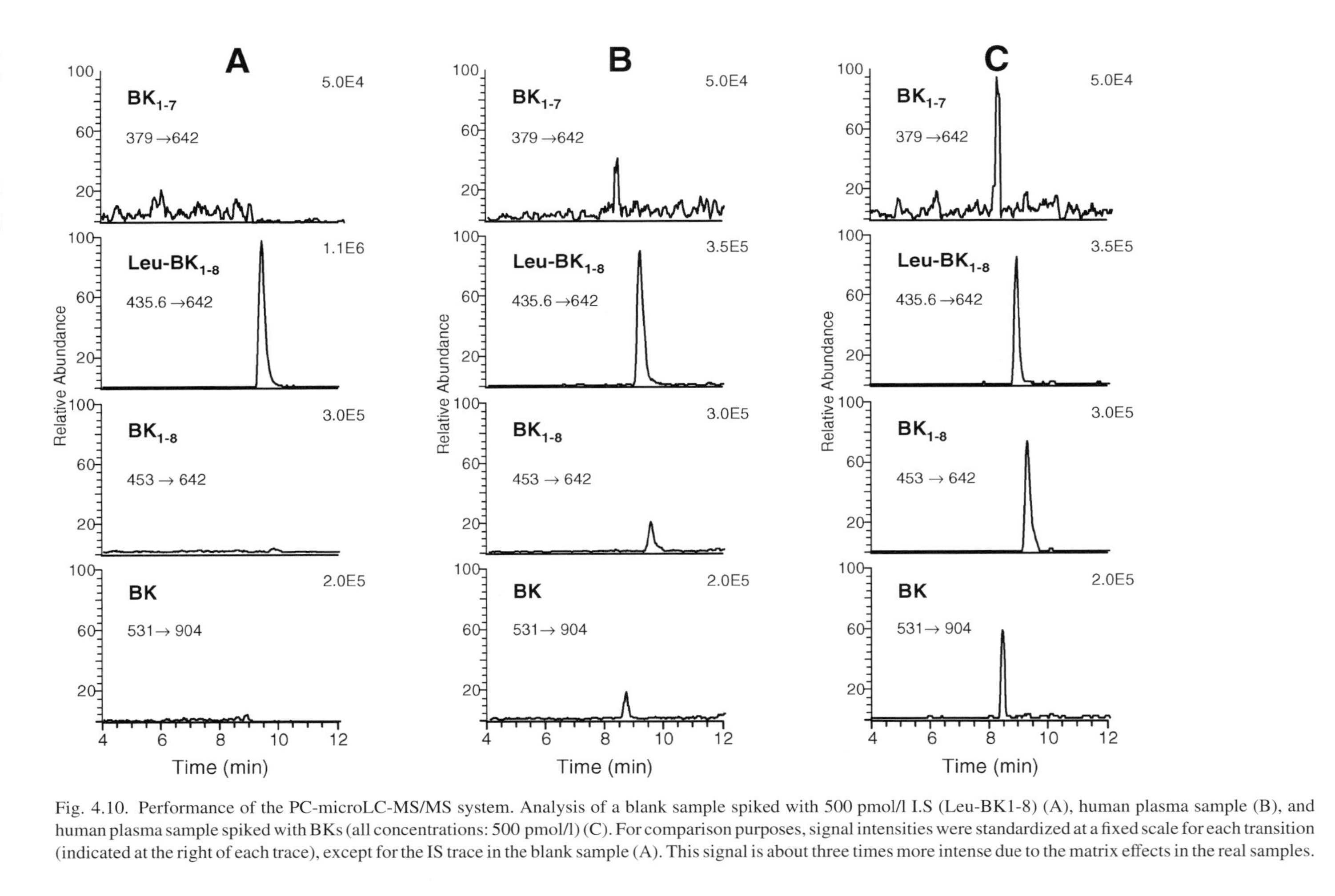

Fig. 4.10. Performance of the PC-microLC-MS/MS system. Analysis of a blank sample spiked with 500 pmol/l I.S (Leu-BK1-8) (A), human plasma sample (B), and human plasma sample spiked with BKs (all concentrations: 500 pmol/l) (C). For comparison purposes, signal intensities were standardized at a fixed scale for each transition (indicated at the right of each trace), except for the IS trace in the blank sample (A). This signal is about three times more intense due to the matrix effects in the real samples.

4.5. ACKNOWLEDGEMENTS

This work was supported by grants SAF2000-0131-C01-02 (CICYT) and PB98-1180 (DGES).

4.6. REFERENCES

1. B.L. Ferraiolo and M.A. Mohler, in: B.L. Ferraolo (Ed.), Protein Pharmacokinetics and Metabolism, Plenum Press, New York, 1992, Chapter 1
2. E. Gelpi, Trends Anal. Chem., 4 (1985) 3
3. E. Gelpi, I. Ramis, G. Hotter, G. Bioque, O. Bulbena and J. Roselló, J. Chromatogr., 492 (1990) 223
4. K.J. Miller and A.C. Herman, Anal. Chem., 68 (1996) 3077
5. A.J. Oosterkamp, H. Irth, L. Heintz, G. Marko-Varga, U. Tjaden and J. Van der Greef, Anal. Chem., 68 (1996) 4101
6. F.E. Karet and A.P. Davenport, J. Cardiovasc. Pharmacol., 22 (1993) S29
7. E.H. van den Burg, J.R. Metz, R.J. Arends, B. Devreese, I. Vandenberghe, J. Van Beeumen, S.E.W. Bonga and G. Flik, J. Endocrinol., 169 (2001) 271
8. X. Liang, D. Lubman, D.T. Rossi, G.D. Nordblom and C.M. Barksdale, Anal. Chem., 70 (1998) 498
9. R. Wang, D. Sweeney, S.E. Gandy and S.S. Sisodia, J. Biol. Chem., 271 (1996) 31894
10. D.S. Stegehuis, U.R. Tjaden, C.M.B. van den Beld and J. Van der Greef, J. Chromatogr., 549 (1991) 185
11. J.J. Kusmierz, R. Sumrada and D.M. Desiderio, Anal. Chem., 62 (1990) 2395
12. C.D. Marquez, M.-L. Lee, S.T. Weintraub and P.C. Smith, J. Chromatogr. B, 700 (1997) 9
13. O.O. Grigorians, J.-L. Tseng, R.R. Becklin and D.M. Desiderio, J. Chromatogr., 695 (1997) 287
14. D.C. Muddiman, A.I. Gusev, A. Proctor, D.M. Hercules, R. Venkataramanan and W. Diven, Anal. Chem., 66 (1994) 2362
15. M. Carrascal, K. Schneider, R.E. Calaf, S. van Leeuwen, D. Canosa, E. Gelpi and J. Abian, J. Pharm. Biomed. Anal., 17 (1998) 1129
16. J. Crowther, V. Adusumalli, T. Mukherjee, K. Jordan, P. Abuaf, N. Corkum, G. Goldstein and J. Tolan, Anal. Chem., 66 (1994) 2356
17. S.M. Darby, M.L. Miller, R.O. Ollen and M. LeBeau, J. Anal. Toxicol., 25 (2001) 8
18. C. Fierens, L.M.R. Thienpont, D. Stockl, E. Willekens and A.P. De Leenheer, J. Chromatogr., 896 (2000) 275
19. C.D. Marquez, S.T. Weintraub and P.C. Smith, J. Chromatogr. B, 694 (1997) 21
20. L.J. Murphey, D.L. Hachey, D.E. Vaughan, N.J. Brown and J.D. Morrow, Anal. Biochem., 292 (2001) 87
21. S. Bobin, M.A. Popot, Y. Bonnaire and J.C. Tabet, Analyst, 126 (2001) 1996
22. G.I. Kirchner, Ch. Vidal, W. Jacobsen, A. Franzke, K. Hallensleben, U. Christians and K.-F. Sewing, J. Chromatogr. B, 721 (1999) 285
23. M. Niwa, K. Enomoto and K. Yamashita, J. Chromatogr. B, 729 (1999) 245
24. N. Brignol, L.M. McMahon, S. Luo and F.L.S. Tse, Rapid Commun. Mass Spectrom., 15 (2001) 898
25. F. Magni, S. Pereira, M. Leoni, G. Grisenti and M. Galli Kienle, J. Mass Spectrom., 36 (2001) 670
26. P. Suder, J. Kotlinska, A. Legowska, M. Smoluch, G. Hohne, J.-P. Chervet, K. Rolka and J. Silberring, Brain Res. Protocols, 6 (2000) 40
27. K. Yamaguchi, M. Takashima, T. Uchimura and S. Kobayashi, Biomed. Chromatogr., 14 (2000) 77
28. J.E. Battersby, V.R. Mukku, R.G. Clark and W.S. Hancock, Anal. Chem., 67 (1995) 447
29. W. Potts, R. Van Horn, K. Anderson, T. Blake, E. Garver, G. Joseph, G. Dreyer, A. Shu, R. Heys and K.-L. Fong, Drug. Metab. Dispos., 23 (1995) 799
30. B. Kaye, M.W.H. Clark, N.J. Cussan, P.V. Macrea and D. Stopher, Biol. Mass Spectrom., 21 (1992) 585
31. H. Fouda, M. Nocerini, R. Schneider and C. Gedutis, J. Am. Soc. Mass Spectrom., 2 (1991) 164
32. J. Abian, A.J. Oosterkamp and E. Gelpi, J. Mass Spectrom., 34 (1999) 244
33. Ch. Dass, J.J. Kusmierz, D.M. Desiderio, S.A. Jarvis and B.N. Green, J. Am. Soc. Mass Spectrom., 2 (1991) 149

34. A.D. Kippen, F. Cerini, L. Vadas, R. Stocklin, L. Vu, R.E. Offord and K. Rose, J. Biol. Chem., 272 (1997) 12513
35. M.S. Wilm and M. Mann, Int. J. Mass Spectrom. Ion Processes, 136 (1994) 167
36. M.S. Wilm, G. Neubauer and M. Mann, Anal. Chem., 68 (1996) 527
37. M.S. Wilm and M. Mann, Anal. Chem., 68 (1996) 1
38. M.R. Emmett and R.M. Caprioli, J. Am. Soc. Mass Spectrom., 5 (1994) 605
39. G.A. Valaskovic, H.D. Keheller, P. Little, D.J. Aaserud and F.W. McLafferty, Anal. Chem., 67 (1995) 3802
40. A. Shevchenko, O.L. Jensen, A.V. Podtelejnikov, F. Sgliocco, M. Wilm, O. Vorm, P. Mortensen, A. Shevchenko, H. Boucherie and M. Mann, Proc. Natl. Acad. Sci. USA, 93 (1996) 14440
41. G. Neubauer, A. Gottschalk, P. Fabrizio, B. Seraphin, R. Lurhrmann and M. Mann, Proc. Natl. Acad. Sci. USA, 94 (1997) 385
42. L.M. Zugaro, G.E. Reid, H. Ji, J.S. Eddes, A.C. Murphy, A.W. Burgess and R.J. Simpson, Electrophoresis, 19 (1998) 867
43. C.L. Gatlin, J.K. Eng, S.T. Cross, J.C. Detter and J.R. Yates III, Anal. Chem., 72 (2000) 757
44. J.F. Kelly, L. Ramaley and R. Thibault, Anal. Chem., 69 (1997) 51
45. P. Cao and M. Moini, J. Am. Soc. Mass Spectrom., 8 (1997) 561
46. M. Mazereeuw, A.J.P. Hofte, U.R. Tjaden and J. van der Greef, Rapid Commun. Mass Spectrom., 11 (1997) 981
47. R.H. Robins and J.E. Guido, Rapid Commun. Mass Spectrom., 11 (1997) 1661
48. D. Figeys, A. Ducret, J.R. Yates III and R. Aebershold, Nature Biotech., 14 (1996) 1579
49. K. Vanhoutte, I. Hoes, F. Lemiere, W. van Dongen, E.L. Esmans, H. van Onckelen, E. van den Eeckhout, R.E.M. van Soest and A.J. Hudon, Anal. Chem., 69 (1997) 3161
50. P.E. Andren and R.M. Caprioli, J. Mass Spectrom., 30 (1995) 817
51. C.L. Gatlin, G.R. Kleemann, L.G. Hays, A.J. Link and J.R. Yates III, Anal. Biochem., 263 (1998) 93
52. M. Zell, C. Husser and G. Hopfgartner, Rapid Commun. Mass Spectrom., 11 (1997) 1107
53. J. Cai and J. Hennion, Anal. Chem., 68 (1996) 72
54. J.P. Chervet, M. Ursem and J. Salzmann, Anal. Chem., 68 (1996) 1507
55. K. Vanhoutte, W. van Dongen and E.L. Esmans, Rapid Commun. Mass Spectrom., 12 (1998) 15
56. A.J. Oosterkamp, E. Gelpi and J. Abian, J. Mass Spectrom., 33 (1998) 976
57. K.B. Tomer, M.A. Moseley, L.J. Deterding and C.E. Parker, Mass Spectrom. Rev., 13 (1994) 431
58. J.P.C. Vissers, H.A. Claessens and C.A. Cramers, J. Chromatogr. A., 779 (1997) 1
59. I.S. Krull, A., Sebag and R. Stevenson, J. Chromatogr. A., 887 (2000) 137
60. C.G. Horvarth, B.A. Preiss and S.R. Lipski, Anal. Chem., 39 (1967) 1422
61. R.P.W. Scott and P. Kucera, J. Chromatogr., 169 (1979) 51
62. Y. Hirata and M. Novotny, J. Chromatogr., 186 (1979) 521
63. D. Ishii, K. Asai, K. Hibi, T. Jonokuchi and M. Nagaya, J. Liq. Chromatogr., 144 (1977) 157
64. D. Ishii, T. Tsuda and T. Takeuchi, J. Chromatogr., 185 (1979) 73
65. R.T. Kennedy and J.W. Jorgenson, Anal. Chem., 61 (1989) 436
66. P.R. Griffin, L.E. Hood and J.R. Yates, The 39th ASMS Conference on Mass Spectrometry and Allied Topics, p. 1157
67. L.N. Tyrefors, R.X. Moulder and K.E. Markides, Anal. Chem., 65 (1993) 2835
68. A. Basey and R.W.A. Oliver, J. Chromatogr., 251 (1982) 265
69. R.T. Kennedy and J.W. Jorgenson, Anal. Chem., 61 (1989) 1128
70. S. Hsieh and J.W. Jorgenson, Anal. Chem., 68 (1996) 1212
71. V. Spikmans, S.J. Lane, U.R. Tjaden and J. van der Greef, Rapid Commun. Mass Spectrom., 13 (1999) 141
72. M. Gucek, M. Gaspari, K. Walhagen, R.J. Vreeken, E.R. Verheij and J. van der Greef, Rapid Commun. Mass Spectrom., 14 (2000) 1448
73. L.A. Colon, T.D. Maloney and A.M. Fermier, J. Chromatogr. A., 887 (2000) 43
74. P.D.A. Angus, C.W. Demarest, T. Catalano and J.F. Stobaugh, J. Chromatogr. A, 887 (2000) 347
75. U. Pyell, J. Chromatogr. A., 892 (2000) 257
76. V.L. Tal'roze, V.E. Skurat and G.V. Karpov, J. Phys. Chem. (Moscow), 43 (1969) 241
77. R.P.W. Scott, C.G. Scott, M. Munroe and J. Hess, J. Chromatogr., 99 (1974) 394
78. W.H. McFadden, H.L. Schwartz and S. Evans, J. Chromatogr., 122 (1976) 389

79. M.A. Baldwin and F.W. McLafferty, Org. Mass Spectrom., 7 (1973) 1111
80. P.J. Arpino, M.A. Baldwin and F.W. McLafferty, Biomed. Mass Spectrom., 1 (1974) 80
81. R.M. Caprioli, T. Fan and J.S. Cotrell, Anal. Chem., 58 (1986) 2949
82. A.C. Barefoot and R.W. Reiser, J. Chromatogr., 398 (1987) 217
83. H. Alborn and G. Stenhagen, J. Chromatogr., 47 (1985) 323
84. H. Alborn and G. Stenhagen, J. Chromatogr., 35 (1987) 394
85. J.E. Coutant, T.-M. Chen and B.L. Ackermann, J. Chromatogr., 529 (1990) 265
86. M.A. Moseley, L.J. Deterding, K.B. Tomer and J.W. Jorgenson, Anal. Chem., 63 (1991) 1467
87. M.L. Vestal, Mass Spectrom. Rev., 2 (1983) 447
88. P.C. Winkler, D.D. Perkins, D.K. Williams and R.F. Browner, Anal. Chem., 60 (1988) 489
89. J. Abian, A. Stone, M.G. Morrow, M.H. Creer, L.M. Fink and J.O. Lay Jr., Rapid Commun. Mass Spectrom., 6 (1992) 684
90. A.P. Bruins, Mass Spectrom. Rev., 10 (1991) 53
91. W.M.A. Niessen, J. Chromatogr., 794 (1998) 407
92. M. Yamashita and J.B. Fenn, J. Phys. Chem., 88 (1984) 4451
93. M.L. Aleksandrov, G.I. Barama, L.M. Gall, N.V. Krasnov, Y.S. Kusner, O.A. Mirgorodskaya, V.I. Nikolaiev and V.A. Shkurov, Bioorg. Khim., 11 (1985) 700
94. E.C. Horning, M.G. Horning, D.I. Carroll, I. Dzidic and R.N. Stillwell, Anal. Chem., 45 (1973) 936
95. T.R. Covey, E.D. Lee, A.P. Bruins and H.D. Henion, Anal. Chem., 58 (1986) 1451A
96. M. Sakairi and H. Kambara, Anal. Chem., 60 (1988) 774
97. C.R. Whitehouse, R.N. Dreyer, M. Yamashita and J.B. Fenn, Anal. Chem., 57 (1985) 675
98. J.B. Fenn, M. Mann, C.K. Meng, S.F. Wong and C.M. Whitehouse, Science, 246 (1989) 64
99. A.P. Bruins, T.R. Covey and J.D. Henion, Anal. Chem., 59 (1987) 2642
100. S. Shen, C. Whitehouse, F. Banks and J.B. Fenn, Proceedings of the 40th ASMS Conference on Mass Spectrometry and Allied Topics, Washington, DC. May 31, 1992
101. R.D. Smith, C.J. Barinaga and H.R. Udseth, Anal. Chem., 60 (1988) 1948
102. E.D. Lee, W. Muck, J.D. Henion and T.R. Covey, J. Chromatogr., 458 (1988) 313
103. C. Stenfors, U. Hellman and J. Silbering, J. Biol. Chem., 272 (1997) 5747
104. J.C. Bett, W.P. Blackstock, M.A. Ward and B.H. Anderton, J. Biol. Chem., 272 (1997) 12922
105. D.F. Hunt, R.A. Henderson, J. Shabanowitz, K. Sakaguchi, H. Michel, N. Sevilir, A.L. Cox, E. Appella and V.H. Engelhard, Science, 255 (1992) 1261
106. R.A. Henderson, H. Michel, K. Sakaguchi, J. Shabanowitz, E. Appella, D.F. Hunt and V.H. Engelhard, Science, 255 (1992) 1264
107. B. Devreese, F. Vanrobaeys and J.V. Beeumen, Rapid Commun. Mass Spectrom., 15 (2001) 50
108. M. Carrascal, S. Carujo, O. Bachs and J. Abian, Proteomics, 2 (2002) 455–468
109. M.P. Washburn, D. Wolters and J.R. Yates III, Nature Biotech., 19 (2001) 242
110. G. Hopfgartner, T. Wachs, K. Bean and J. Henion, Anal. Chem., 65 (1993) 439
111. E.C. Huang and J.D. Henion, Anal. Chem., 63 (1991) 732
112. M.A. Moseley and S.E. Unger, J. Microcol. Sep, 4 (1992) 393
113. J.R. Perkins, C.E. Parker and K.B. Tomer, J. Am. Soc. Mass Spectrom., 3 (1992) 139
114. C.E. Parker, J.R. Perkins, K.B. Tomer, Y. Shida, K. O'Hara and M. Kono, J. Am. Soc. Mass Spectrom., 3 (1992) 563
115. J.H. Purnell, Gas Chromatography, John Wiley and Sons, New York, 1962
116. J.F. Banks, J. Chromatogr. A., 743 (1996) 99
117. G. Hopfgartner, K. Bean, J. Henion and R. Henry, J. Chromatogr., 647 (1993) 51
118. R.D. Smith, J.A. Loo, Ch.G. Edmonds, Ch.J. Barinaga and H. Udseth, Anal. Chem., 62 (1990) 882
119. P. Kebarle and L. Tang, Anal. Chem., 65 (1993) 972A
120. P.E. Andren, M.R. Emmett and R.M. Caprioli, J. Am. Soc. Mass Spectrom., 5 (1994) 867
121. N.J. Clarke, F.W. Crow, S. Younkin and S. Naylor, Anal. Biochem., 298 (2001) 32
122. J.P.C. Vissers, J.-P. Chervet and J.-P. Salzmann, J. Mass Spectrom., 31 (1996) 1021
123. L.A. Holland and J.W. Jorgenson, J. Microcol. Sep., 12 (2000) 371
124. R.B. White, T.D. Oglesby and P.A. Taylor, The 43rd ASMS Conference on Mass Spectrometry and Allied Topics, p. 276

125. N.J. Thatcher and S. Murray, Biomed. Chromatogr., 15 (2001) 374
126. A. Oosterkamp, M. Carrascal, D. Closa, G. Escolar, E. Gelpi and J. Abian, J. Microcol. Sep., 13 (2001) 265
127. M.T. Davis, D.C. Stahl and T.D. Lee, J. Am. Soc. Mass Spectrom., 6 (1995) 571 M.T. Davis, D.C. Stahl, S.A. Hafta and T.D. Lee, Anal. Chem., 67 (1995) 4549
128. M.T. Davis and T.D. Lee, J. Am. Soc. Mass Spectrom., 8 (1997) 1059
129. M.T. Davis and T.D. Lee, J. Am. Soc. Mass Spectrom., 9 (1998) 194
130. M.T. Davis and T.D. Lee, Protein Sci., 1 (1992) 935
131. R. Grimm, M. Serwe and J.-P. Chervet, LC/GC, 15 (1997) 960
132. M. Carrascal and J. Abian, in M-I Aguilar (ed), HPLC of Peptides and Proteins. Methods and Protocols. Method in Molecular Biology Series, Humana Press, NJ, USA, 2003 (In Press)
133. F.J. Yang, J. Chromatogr., 236 (1982) 265
134. V.L. McGuffin and M. Novotny, J. Chromatogr., 255 (1983) 381
135. K.E. Karlsson and M. Novotny, Anal. Chem., 60 (1988) 1662
136. R.T. Kennedy and J.W. Jorgenson, J. Microcol. Sep., 2 (1990) 120
137. T. Takeuchi and D. Ishii, J. Chromatogr., 238 (1982) 409
138. K.P. Bateman, R.L. White and P. Thibault, J. Mass Spectrom., 33 (1998) 1109
139. A. Cappiello, P. Palma and F. Mangani, Chromatographia, 32 (1991) 389
140. D. Roulin, R. Smoch, R. Carney, K.D. Bartlee, P. Myers, M.R. Euerby and C. Johnson, J. Chromatogr. A., 887 (2000) 307
141. T. Takeuchi and D. Ishii, J. Chromatogr., 21 (1981) 25
142. J.C. Gluckman, A. Hirose, V.L. McGuffin and M. Novotny, Chromatographia, 17 (1983) 303
143. S. Hoffman and L. Blomberg, Chromatographia, 24 (1987) 416
144. G. Crescentini, F. Bruner, F. Mangani and G. Yafeng, Anal. Chem., 60 (1988) 1659
145. J.-T. Wu, P. Huang, M.X. Li and D.M. Lubman, Anal. Chem., 69 (1997) 2908
146. T.M. Zimina, R.M. Smith, P. Myers and B.W. King, Chromatographia, 40 (1995) 662
147. A. Malik, W. Li and M.L. Lee, J. Microcol. Sep., 5 (1993) 361
148. D. Tong, K.D. Bartle, A.A. Clifford and A.M. Edge, J. Microcol. Sep., 7(3) (1995) 265
149. Y. Guan, L. Zhou and Z. Shang, J. High Resolut. Chromatogr., 15 (1992) 434
150. J.E. MacNair, K.D. Patel and J.W. Jorgenson, Anal. Chem., 71 (1999) 700
151. N. Wu, D.C. Collins, J.A. Lippert, Y. Xiang and M.L. Lee, J. Microcol. Sep., 12 (2000) 462
152. J.E. MacNair, G.J. Opitek, J.W. Jorgenson and M.A. Moseley III, Rapid Commun. Mass Spectrom., 11 (1997) 1279
153. G.S. Chirica and V. Remcho, Anal. Chem., 72 (2000) 3605
154. I. Gusev, X. Huang and C. Horváth, J. Chromatogr. A., 855 (1999) 273
155. J.D. Hayes and A. Malik, Anal. Chem., 72 (2000) 4090
156. S.K. Chowdhury and B.T. Chait, Anal. Chem., 63 (1991) 1660
157. J.C. Hannis and D.C. Muddiman, Rapid Commun. Mass Spectrom., 12 (1998) 443
158. M.S. Kriger, K.D. Cook and R.S. Ramsey, Anal. Chem., 67 (1995) 385
159. J.-T. Wu, M.G. Qian, M.X. Li, L. Liu and D.M. Lubman, Anal. Chem., 68 (1996) 3388
160. L. Fang, R. Zhang, E.R. Williams and R.D. Zare, Anal. Chem., 66 (1994) 3696
161. K. Schmeer, B. Behnke and E. Bayer, Anal. Chem., 67 (1995) 3656
162. S.E.G. Dekkers, U.R. Tjaden and J. van der Greef, J. Chromatogr., 712 (1995) 201
163. J.-T. Wu, P. Huang, M.X. Li, M.G. Qian and D.M. Lubman, Anal. Chem., 69 (1997) 320
164. G.M. Rubanyi and M.A. Polokoff, J. Biol. Chem., 46 (1994) 325
165. K. Schneider, S. Van Leeuwen, C. Peralta, J. Roselló-Catafau, E. Gelpi and J. Abián, 44th ASMS Conference, Portland, Oregon, 1996
166. M.J. Ashby, C. Plumpton, P. Teale, R.E. Kuc, E. Houghton and A. Davenport, J. Cardiovasc. Pharmacol., 26 (1995) S247
167. T. Eimer, K.K. Unger and J. van der Greef, Trends Anal. Chem., 15 (1996) 463

G.A. Marko-Varga and P.L. Oroszlan (Editors)
Emerging Technologies in Protein and Genomic Material Analysis
Journal of Chromatography Library, Vol. 68

CHAPTER 5

On-line Continuous-flow Multi-protein Biochemical Assays for the Characterization of Bio-active Compounds Using Electrospray Quadrupole Time-of-Flight Mass Spectrometry

R.J.E. DERKS, A.C. HOGENBOOM, G. VAN DER ZWAN and H. IRTH*

Department of Analytical Chemistry and Applied Spectroscopy, Division of Chemistry, Faculty of Sciences, vrije Universiteit amsterdam, De Boelelaan 1083, 1081 HV Amsterdam, The Netherlands

Abstract

The applicability of homogeneous on-line continuous-flow, multi-protein biochemical assays was demonstrated for the interaction between fluorescein-biotin and streptavidin and for digoxin and anti-digoxigenin using electrospray quadrupole time-of-flight mass spectrometry (Q-TOF MS). In the on-line continuous-flow biochemical MS-based system, several receptors (e.g. streptavidin and anti-digoxigenin, respectively) were allowed to react with corresponding reporter ligands (e.g. fluorescein-biotin and digoxin, respectively). The methodology presented allows the simultaneous measurement of affinity and molecular mass of an active compound. By using automated MS and MS-MS switching functions of the Q-TOF structure, information is obtained allowing the characterization of bioactive compounds. No cross-reactivities were observed between the two model systems fluorescein-biotin/streptavidin and digoxin/anti-digoxigenin.

5.1. INTRODUCTION

The measurement of biochemical interactions is becoming increasingly important in various areas of analytical chemistry. The detection of interactions of small or large molecules with biomolecular targets of interest generates information on the potential biological activity of these molecules. Such measurements are highly relevant, for

*Corresponding author.

example, in drug discovery, and also in areas such as environmental toxicology and monitoring. The characterization of the interaction of ligands (analytes) with biomolecular targets can be performed with several analytical tools. In general, fluorescence and/or radioactivity detection are the most used detection techniques for measuring biochemical interactions. Recently, mass spectrometry (MS) has become an important detection technique in the measurement of biochemical interactions [1–5]. One of the main difficulties in MS is to preserve the weak noncovalent complexes during the ionization process. Electrospray ionization (ESI) is the most widely used ionization technique for the ionization of weak noncovalent complexes, due to the relatively soft ionization process [6].

In general, biochemical assays are performed in batch, i.e. microtiterplates (HTS). However, in recent years, biochemical assays have been widely implemented in analytical techniques such as affinity electrophoresis and on-line biochemical assays [7–13]. Our concept is derived from similar biochemical assays based on fluorescence detection [14]. In our first study, the concept is developed with two model systems separately, i.e. fluorescein-biotin/streptavidin and digoxin/anti-digoxigenin assay [15]. The assay is carried out in a continuous-flow post-column reaction detection system where the affinity protein (receptor molecule) and reporter ligand are sequentially added to the carrier solution of a flow-injection system or the eluate of an LC separation column. This methodology has the potential for multiplexing, i.e. performing several biochemical assays in parallel. In the present chapter, both model systems are combined into an on-line continuous-flow multi-protein biochemical assay. By employing MS as a detection method, two types of information can be obtained. First of all, from the MS data, biochemical information (binding affinity) can be obtained by monitoring the reporter ligand. Secondly, by interpreting the MS data, chemical information (molecular mass, MS-MS spectra) can be obtained. With a quadrupole time-of-flight (Q-TOF) mass spectrometer it is possible to acquire the MS data in 'full scan' to obtain the accurate molecular mass of the active ligand. In addition, by using automated MS-MS switching, MS-MS spectra of the active ligand can be obtained in a single run. This will provide specific fragments to elucidate the structure of the active ligand. By employing MS as a detection technique for biochemical assays it may substantially reduce the time it takes to discover and characterize biologically active compounds in areas such as combinatorial chemistry and natural product screening.

5.2. EXPERIMENTAL

5.2.1. Chemicals

Streptavidin, biotin, fluorescein-biotin, digoxin, digoxigenin and Leu-enkephalin were obtained from Sigma (St. Louis, MO, USA). *N*-biotinyl-6-aminocaproic hydrazide, *N*-biotin-hydrazide, *N*-biotinyl-L-lysine, and biotin-*N*-succinimidylester were purchased from Fluka Chemie (Buchs, Germany). Anti-digoxigenin Fab-fragments were purchased from Boehringer Mannheim (Mannheim, Germany). Methanol (HPLC-grade) was obtained from Merck (Darmstadt, Germany) and ammonium formate was purchased from Aldrich (Zwijndrecht, The Netherlands). All aqueous solutions were prepared with

Milli-Q water (Milli-Q Academic system) from Millipore (Bedford, MA, USA). The binding solution consisted of either ammonium acetate or ammonium formate (10 mmol/l, pH 7.5). The carrier solution was the binding solution/methanol (90:10% v/v). All the stock solutions of biotin, fluorescein-biotin, digoxin and digoxigenin were prepared in methanol. Dilutions were made in the carrier solution. Stock solutions of streptavidin and anti-digoxigenin were prepared in the binding solvent.

5.2.2. Continuous-flow set-up

The continuous-flow system (Fig. 5.1) consisted of three LC-10Ai solvent delivery modules from Shimadzu Europe (Duisburg, Germany). The first pump (carrier solution) was connected to a Gilson 234 autoinjector (Villiers-Le-Bel, France). A Rheodyne six-port injection valve with a 10 μl injection loop was implemented in the autosampler. The second pump (receptor pump) was connected with the carrier solution pump via an inverted Y-piece (home-made) followed by reaction coil 1, which was made from knitted poly(tetrafluoroethylene) tubing (33 μl, 300 μm i.d.). The third pump (reporter pump) was connected with reaction coil 1 via an inverted Y-piece followed by reaction coil 2 (knitted poly(tetrafluoroethylene) tubing, 17 μl, 300 μm i.d.), which was connected to the next inverted Y-piece. The first pump was used for pumping the eluent, 10 mmol/l ammonium formate:methanol (90:10% v/v) at pH 7.5, at a flowrate of 25 μl/min. The second pump delivered the receptors, dissolved in 10 mmol/l ammonium formate at pH 7.5, at a flowrate of 25 μl/min. The reporter molecules, dissolved in 10 mmol/l ammonium formate:methanol (90:10% v/v) at pH 7.5, were delivered by the last pump at a flowrate of 50 μl/min. The last inverted Y-piece, after reaction coil 2, was used as a flow splitter directing 10 μl/min to the MS and 90 μl/min to the waste. A Micromass (Wythenshawe, Manchester, UK) Q-TOF2 mass spectrometer equipped with a Micromass Z-spray electrospray (ESI) source was used for detection. Masslynx software (version 3.5) running under Windows NT was

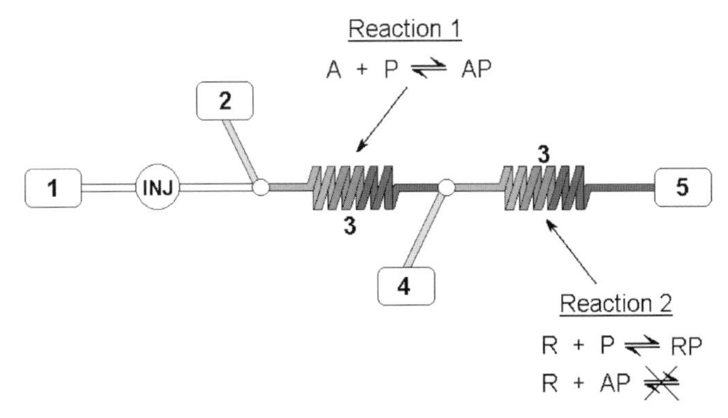

Fig. 5.1. Scheme of the on-line continuous-flow multi-protein biochemical assay. 1 = carrier solution (10 mM NH_4 formate + 10% MeOH, pH 7.5, 25 μl/min), 2 = reagent pump with affinity receptors (P) in binding solvent (10 mM NH_4 formate + 10% MeOH, pH = 7.5, 25 μl/min), 3 = reaction coil (33 and 17 μl, respectively), 4 = reagent pump with reporter molecule (R) in the carrier solution (10 mM NH_4 formate + 10% MeOH, pH 7.5, 50 μl/min), 5 = Micromass Q-TOF2 mass spectrometer (10 μl/min split into MS).

used to control the system and data acquisition. Mass spectrometry was operated at a 20 kHz frequency with a spectrum integration time of 3 and 2 s (for digoxin and fluorescein-biotin, respectively) in 'full scan' MS in the positive-ion mode (PI) in the range $m/z = 200-850$ ('interscan' time, 0.1 s). The electrospray source conditions for the multi-protein biochemical assay were: source temperature, 80 °C; desolvation temperature, 150 °C; capillary voltage, 2.4 kV. The sampling cone voltages were 23 and 65 V for digoxin and fluorescein-biotin, respectively, with an extraction cone voltage of 2 V. Nitrogen (99.999% purity; Praxair, Oevel, Belgium) was used with flowrates of 20 l/h for nebulization, 50 l/h for cone-gas and 350 l/h for desolvation. Argon (99.9995% purity; Praxair) was used for the MS-MS experiments.

5.2.3. MUX-technology set-up

Both the fluorescein-biotin/streptavidin and the digoxin/anti-digoxigenin on-line continuous-flow biochemical assays were performed on the MUX. The bioassay was split into four separate lines, each connected to the one of four sprayers of the MUX (Fig. 5.2).

For the next experiments with the MUX another set-up was developed (see Fig. 5.3). Probes 1 and 2 were used for the bioassay (waste assay + assay, respectively). Probe 3 sprayed the eluent directly after the injection. Probe 4 sprayed the lock mass. Here, 10 μl/min of 1 μmol/l of leucine-enkephalin was infused directly into the MS. Unfortunately, it is not yet possible for the Masslynx software to calculate the accurate mass of a compound if the lock mass is in a separate data file.

5.2.4. Batch experiments set-up

Batch incubate samples were injected in a modified continuous-flow set-up, i.e. the flow injection (FI) set-up. The system consisted of only one LC-10A insolvent delivery module. The pump was connected to a Gilson 234 autoinjector implemented with a Rheodyne six-port injection valve (10 μl injection loop). An inverted Y-piece was used as a flow splitter directing 10 μl/min to the MS detector and 90 μl/min to the waste. A Micromass Quattro II triple–quadrupole mass spectrometer (MS-MS) equipped with an ESI source was used for detection. Masslynx software (version 2.22) running under Windows NT was used for control of the system and data acquisition. The electrospray source conditions for the batch

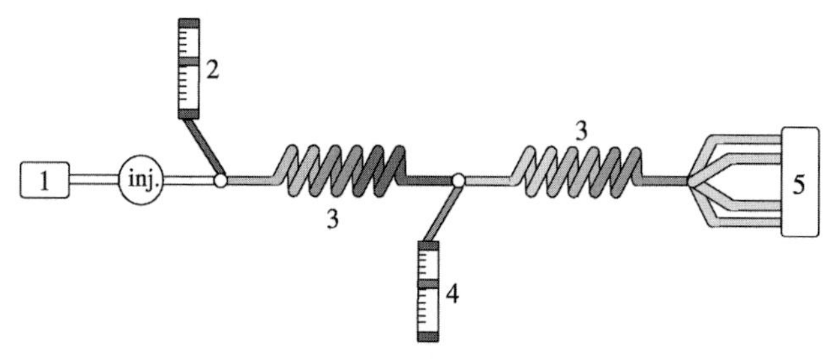

Fig. 5.2. On-line continuous flow biochemical assay set-up with the MUX technology.

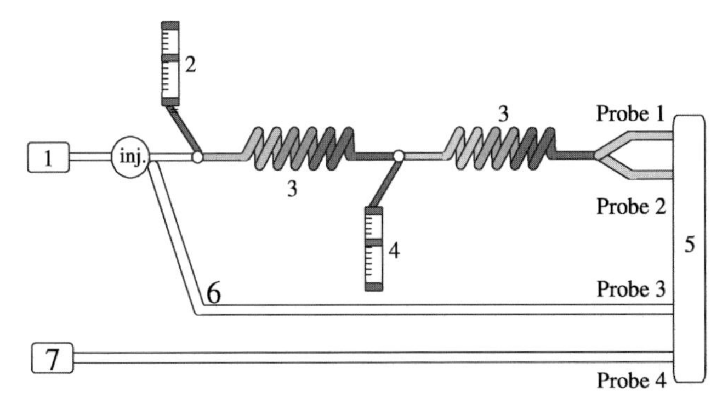

Fig. 5.3. Scheme of the new on-line continuous-flow biochemical assay with the MUX. 1–4 = the on-line continuous-flow biochemical assay described in Fig. 5.1, 5 = Micromass Q-TOF2 mass spectrometer with MUX (approximately 10 μl/min into MS, 6 = split direct after injection (10 mM NH_4 formate + 10% MeOH, pH 7.5, 10 μl/min), 7 = lock mass (1 μmol/l leucine-enkephalin in 10 mM NH_4 formate + 10% MeOH, pH 7.5, 10 μl/min).

experiments were: temperature 80 °C; capillary voltage, 3.5 kV. The sample cone voltages were 60 and 20 V for fluorescein-biotin and digoxin, respectively. Sample cone voltage for digoxigenin and biotin was 25 V. Nitrogen (99.999% purity; Praxair, Oevel, Belgium) was used with flowrates of 20 l/h for nebulization and 350 l/h for desolvation.

5.3. RESULTS AND DISCUSSION

Recently, an on-line continuous-flow analytical screening system using ESI-MS to measure the interaction of biologically active compounds with soluble affinity proteins was presented by our group [15]. Similar to other biochemical assays, a known ligand is used as a reporter molecule to detect the interaction of unknown ligands with the biomolecular target. The system set-up using a two-step reaction sequence is shown in Fig. 5.1. In the first step, the compound(s) to be screened is allowed to react with the biomolecular target. The concentration of free binding sites after reaction 1 is measured by the addition of MS reporter molecules in the second step. The entire reaction mixture is then introduced into the mass spectrometer. By monitoring the unbound concentration of the MS reporter molecule at its specific m/z, the interaction of active compounds with the biomolecular target can be detected. An increase of the active ligand concentration in the sample leads to a decrease of the bound concentration and an increase of the free reporter molecule concentration.

A mathematical treatment of the reaction kinetics of on-line assays is described in the appendix. As can be expected, the amount of complex formed depends mainly on the reaction time and the reaction rate constants of the association and dissociation reactions involved. For both interactions studied in this paper, more than 80% of the ligands involved in the reaction are bound to the biomolecular target. Although the reaction time is too short to reach an equilibrium, sufficient complex formation occurs to allow the detection of active ligands.

5.3.1. Optimization of MS conditions: buffers and organic modifier

Biochemical interaction experiments were performed to study the behavior of the interaction in different MS compatible carrier solutions, i.e. ammonium acetate and ammonium formate (10 mmol/l, pH 7.5). Two concentration series' of digoxigenin of 0, 50, 100, 200 and 300 nmol/l, one without anti-digoxigenin and one in 200 nmol/l anti-digoxigenin, were flow-injected into MS in both the carrier solutions. Biochemical interaction between anti-digoxigenin and digoxin could be observed in both carrier solutions. Slight dissociation of the complex was observed. However, the MS response was higher with ammonium formate than ammonium acetate (data not shown). All further experiments were performed in 10 mmol/l ammonium formate at pH 7.5. Similar results were obtained for the interaction of fluorescein-biotin/streptavidin earlier [15]. No additional blocking reagents were used in these experiments. One has to keep in mind that this may cause non-specific binding of proteins and protein complexes to the surface of the reactor tubing.

5.3.2. MS-based bioassay: optimization with the use of batch experiments

To study the feasibility of combining the two biochemical assays in an on-line continuous-flow set-up, a number of batch experiments were performed in order to examine the cross-reactivities of the two biomolecular targets [15]. The assay format presented in Ref. [15] has the potential for multiplexing, i.e. performing several assays in parallel, by pumping mixtures of receptors, i.e. streptavidin and anti-digoxigenin, and reporter ligands, i.e. fluorescein-biotin and digoxin. Clearly this approach will only be feasible for those assays where no cross-reactivity exists between receptors and reporter ligands. In order to study the feasibility of a multi-protein bioassay, several batch-incubated samples were flow-injected into the MS. The different experiments performed are presented in Table 5.1.

The results of the batch experiments are shown in Fig. 5.4. The *m/z* values correspond to the masses used for monitoring the different ligands in these experiments. Peak heights are compared with standard injections of the same concentration (data not shown). The first experiment is performed to study whether an interaction between biotin/fluorescein-biotin and anti-digoxigenin will occur. The response of fluorescein-biotin and biotin in a solution

TABLE 5.1
BATCH EXPERIMENTS PARALLEL BIOASSAY

Experiment	Solutions	After 1 h incubation
1	1 part fluorescein-biotin + 1 part biotin + 1 part anti-digoxigenin	–
2	1 part digoxin + 1 part digoxigenin + 1 part streptavidin	–
3	1 part fluorescein-biotin + 1 part streptavidin	1 part digoxin (excess)
4	1 part digoxin + 1 part anti-digoxigenin	1 part biotin (excess)
5	1 part fluorescein-biotin + 1 part streptavidin	1 part anti-digoxigenin
6	1 part digoxin + 1 part anti-digoxigenin	1 part streptavidin

1 part is 300 μl of fluorescein-biotin, biotin, digoxin, digoxigenin or anti-digoxigenin in 1000 nmol/l. Streptavidin 250 nmol/l. Excess digoxin and biotin 10 000 nmol/l.

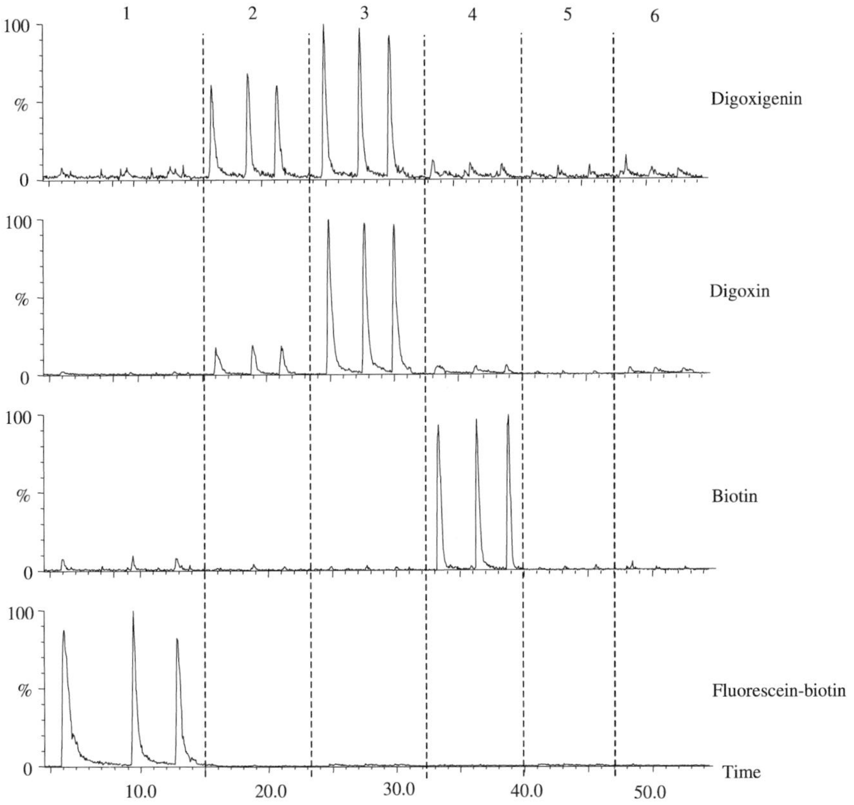

Fig. 5.4. Ion extracted chromatogram of digoxigenin ($m/z = 390.0$), digoxin ($m/z = 651.2$), biotin ($m/z = 245.1$) and fluorescein-biotin ($m/z = 391.0$).

with anti-digoxigenin is the same as the response of fluorescein-biotin and biotin in a carrier solution, indicating that there is no interaction and/or ion suppression observed. Similar results were obtained in experiment 2 for digoxin and digoxigenin with streptavidin. In experiments 3–6, the binding strength of the reporter molecule with the receptor molecule is studied in the presence of an excess of another ligand or receptor molecule, which is supposed to have no affinity for the receptor. The complex should stay intact when it is subjected to an excess (10 times) of the ligand or receptor. In this way the potential of performing the biochemical assay in parallel is studied.

In experiment 3, no peaks were observed for the fluorescein-biotin, which indicated that all the fluorescein-biotin was still bound to the streptavidin and the excess of digoxin did not influence the binding between fluorescein-biotin and streptavidin. Due to cone voltage fragmentation of digoxin (loss of all three sugar groups) peaks were also observed in the digoxigenin trace. Experiment 4 shows only small peaks of digoxin indicating that a high concentration of biotin had no significant influence on the binding of digoxin and anti-digoxigenin.

In experiment 5, no peaks were present in the traces, which indicated that adding anti-digoxigenin to fluorescein-biotin and streptavidin had no influence on the complex.

Similar results were observed in experiment 6. No cross-reactivity is observed and therefore it can be concluded that it is possible to combine both separate biochemical assays into one biochemical assay.

5.4. ON-LINE CONTINUOUS-FLOW MS-BASED BIOCHEMICAL INTERACTION

5.4.1. Fluorescein-biotin/streptavidin assay

The fluorescein-biotin/streptavidin assay was developed and optimized on a Micromass triple–quadrupole MS [15]. In this paper, the biochemical assay was performed on a Micromass Q-TOF2, and all experiments were done in 'full scan' MS instead of selected-ion monitoring (SIM). The Q-TOF2 has the same sensitivity in 'full scan' MS as the triple–quadrupole in SIM (50 nmol/l digoxin, 10 μl injection). This is a great advantage of the use of Q-TOF (TOF) instruments over the triple–quadrupole MS in the screening for (un)known (bioactive) compounds. First, the biochemical assays were performed separately. Triplicate injections of a blank, 1 μmol/l digoxigenin and biotin were performed. Digoxigenin was injected to study the cross-reactivity. No response was observed for digoxigenin in the fluorescein-biotin/streptavidin biochemical assay.

5.4.2. Digoxin/anti-digoxigenin assay

Similar experiments were performed to monitor the interaction between digoxin/anti-digoxigenin and biotinylated ligands. In the digoxin/anti-digoxigenin assay, 100% binding was not achieved, due to the relatively low concentration levels of both receptor and ligand reporter levels. In the digoxin/anti-digoxigenin assay, the digoxin signal did not decrease completely. At the concentration level used in the batch experiments about 90% of the complex could be formed. In the continuous-flow bio-assay the concentrations are lower and only up to approximately 80% of the complex can be formed. If only approximately 80% of the complex can be formed, the signal of the reporter molecule digoxin will not decrease completely, upon addition of anti-digoxigenin, as was observed for the fluorescein-biotin/streptavidin biochemical assay. In order to study cross-reactivity in the continuous-flow biochemical assay, triplicate injections of blank, 1 μmol/l digoxigenin and biotin were performed. No response on the digoxin trace was observed for biotin indicating no cross-reactivity between biotin and anti-digoxigenin (data not shown).

5.4.3. Parallel biochemical assay

The parallel biochemical assay was performed by dissolving both receptors in one solution and the two reporter ligands in another solution. Both reporter ligands (fluorescein-biotin and digoxin) were pumped together with one (reporter) pump. The receptors (streptavidin and anti-digoxigenin) were pumped with another pump. In Fig. 5.5, the extracted-ion chromatograms for both reporter molecules are shown. The two lower traces represent the reporter molecule digoxin ($m/z = 798.5$) and the reporter molecule

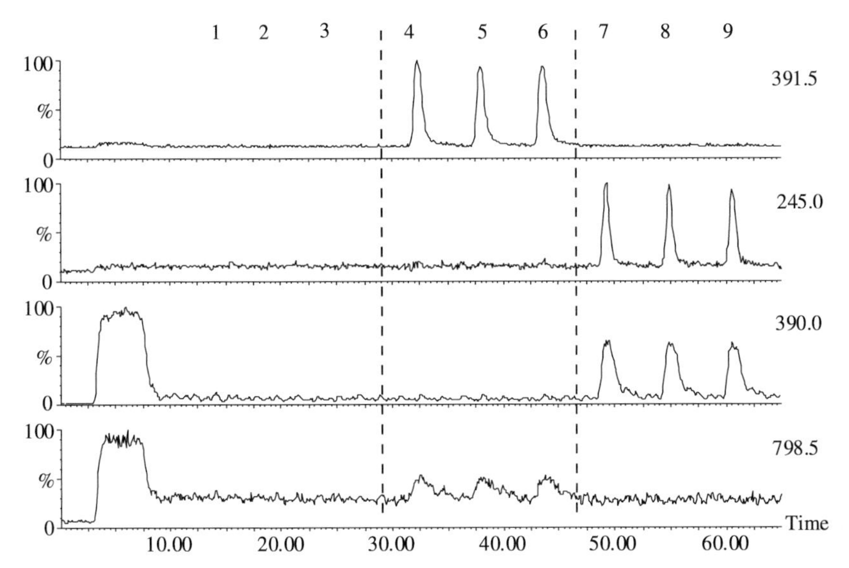

Fig. 5.5. On-line continuous-flow, multi-protein biochemical assay. The lower two extracted-ion chromatograms represent the biochemical assays of digoxin/anti-digoxigenin ($m/z = 798.5$) and fluorescein-biotin/streptavidin ($m/z = 390.0$), respectively. The upper two extracted-ion chromatograms are biotin ($m/z = 245.0$) and digoxigenin ($m/z = 391.5$), respectively.

fluorescein-biotin ($m/z = 390.0$), respectively. Triplicate injections of blank, 1 μmol/l digoxigenin and 1 μmol/l biotin were performed. As expected, the injection of an active compound resulted in an increase in the concentration of the respective unbound reporter molecules. In addition, peaks were observed in the extracted-ion chromatograms of biotin and digoxigenin. This is a result of the fact that the concentration injected into the carrier solution is a large excess compared with the concentration of receptor present in the carrier solution.

5.4.4. MS and MS-MS switching for identification of (bio-)active compounds in a continuous flow biochemical assay

For the characterization of active compounds, structure information needs to be obtained, e.g. through fragmentation of the compound in MS-MS experiments. With the Q-TOF2 it is possible to perform a 'survey scan', i.e. MS and MS-MS switching in an automated alternating manner. This option was also used in the continuous-flow fluorescein-biotin/streptavidin bioassay (Fig. 5.6). The scan method was set up in such a way has to obtain an MS-MS spectrum of the first most intense ion recorded in the previous MS scan. Background ions, including reporter molecules, were specified prior to analysis and automatically ignored by the Masslynx software. Firstly, the fluorescein-biotin/streptavidin assay ($m/z = 390$) was chosen to monitor the reporter molecule. However, when performing the survey scan this was not possible because a high cone voltage generates fragmentation of the injected ligands. At a lower cone voltage ($m/z = 416$), this is more clearly present and less fragmentation of the injected ligands occurs. Triplicate

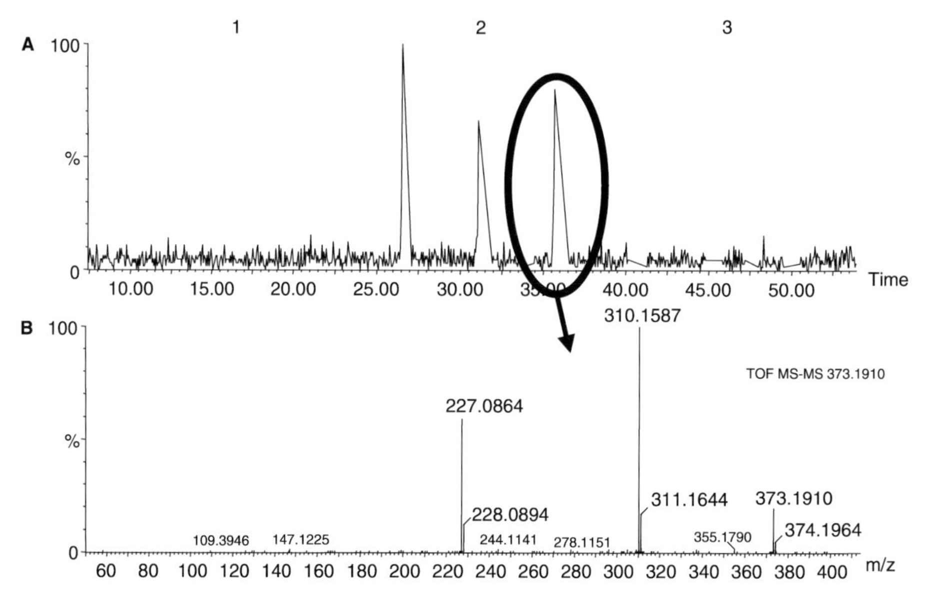

Fig. 5.6. (A) Extracted-ion chromatogram of fluorescein-biotin from the fluorescein-biotin/streptavidin assay. Triplicate injections of blank (1), *N*-biotinyl-lysine (2) and digoxigenin (3) were performed. (B) MS-MS spectrum of *N*-biotinyl-L-lysine found with survey scan.

injections of blank (1), 1 μmol/l biotinhydrazide (2) and 1 μmol/l *N*-biotinyl-L-lysine (3) were performed. The Q-TOF2 MS-MS was performed on the protonated molecule of digoxigenin (i.e. $m/z = 391.24$) and *N*-biotinyl-L-lysine (i.e. $m/z = 373.19$). The extracted-ion chromatogram of fluorescein-biotin shows that a compound with affinity for streptavidin was injected.

From the MS data the accurate mass and isotopic pattern can be determined. Leucine-enkephalin was used as a reference mass for the accurate mass measurements. After correction the protonated molecule at $m/z = 556.2770$ was set as the lock mass. The m/z value of the protonated molecule of *N*-biotinyl-L-lysine was determined to be 373.1904. From the accurate mass, possible elemental composition can be calculated and from the isotopic pattern it was determined that no Cl or Br was present in the compound. The calculation of the elemental composition yielded 25 solutions (see Table 5.2).

To reduce the number of solutions the double-bond equivalent was calculated. The injected compound gave a response on the fluorescein-biotin/streptavidin biochemical assay; therefore, it was assumed that it was a biotinylated compound. The DBE of biotin is 4 or 3.5 if the molecule is protonated. If the DBE of biotin was taken into account the number of solutions was reduced to 13. If a compound contains biotin it also contains at least one sulfur atom, one or two oxygen atoms and two nitrogen atoms reducing the number of solutions to 1 (see Table 5.2, calculated masses in bold). The m/z value is than 373.1910, which is the monoisotopic mass of protonated *N*-biotinyl-L-lysine.

This was also confirmed by comparing the MS-MS spectrum, found with the survey scan, with an MS-MS spectrum obtained from an infusion of *N*-biotinyl-L-lysine. Figure 5.7 shows the fragmentation pathway of *N*-biotinyl-L-lysine derived from an MS-MS spectrum.

TABLE 5.2

SOLUTIONS OF THE CALCULATION OF AN ELEMENTAL COMPOSITION BY THE ELECOMP PROGRAM (MICROMASS)

M	$M_{calculated}$	mDa	ppm	DBE	Formula
373.1904	373.1904	0.0	0.1	0.5	C14 H31 N2 O6 F P
	373.1903	0.1	0.3	6.5	C19 H25 N2 O F4
	373.1906	−0.2	−0.5	9.5	C18 H26 N6 O P
	373.1901	0.3	0.8	3.5	C18 H34 N2 P S2
	373.1907	−0.3	−0.9	0.5	C14 H31 N4 O F2 S2
	373.1908	−0.4	−1.1	5.5	C20 H29 O F3 P
	373.1909	−0.5	−1.3	−0.5	C15 H35 O6 P2
	373.1899	0.5	1.4	1.5	C13 H27 N4 O6 F2
	373.1910	**− 0.6**	**− 1.5**	**4.5**	**C16 H29 N4 O4 S**
	373.1898	0.6	1.7	1.5	C14 H30 N4 F3 P2
	373.1897	0.7	2.0	9.5	C23 H28 F2 P
	373.1896	0.8	2.2	4.5	C17 H30 N4 F S2
	373.1912	−0.8	−2.3	−0.5	C15 H35 N2 O F P S2
	373.1895	0.9	2.3	−0.5	C16 H36 O2 F P2 S
	373.1914	−1.0	−2.8	2.5	C16 H26 N2 O2 F5
	373.1893	1.1	3.0	2.5	C13 H26 N6 F4 P
	373.1892	1.2	3.1	4.5	C17 H30 N2 O5 P
	373.1916	−1.2	−3.2	13.5	C24 H25 N2 O2
	373.1892	1.2	3.3	10.5	C22 H24 N2 F3
	373.1917	−1.3	−3.5	5.5	C15 H27 N6 O2 F P
	373.1890	1.4	3.7	0.5	C15 H32 N2 O2 F2 P S
	373.1920	−1.6	−4.2	1.5	C17 H30 O2 F4 P
	373.1887	1.7	4.5	5.5	C16 H26 N4 O5 F
	373.1921	−1.7	−4.5	0.5	C13 H30 N4 O5 F S
	373.1922	−1.8	−4.9	4.5	C16 H31 N4 O2 P2

Calculation performed for *N*-biotinyl-L-lysine ($m/z = 373.1910$). Parameter settings: C, 0–35 atoms; H, 0–70 atoms; N, 0–6 atoms; P, 0–2 atoms; S, 0–4 atoms; O, 0–6 atoms; and F, 0–6 atoms. Even-electron ions, DBE, −0.5 to 50. Search window: 5 ppm.

The *m/z* values obtained from the MS-MS spectra can be used to confirm the fragmentation pathway. From the accurate masses, possible elemental compositions can be calculated. The calculated solutions can be compared with the MS-MS spectra.

5.4.5. MUX technology

The MUX technology of Micromass with four inlets into the ion source offers the possibility of detecting four different traces in parallel for four different applications. In this paper some preliminary experiments were performed to investigate the feasibility of the on-line continuous-flow biochemical assays in combination with the MUX-technology. Both the fluorescein-biotin/streptavidin and the digoxin/anti-digoxigenin bioassay were tested with this new set-up. For the fluorescein-biotin/streptavidin assay, triplicate injections of four different concentrations (1, 2, 5 and 10 μmol/l) of biotin and digoxin

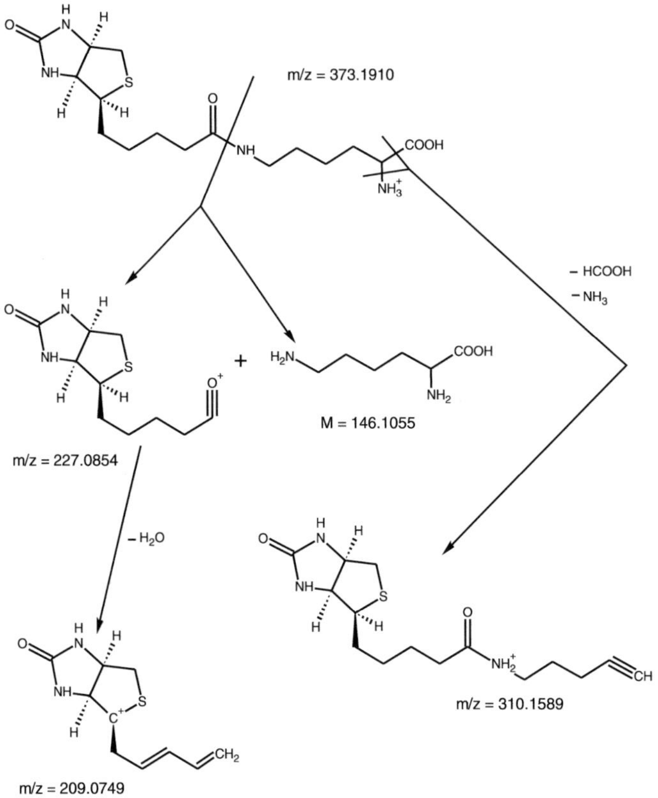

Fig. 5.7. Fragmentation pathway of *N*-biotinyl-L-lysine. The *m/z* values are calculated with the isotopic masses of the atoms.

were given and for the digoxin/anti-digoxigenin assay triplicate injections of four different concentrations (1, 2, 5 and 10 μmol/l) of biotin and digoxigenin were given.

Figure 5.8 shows the different traces/chromatograms of the different probes of the MUX. The upper trace represents the extracted ion chromatogram ($m/z = 556.2$) of the lock mass (4th probe). The second and third traces represent the extracted-ion chromatograms ($m/z = 416.2$) of fluorescein-biotin (1st and 2nd probes, respectively). These two traces represent the fluorescein-biotin/streptavidin assay. The lower trace represents the TIC of the line that was directly split after injection (3rd probe). In this line no reporter or receptor molecules were present, just the eluent and the injected compounds.

For both the assay traces similar peaks were observed for the biotin injections (2 μmol/l and above). No peaks were observed for the blank or digoxin injections. Even with the high concentrations of digoxin no response was seen in the bioassay. If this new set-up is compared with the first on-line continuous-flow bioassay set-up then higher concentrations of biotin need to be injected. This is mainly caused by the 'scan time' of each sprayer of the MUX since the MS samples from each probe individually and sequentially. The MS monitors the first three probes for 3 s and the fourth for 1 s resulting in a cycle time of

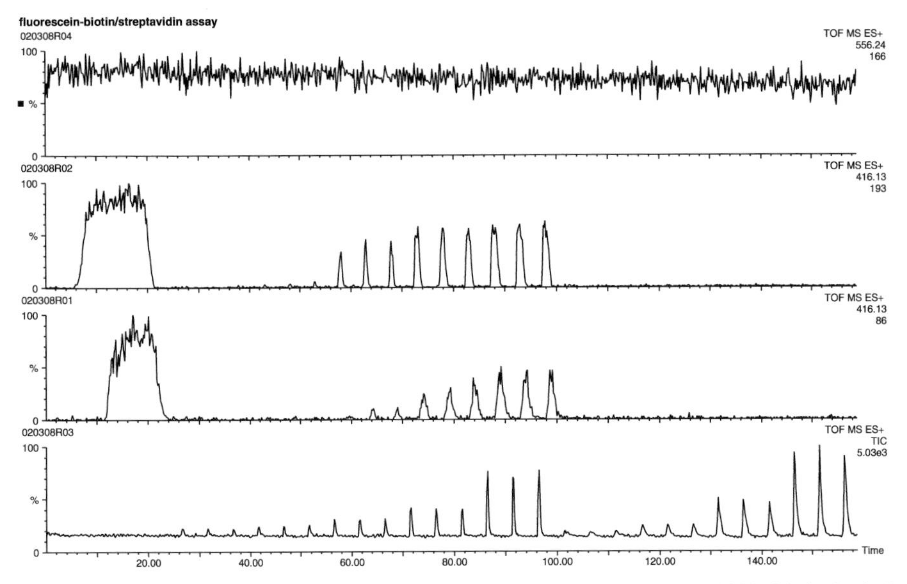

Fig. 5.8. Fluorescein-biotin/streptavidin assay on the MUX. The upper trace (020308R04) is the lock mass sprayer, which sprays 1 μmol/l leucine-enkephalin in eluent at 10 μl/min. The second trace (020308R02) is the assay sprayer (± 10 μl/min). The third trace (020308R01) is the waste of the assay (± 10 μl/min). The fourth trace (020308R03) is the line that was directly split after the injection.

10.4 s ('interscan' time = 100 ms). If the peak is too small then not enough points are acquired over the peak to enable adequate detection.

By analyzing the fourth trace (TIC) for all injections we found that the injected analyte could be detected more sensitively when not in the presence of the bioassay. When comparing the extracted-ion chromatograms the unbound analyte could be observed 5–10 times better than in the bioassay trace. Summing the spectra of these peaks gave the MS-spectrum of the injected analytes (data not shown). This is a major advantage compared with the first system. Unfortunately, it is not yet possible to perform a survey scan in combination with the MUX.

Figure 5.9 shows the different traces/chromatograms of the different probes. The upper trace represents the extracted-ion chromatogram ($m/z = 556.2$) of the lock mass. The second and third traces represent the extracted-ion chromatograms ($m/z = 651.2$) of digoxigenin (2nd and 3rd sprayers, respectively). These two traces represent the digoxin/anti-digoxigenin assay. The lower trace represents the TIC of the line that was directly split after injection (3rd sprayer). The bioassay did not respond to the blank or biotin injections. Even for the high concentration of biotin (10 μmol/l) no peak was observed in the bioassay. Peaks were observed for the complete series of digoxin in both bioassay traces. Digoxigenin and biotin can be detected in the lowest trace and summing the spectra of these peaks gave the MS-spectrum of the injected analytes (data not shown). These results show that the potential of using the MUX on the Q-TOF for the analysis of biochemical interactions works. The feasibility was demonstrated by using the fluorescein-biotin/streptavidin and digoxin/anti-digoxigenin bioassay as a model system.

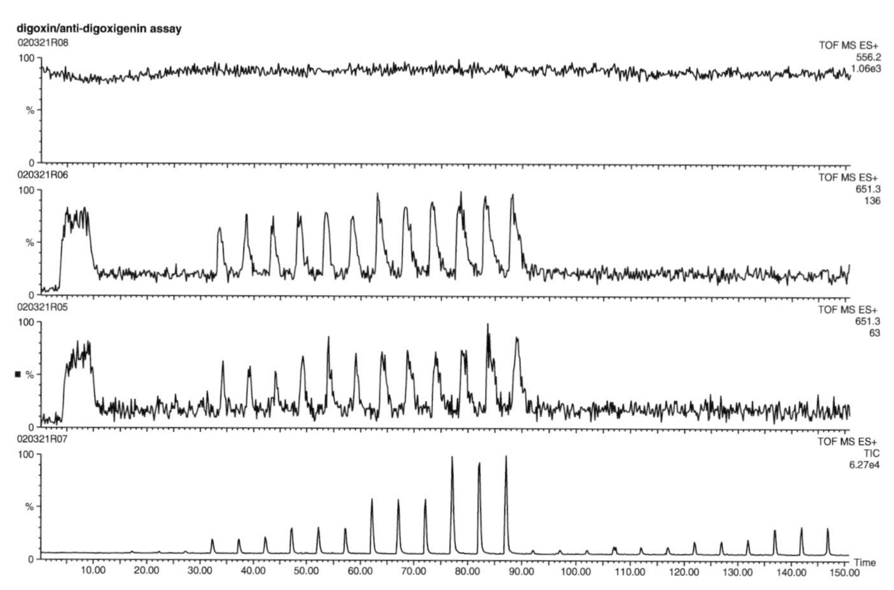

Fig. 5.9. Digoxin/anti-digoxigenin assay on the MUX. The upper trace (020321R08) is the lock mass sprayer, which sprays 1 μmol/l leucine-enkephalin in eluent at 10 μl/min. The second trace (020321R06) is the assay sprayer (± 10 μl/min). The third trace (020321R05) is the waste of the assay (± 10 μl/min). The fourth trace (020321R07) is the line that was directly split after the injection, so no assay.

5.5. CONCLUSIONS

The applicability of homogeneous on-line continuous-flow multi-protein bio-chemical assays was demonstrated for the interaction between fluorescein-biotin and streptavidin and for digoxin and anti-digoxigenin. No significant cross-reactivity was observed between the two biochemical assays in the batch and on-line experiments. Principally, for the development of a new assay, known high-affinity ligands for the receptor molecule have to be screened. The ligand with the highest sensitivity can be used as the reporter molecule for the biochemical assay. With this approach, assay development times can be substantially reduced. Moreover, the development of biochemical assays becomes possible for those receptors, where labeling of a known ligand with a fluorophor or an enzyme results in a dramatic decrease of binding affinity and, consequently, of detection sensitivity. An important requirement for the biochemical assay is that the binding between the reporter and receptor molecules can take place adequately under MS-compatible conditions. Volatile buffers such as ammonium acetate or formate have to be used rather than phosphate buffers. Moreover, salts such as sodium chloride, which are often added to adjust the background ionic strength of the assay, have to be avoided. The biochemical assay is limited by the binding strength between the reporter and receptor molecules. The binding strength between fluorescein-biotin/streptavidin is 10^{-15} mol/l and digoxin/anti-digoxigenin is 10^{-9} mol/l. In the fluorescein-biotin/streptavidin assay 100% of the complex can be formed and the signal of the reporter molecule fluorescein-biotin decreases completely. In the digoxin/anti-digoxigenin assay only up to 80% of the complex can be formed. With (automated) MS-MS switching the Q-TOF2 was able to acquire the MS-MS spectra of the injected analytes. In this way it is possible to obtain biochemical (binding affinity) and structural information in a single run. From accurate mass measurements the elemental composition of the analyte can be calculated. Combining this information with the information obtained in the MS-MS experiment, it was possible to confirm and identify (un)knowns. An internal standard (lock mass) needs to be added to the carrier solution in order to perform accurate mass measurements. This compound might influence the binding between the reporter and receptor molecules or the analyte and receptor molecules. With the MUX technology four separate probes are available, where one inlet was used for the lock mass. Another inlet was used for the detection of the unbound analyte. It was possible to detect the unbound analyte much better than in the presence of the bioassay. Peaks for the unbound analyte were 5–10 times higher compared with the peaks of the unbound analyte in the bioassay traces.

5.6. REFERENCES

1. B. Ganem, Y. Li and J. Henion, Observation of noncovalent enzyme–substrate and enzyme–product complexes by ion-spray mass spectrometry. J. Am. Chem. Soc., 113 (1991) 7818–7819
2. B. Pramanik, P. Bartner, U. Mirza, Y. Liu and A. Ganguly, Electrospray ionization mass spectrometry for the study of non-covalent complexes: an emerging technology. J. Mass Spectr., 33 (1998) 911–920
3. J. Loo, Studying noncovalent protein complexes by electrospray ionization mass spectrometry. Mass Spectr. Rev., 16 (1997) 1–23
4. B. Ganem, Y. Li and J. Henion, Detection of noncovalent receptor–ligand complexes by mass spectrometry. J. Am. Chem. Soc., 113 (1991) 6294–6296
5. T. Veenstra, Electrospray ionization mass spectrometry in the study of biomolecular non-covalent interactions. Biophys. Chem., 79 (2) (1999) 63–79
6. R. Smith and K. Light-Wahl, The observation of non-covalent interactions in solution by electrospray ionization mass spectrometry: promise, pitfalls and prognosis. Biol. Mass Spectr., 22 (1993) 493–501
7. Y. Zhang and F. Gomez, Multiple-step ligand injection affinity capillary electrophoresis for determining binding constants of ligands to receptors. J. Chromatogr. A, 897 (1–2) (2000) 339–347
8. J. Heintz, M. Hernandez and F. Gomez, Use of a partial-filling technique in affinity capillary electrophoresis for determining binding constants of ligands to receptors. J. Chromatogr. A, 840 (11) (1999) 261–268
9. N. Heegaard, S. Nilsson and N. Guzman, Affinity capillary electrophoresis: important application areas and some recent developments. Review, J. Chromatogr. B, 715 (1998) 29–54
10. M. Van Bommel, A. De Jong, U. Tjaden, H. Irth and J. Van der Greef, High-performance liquid chromatography coupled to enzyme-amplified biochemical detection for the analysis of hemoglobin after pre-column biotinylation. J. Chromatogr. A, 886 (11) (2000) 19–29
11. J. Emnus and G. Marko-Varga, Biospecific detection in liquid chromatography. J. Chromatogr. A, 703 (1–2) (1995) 191–243, review
12. P. Onnerfjord, S. Eremin, J. Emnus and G. Marko-Varga, High sample throughput flow immunoassay utilising restricted access columns for the separation of bound and free label. J. Chromatogr. A, 800 (15) (1998) 219–230
13. K. Graefe, Z. Tang and H. Thomas Karnes, High-performance liquid chromatography with on-line post-column immunoreaction detection of digoxin and its metabolites based on fluorescence energy transfer in the far-red spectral region. J. Chromatogr. B, 745 (2) (2000) 305–314
14. H. Irth, A. Oosterkamp, U. Tjaden and J. Van der Greef, Strategies for on-line coupling of immunoassays to high-performance liquid chromatography. Trends Anal. Chem., 14 (7) (1995) 355–361

15. A. Hogenboom, A. De Boer, R. Derks and H. Irth, Continuous-flow, on-line monitoring of biospecific interactions using electrospray mass spectrometry. Anal. Chem., 73 (16) (2001) 3816–3823
16. A. Oosterkamp, H. Irth, U. Tjaden and J. Van der Greef, On-line coupling of liquid chromatography to biochemical assays based on fluorescent-labeled ligands. Anal. Chem., 66 (23) (1994) 4295–4301

5.7. APPENDIX

5.7.1. Theory

The most important difference between digoxin/anti-digoxigenin (or digoxigenin/anti-digoxigenin) and fluorescein-biotin/streptavidin assays is the binding strength between the reporter molecule (digoxin/digoxigenin and fluorescein-biotin, respectively) and the receptor molecule (anti-digoxigenin and streptavidin, respectively). The dissociation constant (K_D) for fluorescein-biotin/streptavidin is about 10^{-15} M and the dissociation constant for digoxin/anti-digoxigenin is about 10^{-9} M. Dissociation can occur if the reporter and receptor concentrations are in the same range as the dissociation constant. An attempt is made to explain this effect with a theoretical calculation of 'a simple chemical rate problem'. The simplest chemical reaction is where particle A plus particle B becomes particle AB.

$$\mathrm{A} + \mathrm{B} \leftrightarrow \mathrm{AB}$$

The following dynamical equations can be given for this reaction:

$$\frac{\mathrm{d}n_\mathrm{A}}{\mathrm{d}t} = -k_\mathrm{f} n_\mathrm{A} n_\mathrm{B} + k_\mathrm{r} n_\mathrm{AB} \tag{1}$$

$$\frac{\mathrm{d}n_\mathrm{B}}{\mathrm{d}t} = -k_\mathrm{f} n_\mathrm{A} n_\mathrm{B} + k_\mathrm{r} n_\mathrm{AB} \tag{2}$$

$$\frac{\mathrm{d}n_\mathrm{AB}}{\mathrm{d}t} = k_\mathrm{f} n_\mathrm{A} n_\mathrm{B} + k_\mathrm{r} n_\mathrm{AB} \tag{3}$$

where n_A, n_B and n_AB is the number of particles of type A, B or AB, respectively and k_f and k_r are the forward and backward rate constants, respectively. The starting conditions for A, B and AB are given by n_A^0, n_B^0 and n_AB^0, respectively. For the loss of every A and B particle one AB particle is created, giving the following equation:

$$n_\mathrm{A} + n_\mathrm{B} + 2n_\mathrm{AB} = \text{constant} = n_\mathrm{A}^0 + n_\mathrm{B}^0 + 2n_\mathrm{AB}^0 = n_\mathrm{A}^\mathrm{e} + n_\mathrm{B}^\mathrm{e} + 2n_\mathrm{AB}^\mathrm{e} \tag{4}$$

where n_i^e refers to the equilibrium solution. In addition, there is another relationship, i.e. taking away one A particle means taking away one B particle, so that:

$$n_\mathrm{A} + n_\mathrm{B} = \text{constant} = n_\mathrm{A}^0 + n_\mathrm{B}^0 \tag{5}$$

n_A and n_B can be expressed using Eqs. 4 and 5.

$$n_\mathrm{A} = n_\mathrm{A}^0 + n_\mathrm{AB}^0 - n_\mathrm{AB} \tag{6}$$

and

$$n_{\mathrm{B}} = n_{\mathrm{B}}^{0} + n_{\mathrm{AB}}^{0} - n_{\mathrm{AB}} \tag{7}$$

The number of A particles plus the number of AB particles is constant, since for the loss of every A, an AB is created. The same is true for $n_{\mathrm{B}} + n_{\mathrm{AB}}$. Both equations can be substituted in Eq. 4 to give, after some rearrangement:

$$\frac{\mathrm{d}n_{\mathrm{AB}}}{\mathrm{d}t} = k_{\mathrm{f}}[n_{\mathrm{A}}^{0} + n_{\mathrm{AB}}^{0}][n_{\mathrm{B}}^{0} + n_{\mathrm{AB}}^{0}] - [k_{\mathrm{f}}N + k_{\mathrm{r}}]n_{\mathrm{AB}} + k_{\mathrm{f}}n_{\mathrm{AB}}^{2} \tag{8}$$

The equilibrium concentrations are related to the rate constants by:

$$\frac{n_{\mathrm{AB}}^{\mathrm{e}}}{n_{\mathrm{A}}^{\mathrm{e}}n_{\mathrm{B}}^{\mathrm{e}}} = \frac{k_{\mathrm{f}}}{k_{\mathrm{r}}} = K_{\mathrm{A}} = \frac{1}{K_{\mathrm{D}}} \tag{9}$$

where K_{A} and K_{D} are the association and dissociation constants, respectively. Substitution in Eq. 8, and some straightforward algebra shows that the equilibrium condition is still satisfied. The right-hand side is a quadratic equation in n_{AB}, which can be factored:

$$k_{\mathrm{f}}[n_{\mathrm{A}}^{0} + n_{\mathrm{AB}}^{0}] - [k_{\mathrm{f}}N + k_{\mathrm{r}}]n_{\mathrm{AB}} + k_{\mathrm{f}}n_{\mathrm{AB}}^{2} = k_{\mathrm{f}}[n_{\mathrm{AB}} - \lambda_1][n_{\mathrm{AB}} - \lambda_2] \tag{10}$$

with

$$\lambda_{1,2} = \frac{N + k_{\mathrm{r}}/k_{\mathrm{f}} \pm \sqrt{(N + k_{\mathrm{r}}/k_{\mathrm{f}})^2 - 4(n_{\mathrm{A}}^{0} + n_{\mathrm{AB}}^{0})(n_{\mathrm{B}}^{0} + n_{\mathrm{AB}}^{0})}}{2} \tag{11}$$

This means Eq. 8 can be rewritten as:

$$\frac{\mathrm{d}n_{\mathrm{AB}}}{[n_{\mathrm{AB}} - \lambda_1][n_{\mathrm{AB}} - \lambda_2]} = k_{\mathrm{f}}\,\mathrm{d}t \tag{12}$$

After some more complicated algebra n_{AB} is solved, which yields:

$$n_{\mathrm{AB}}(t) = \frac{\lambda_1(n_{\mathrm{AB}}^{0} - \lambda_2) - \lambda_2(n_{\mathrm{AB}}^{0} - \lambda_1)\mathrm{e}^{(\lambda_1 - \lambda_2)k_{\mathrm{f}}t}}{n_{\mathrm{AB}}^{0} - \lambda_2 - (n_{\mathrm{AB}}^{0} - \lambda_1)\mathrm{e}^{(\lambda_1 - \lambda_2)k_{\mathrm{f}}t}} \tag{13}$$

With Eq. 13, it is possible to calculate the amount of AB particles in time.

The reaction between fluorescein-biotin and streptavidin is a 4:1 (fluorescein-biotin:streptavidin) reaction. To keep the calculation as simple as possible a 1:1 reaction is assumed. The concentrations in reaction coil 2 are:

- Fluorescein-biotin = [A] = 24 nmol/l; streptavidin = [B] = 24 nmol/l
- $k_{\mathrm{f}} = 2.4 \times 10^{+7}\ \mathrm{l\ mol^{-1}\ s^{-1}}$ [16]
- $k_{\mathrm{r}} = 0.4 \times 10^{-7}\ \mathrm{s^{-1}}$
- $K_{\mathrm{A}} = k_{\mathrm{f}}/k_{\mathrm{r}} = (2.4 \times 10^{+7})/(0.4 \times 10^{-7}) = 6 \times 10^{+14}\ \mathrm{M^{-1}}$
- $K_{\mathrm{D}} = 1/K_{\mathrm{A}} = 1/(6 \times 10^{+14}) = 1.67 \times 10^{-15}\ \mathrm{M}$

The graph in Fig. 5.10 shows that at concentrations of 24 nmol/l fluorescein-biotin and streptavidin, with a $K_{\mathrm{D}} \approx 1.67 \times 10^{-15}$ M, it is possible to obtain almost 100% complex.

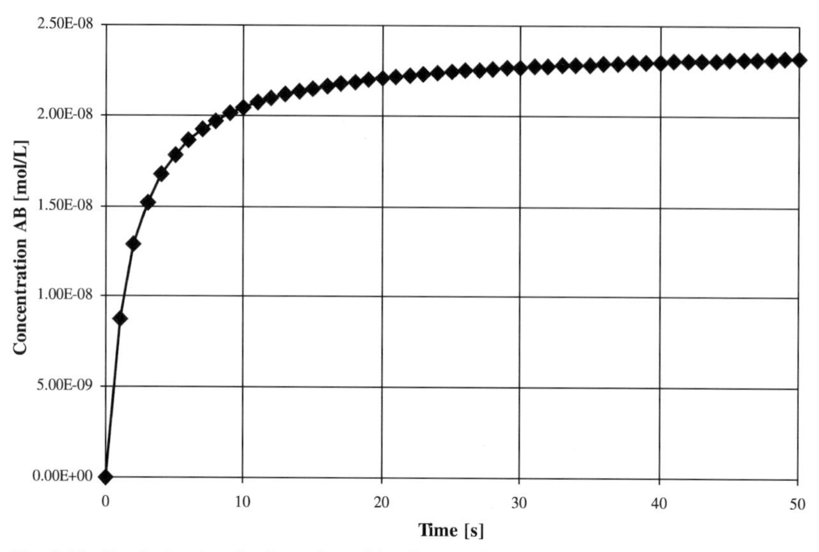

Fig. 5.10. Graph showing the formation of the fluorescein-biotin/streptavidin complex (AB) over time.

After 50 s, 23 nmol/l (approximately 96%) of complex can be formed. The reaction between digoxin and anti-digoxigenin is a 1:1 (digoxin:anti-digoxigenin) reaction. The concentrations in reaction coil 2 are:

- Digoxin = [A] = 50 nmol/l; anti-digoxigenin = [B] = 50 nmol/l
- $k_f = 2.4 \times 10^{+7}$ l mol^{-1} s^{-1} (assumption, derived from literature)

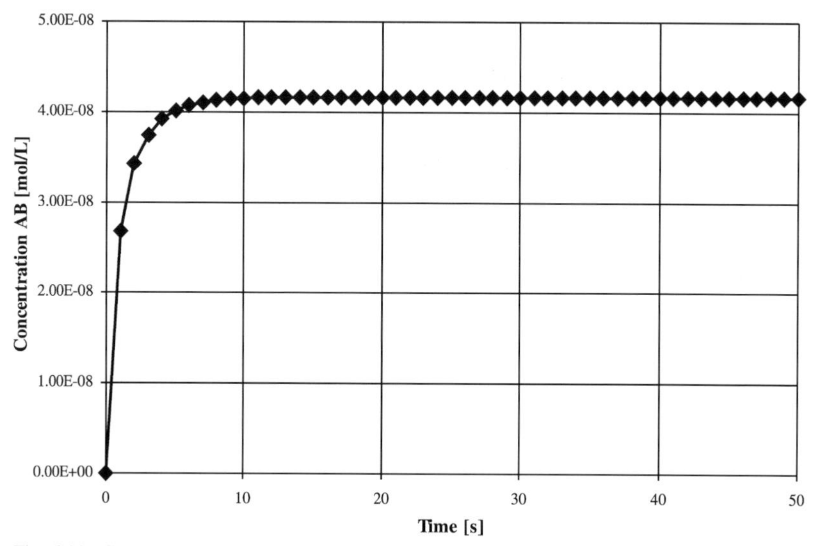

Fig. 5.11. Graph showing the formation of the digoxin/anti-digoxigenin complex (AB) over time.

- $k_r = 0.4 \times 10^{-1}\ s^{-1}$ (assumption, derived from literature)
- $K_A = k_f/k_r = (2.4 \times 10^{+7})/(0.4 \times 10^{-1}) = 6 \times 10^{+8}\ M^{-1}$
- $K_D = 1/K_A = 1/(6 \times 10^{+8}) = 1.67 \times 10^{-9}\ M$

The graph in Fig. 5.11 shows that at concentrations of 50 nmol/l digoxin and anti-digoxigenin, with a $K_D \approx 1.67 \times 10^{-9}$ M, it is not possible to obtain 100% complex. After 50 s, 42 nmol/l (approximately 83%) of complex can be formed.

G.A. Marko-Varga and P.L. Oroszlan (Editors)
Emerging Technologies in Protein and Genomic Material Analysis
Journal of Chromatography Library, Vol. 68

CHAPTER 6

Capillary Isoelectric Focusing Developments in Protein Analysis

A. PALM*

Department of Chemistry, Indiana University, Bloomington, IN 47405, USA

6.1. INTRODUCTION

The following chapter is a review of recent literature reports, from 1999 to 2001, in the field of capillary isoelectric focusing (CIEF) and protein analysis. Included are CIEF applications to recombinant proteins, glycoproteins, antibodies, hemoglobins, peptides, microorganisms, protein complexes, and protein mixtures from cell lysates. Also presented are novel, or extension of earlier, detection methods, capillary surface coatings, micropreparative CIEF, CIEF without carrier ampholytes, isoelectric point markers and two-dimensional chromatography-CIEF. Recent reports on CIEF coupled to mass spectrometry and the more novel field of miniaturized CIEF and microfabricated chip-IEF are also included.

For the reader less familiar with CIEF several handbooks in capillary electrophoresis can be of interest where basic concepts and principles are explained, see e.g. Ref. [1]. The latest CIEF review covers some literature reports up to the year 1999 [2].

The following review is, among other things, based on a database search using '(capillary) isoelectric focusing' as key search words in the title, and by the following life sciences databases: BIOSIS, CABA, CAplus, EMBASE, ESBIOBASE, KOSMET, MEDLINE, and SCISEARCH.

Since the chapter primarily addresses CIEF of proteins (with few exceptions) some recent CIEF reports are not covered herein, and can be found in Ref. [3] ('analytical- and preparative-scale isoelectric focusing separation of dansyl phenyl (DNS-Phe) enantiomers'), Ref. [4] ('enantioseparation of derivatized amino acids by CIEF using cyclodextrin complexation'), Ref. [5] ('dynamics of CIEF in the absence of fluid flow:

Abbreviations: i.d., internal diameter; p*I*, isoelectric point; CE, capillary electrophoresis; CZE, capillary zone electrophoresis; TEMED, tetramethylethylenediamine; MC, methylcellulose; HPC, hydroxypropyl cellulose; HPMC, hydroxypropylmethylcellulose; RSD, relative standard deviation; LC, liquid chromatography; MS, mass spectrometry; ESI, electrospray ionization; FTICR, Fourier transform ion cyclotron resonance.

* *E-mail address:* apalm@indiana.edu

high-resolution computer simulation and experimental validation with whole column optical imaging'), and Ref. [6] ('digitally controlled electrophoretic focusing').

6.2. DETECTION

6.2.1. Imaging CIEF

CIEF with whole column imaging detection requires no mobilization since the focusing process can be followed in real-time and the analysis is completed once the analytes are seen focused. Therefore, the analysis time is shortened and problems sometimes inherited with mobilization such as protein precipitation and pH gradient distortion are alleviated. Moreover, imaged CIEF reveals the dynamics of the focusing process. For a review, see Refs. [7,8]. It is a common phenomenon in CIEF (as in CE in general) that similar analytes dissolved in a matrix containing various amounts of salt will show different separation patterns and if too high a content of salt is present the analysis will fail, see Ref. [1]. The salt affects the CIEF separation negatively by adversely generated joule heat, pH gradient compression, and decreased reproducibility. It is therefore often necessary to work under low-salt conditions; excess salt can be removed off-line or on-line, see e.g. Refs. [9–13].

Mao and Pawliszyn [14] studied the effect of salt concentration on separation pattern in CIEF with whole column imaging detection. The experiments were performed using the iCE280 imaged CE system (Convergent Bioscience) with the absorbance detection mode at 280 nm. The focusing process was monitored every 20 s. A 5 cm × 100 μm i.d. fused-silica capillary coated with fluorocarbon (J&W Scientific) was filled with sample (0.1 mg/ml myoglobin and/or low molecular mass p*I* markers) mixed with 2% carrier ampholytes pH 2–11 (Sigma), 0.15% MC, and 0.1% TEMED. Before use, the capillary was conditioned with methylcellulose for 20 min as to further reduce the EOF. Anolyte was 100 mM H_3PO_4 and catholyte 40 mM NaOH. Focusing was at 1.5 or 3 kV for 3.5 min. Results indicated standard deviations of p*I* markers (10 replicate runs) of less than 0.01 pH units relative to an internal standard. The effect of salt concentration is shown in Fig. 6.1 for a mixture of five p*I* markers. With increasing concentration of phosphate buffered saline (PBS) in the sample, the pH gradient was gradually compressed. In quantitative terms the pH gradient was compressed to a degree of 4.3% at a PBS concentration interval of 10 mM. The separation performance of myoglobin was similarly affected by an increased salt content, decreasing the detectability of the minor myoglobin variant (p*I* 6.8). The authors proposed that by using two p*I* markers the separation pattern in the presence of salts could be corrected relative to a salt-free matrix by simply stretching the electropherogram, thus simplifying p*I* determination of samples with different salt contents.

Huang et al. [15] used CIEF with imaging detection in a study to separate proteins in pure water, i.e. without carrier ampholytes. The natural pH gradient was established by protons and hydroxyl groups produced by electrolysis of water at anode and cathode, respectively, migrating into the capillary (see also Ref. [106]). Two imaging detection modes were used in this study: UV absorbance using the iCE280 imaging system and coated capillary as described above, and an axially illuminated laser-induced

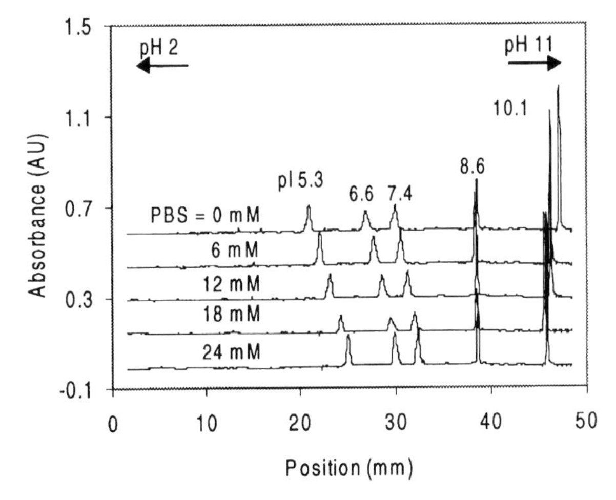

Fig. 6.1. CIEF with absorption imaging detection. Electropherogram shows a mixture of five p*I* markers separated at five different phosphate buffered saline (PBS) concentrations. With increasing PBS concentration the pH gradient gradually compressed. From Ref. [14], with permission.

fluorescence (LIF) imaging detection, as recently reported [16]. In the experiment with LIF detection a separation capillary made of poly(tetrafluoroethylene) was used. This capillary, 6 cm × 200 μm i.d., was treated with HPMC to reduce EOF and protein adsorption. The sample (1.8×10^{-8} M of green fluorescent protein and 3.3×10^{-10} M of R-phycoerythrin) was prepared by dissolving the proteins in 20% glycerol for the LIF experiment or 0.35% HPMC for the UV experiment. The experiment using UV detection is shown in Fig. 6.2 where cytochrome C (6.1×10^{-4} M) and hemoglobin (3.1×10^{-6} M) are seen focused after 120 s at 250 V/cm. An anolyte (100 mM phosphoric acid, H_3PO_4) and catholyte (100 mM sodium hydroxide, NaOH) were used in this experiment but are not necessary to establish the pH gradient, they merely accelerated the focusing process. The authors pointed out several potential possibilities and problems using this approach.

Fang et al. [17] earlier explored the possibilities using temperature sensitive buffers instead of carrier ampholytes in CIEF with imaging detection. A pH gradient in the capillary was developed by changing the temperature of the buffer (large dpH/dT slope). This method, however, suffered from a narrow pH range and/or poor separation performance.

A recent report by Wu et al. [18] applied imaging CIEF to the study of the isoelectric points of peanut allergenic proteins and its antibody, rabbit IgG. Two main proteins were found. Also the immunoreactivity of the allergen proteins with antibody was studied, where one protein showed a larger affinity for binding to the antibody than the other.

Tragas et al. [19] coupled an on-line version of gel filtration chromatography (GFC) to imaged CIEF for protein analysis. Effluent from the GFC column passed through a tee where it was split into two streams. One part was directed to a UV detector to monitor the GFC separation while another part was directed towards a sample preparation device. This preparation device consisted of a hollow fiber membrane (15 cm long × 180 μm i.d.) and provided desalting and mixing of the eluted sample with carrier ampholytes prior to CIEF. A schematic representation of the set-up is shown in Fig. 6.3. Twenty microliters of two

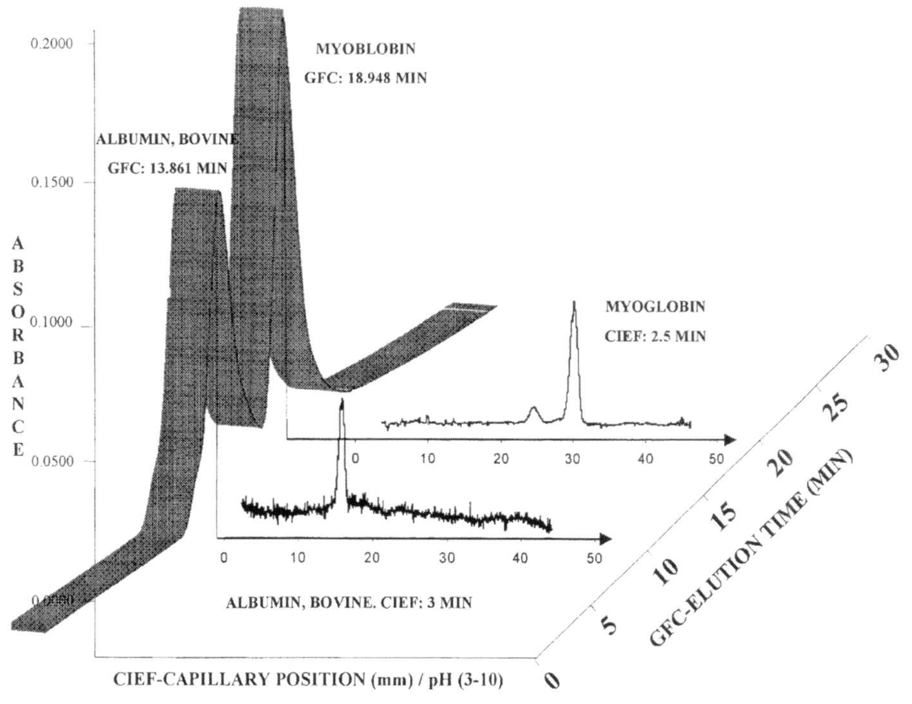

Fig. 6.4. A chromatogram and electropherogram combined in a 3-D display. GFC elution times of two proteins are plotted against CIEF position of respective protein. The two proteins were separated by GFC and further transferred on-line to—and analyzed by—imaging CIEF. From Ref. [19], with permission.

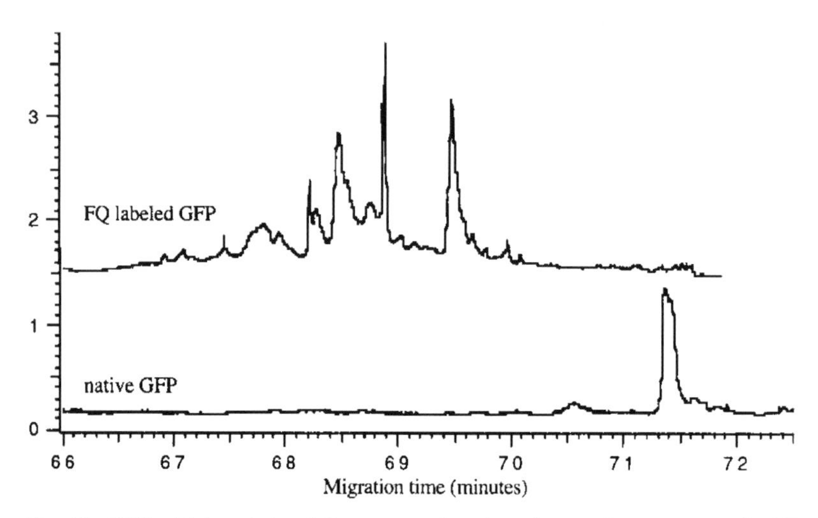

Fig. 6.5. CIEF with laser-induced fluorescence detection of green fluorescent protein (GFP). The effect of labeling GFP with a fluorophore (FQ) is shown in the upper electropherogram while the lower electropherogram shows the native GFP. From Ref. [20], with permission.

achieved (except between the two different myoglobins with p*I* 7.2 and 6.8, respectively). Identity of peaks a and b was not defined. Detection limit for cyt C was 6×10^{-9} M using the 100 μm i.d. capillary (1×10^{-8} M with the 75 μm i.d.) and the CL response was linear for greater than two orders of magnitude (both capillaries).

In an effort to decrease the mobilization time, Hashimoto et al. [25] later presented a similar CIEF system employing pressure mobilization instead of chemical mobilization. An interface device was here made between the separation capillary and a syringe pump for mobilization, to which a platinum wire was inserted as an anodic electrode. This system was shown to be more effective than chemical mobilization.

Verbeck et al. [26] presented a two-wavelength LIF system for CIEF of peptides and proteins. According to the authors, the performance of the CIEF/LIF system compared with a conventional UV absorbance detection resulted in: (i) reduced background, (ii) more sensitive detection, (iii) the two-channel system sample and p*I* standards being more easily identified, even in a noisy spectrum. Standards and unknown were labeled with fluorophores showing different emission wavelengths and the fluorescence signal from different fluorophores was separated by a dichroic mirror, spectrally filtered, and collected in separate channels. CIEF was performed in a laboratory made system.

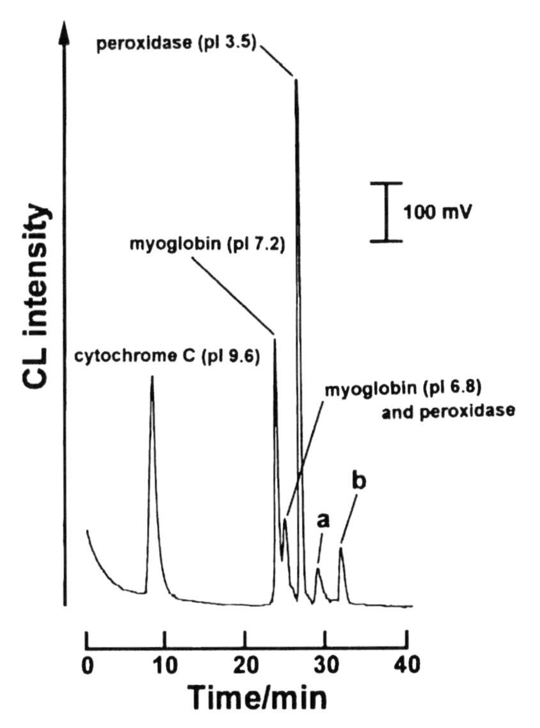

Fig. 6.6. CIEF of heme proteins using chemiluminescence (CL) detection. Identity of peaks a and b was not defined. Chemical (cathodic) mobilization was performed with added luminol and hydrogen peroxide in catholyte for post-column CL detection. From Ref. [24], with permission.

Approximately 10^{-8} M of sample was mixed with 2% Bio-Lyte pH 3–10 and 0.1% Tween and injected into a polyacrylamide coated capillary (35 cm × 75 μm i.d.). Focusing was at 20 kV for 10 min (mobilization not defined). A green helium–neon laser (543.5 nm) was used for the LIF detection. Several fluorophores with amine reactive moieties for protein/peptide labeling and suitable for excitation at 543.5 nm were investigated: tetramethylrodamine isothiocyanate (TRITC), 5-carboxyrhodamine 6G (5-CR), Alexa 532 dye, BODIPY 530/550, in conjunction with Texas Red-X succinimidyl ester (TR-X). TRITC was found to be less suitable. The emission spectrum from Texas Red was collected only in the transmitted channel while the spectrum from, e.g. 5-CR, was collected in both the transmitted and reflected channels. TR-X and 5-CR had similar detection limits. The CIEF analysis of peptides from a protein digest together with p*I* standards is shown in Fig. 6.7. The p*I* standards were five peptides labeled with 5-CR and the protein digest was from β-casein labeled with TR-X (derivatized after digestion). Figure 6.6A shows the transmitted channel (signal from TR-X and 5-CR); peak 1 is free fluorophore while peaks 2–6 are the standard peptides, non-denoted peaks are peptides from β-casein. Figure 6.7B shows the reflected channel (signal from 5-CR only), while Fig. 6.7C shows a subtraction (reflected channel–transmitted channel). Therefore, the peptide peaks from β-casein not seen in the electropherogram shown in Fig. 6.7A (since they were hidden by the standard peptides) could be detected in Fig. 6.7C after the signal from the p*I* standards was subtracted; those peptides are marked with an asterisk.

Strong et al. [27] discussed the advantages and limitations of derivatization of proteins for improved performance and detectability in CIEF. Several small and large proteins were derivatized with 6-aminoquinolyl-*N*-hydroxysuccinimidyl carbamate (AQC) and subjected to CIEF analysis. An ISCO model 3850 CE system with UV detection at 280 nm and a microsil FC or DB-1 coated capillary (J&W Scientific) was employed. The derivatized sample was diluted and mixed with various concentrations of carrier ampholytes pH 2–11 (Sigma) including a FC-PN detergent and 4–5 M urea. Urea was found to be necessary to limit precipitation during focusing. A single-step CIEF protocol was used, i.e. the EOF was strong enough to sweep the focused analytes past the detector. The p*I* reproducibility was estimated to be lower than ±0.1 p*I* unit and even less for the derivatized species, using standard proteins such as p*I* markers, despite a non-linear pH gradient. AQC is a neutral chromophore/fluorophore and reacts with protein amino groups; as a result the derivative exhibits a lower p*I* than the native protein. Interestingly some derivatized proteins were shown to produce a single species, or multiple species, which behaved as a single species, upon CIEF analysis. Improved detectability as well as separation efficiency was achieved when compared with the native species. Limit of detection (LOD) for derivatized insulin was ~24.7 μg/ml while LOD for the native species was ~50 μg/ml using UV detection. Also limit of derivatization for insulin was defined and determined to be 710 μg/ml. The immunorecognition properties were preserved for at least one derivatized protein as shown by antibody recognition. The authors stressed the problem encountered with precipitation during derivatization and focusing (despite the use of urea and detergent), a problem that was more pronounced with basic proteins.

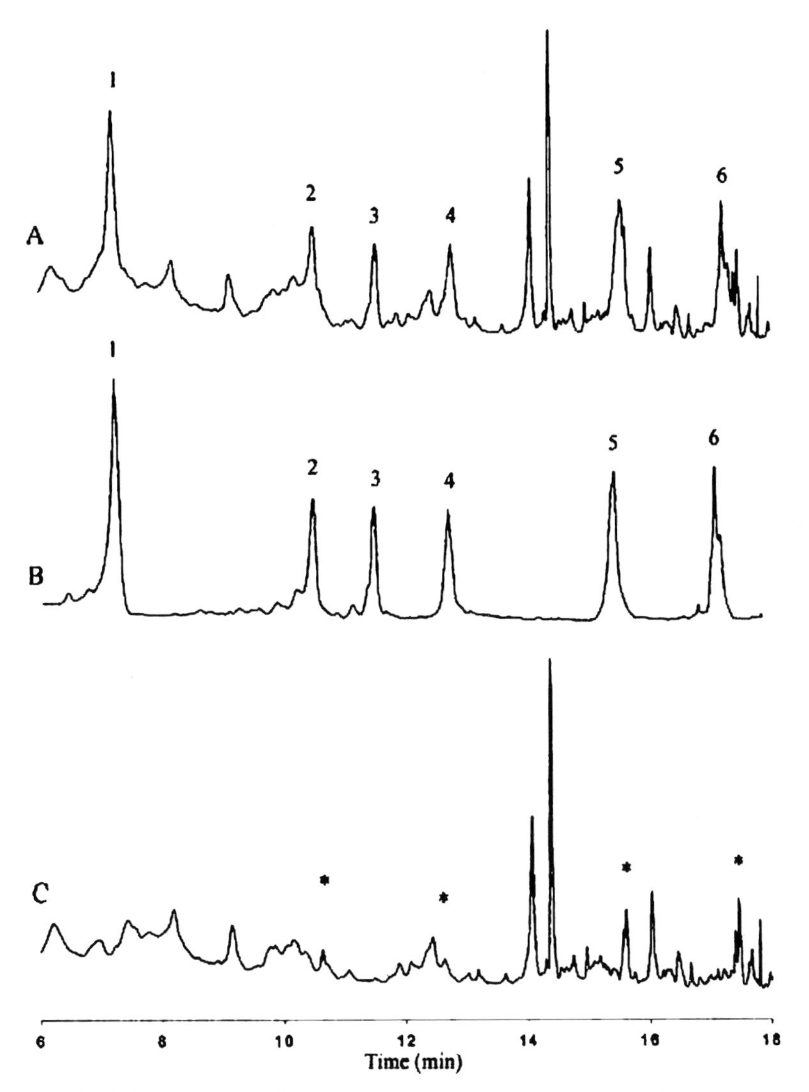

Fig. 6.7. CIEF with two-wavelength laser-induced fluorescence detection of peptides from a protein digest, including peptide p*I* standards. The digest peptides and p*I* markers were labeled with different fluorophores. The emission signal from respective fluorophore was resolved in separate channels whereby identification was possible. See text for details. From Ref. [26], with permission.

6.3. MICROPREPARATIVE CIEF

Minarik et al. [28] described a new micropreparative fraction collection system for CZE and CIEF. A laboratory made system was used. Capillaries with i.d. 75, 100, and 200 μm coated with poly(vinyl alcohol) (PVA) were evaluated. Four standard proteins, 0.1 mg/ml each, were mixed with 2% carrier ampholytes and loaded into the capillary. Phosphoric acid (100 mM) was used as anolyte and 30 mM ammonium hydroxide (catholyte), pH 9.3, as sheat liquid at the exit (fraction collection) side of the capillary. A lower concentration, and therefore pH, of the catholyte prevented degradation of the PVA coating. To suppress siphoning, as well as backflow arising from the sheat flow, the separation system was closed at the inlet by a semipermeable membrane. Joule heating was reduced by initially focusing on a constant current mode (30 μA) until an upper voltage of 15 kV was reached, whereby the voltage was kept constant and current started to decline. A diagram of the CIEF system is shown in Fig. 6.8. Mobilization was by pressure (syringe pump). Two UV detectors near the capillary exit were used to exactly determine the mobilization rate of the standard proteins whereby individual proteins could be selectively fractionated. Fractions were collected in vials containing 20 μl catholyte. The mobilization rate was optimized and set at 0.6 cm/min, larger rates gave decreased resolution. The 100 μm i.d. capillary was chosen as optimal since the 200 μm i.d. exhibited less resolution while the 75 μm i.d. capillary provided less sample load. The sample identity of the collected fraction was confirmed by MALDI mass spectrometry and by re-analysis of part of the sample in the CIEF system.

6.4. INTERNAL STANDARDS

Low-molecular-mass p*I* markers are proposed to be more suitable as internal standards (reference markers) than native- and fluorophore labeled-proteins since they are less prone to precipitate at pH close to p*I*, more stable over time, have a higher degree of purity, and are homogeneously labeled.

Horká et al. [29] investigated the suitability of low-molecular-mass fluorescent compounds as p*I* markers for CIEF with UV photometric and UV excited fluorometric detection. A laboratory made CIEF system was used, employing fused-silica capillaries coated with silane (γ-glycidoxypropyl trimethoxysilane). While silane alone suppressed EOF, 0.25% HPMC was added to the sample ampholyte mixture. Eighteen p*I* markers spanning the p*I* range from 3.5 to 10.3 were separated and characterized together with dansyl (Dns) fluorescent derivatized proteins. The p*I* markers showed good focusing properties-, stability-, and spectroscopic identification. Their optical properties make them detection compatible with the products of protein fluorescent derivatization, e.g. fluorescamine, dansyl chloride, *o*-phthaldialdehyde and coumarin labeling.

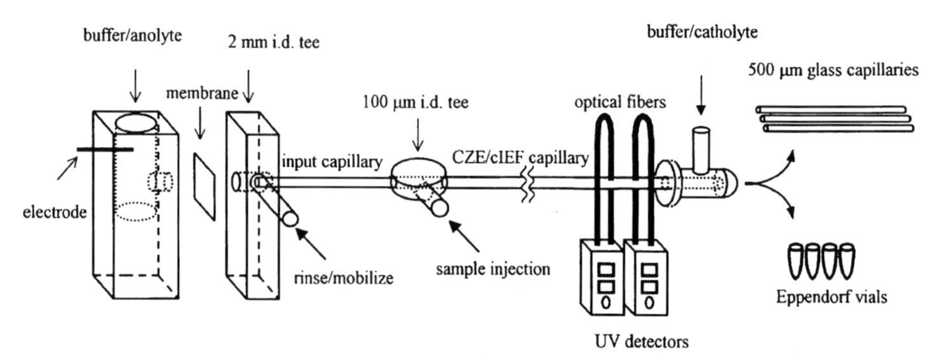

Fig. 6.8. Micropreparative system for CIEF. After focusing proteins were mobilized by pressure towards the cathode. Fraction collection times were determined after the mobilized zones traveled between the two UV detection points. Proteins could therefore be collected in separate vials. From Ref. [28], with permission.

Shimura et al. [30] evaluated the accuracy of isoelectric point determination by CIEF. The p*I* values of nine protein standards frequently used in CIEF were compared with the p*I*s of peptide markers (p*I* ranging from 3.38 to 10.17) recently characterized by the same authors [31]. The experiments were performed using the Beckman P/ACE 2210 system and coated capillaries (eCAP neutral coated capillary (Beckman) and linear polyacrylamide coated (LPA) capillary [23]). Carrier ampholytes were from two different suppliers (Pharmalytes and Servalytes of various pH ranges). Anolyte was 91 mM (eCAP capillary) or 20 mM (LPA capillary) phosphoric acid except for pH range 5–10.5 where glutamic acid (20 mM) was used, and catholyte was 20 mM sodium hydroxide. HPMC (0.2–0.4%) and TEMED (0.1–0.3%) were occasionally included in the sample-ampholyte solution. Focusing was at 500 V/cm for 2–20 min and mobilization by pressure (0.5 psi) maintaining the voltage. The authors did not mention any discrepancy in performance regarding type of coating or anolyte, except that a 15 min rinse period of the LPA coated capillary with water (20 psi) was necessary to achieve a consistent performance. p*I* determination of neutral to acidic proteins using Servalyte pH 3–7 provided better results with a linear pH profile than Pharmalyte pH 2.5–8. When a broad pH range was used, Pharmalyte pH 3–10 showed a better performance for neutral to basic proteins while Servalyte pH 3–10 was more effective for neutral to acidic proteins. The p*I* of trypsin inhibitor (a common p*I* marker) seemed especially difficult to determine. The authors proposed, since the pH profile looked different for various ampholytes and pH ranges when mobilization was performed by hydrodynamic flow, that the number of p*I* markers covering the particular pH range of interest and curve fitting procedures are important for accurate p*I* determination. The set of peptide p*I* markers investigated formed very sharp peaks and allowed protein p*I* determination within an average error of less than 0.1 pH unit.

6.5. CAPILLARY COATINGS

Shen et al. [32] studied the influence of different capillary coatings on CIEF separations of complex protein mixtures from cellular lysates of microorganisms. Linear polyacrylamide (LPA), poly(vinyl alcohol) (PVA), and hydroxypropyl cellulose (HPC) were tested. Separations were performed in a laboratory made system employing UV detection at 280 nm. Capillary length was 65 cm (54 cm to detector) × 50 μm i.d. Focusing was at 20 kV and mobilization by gravity (hydrodynamic flow) with maintained voltage. Ammonium hydroxide, 1% and acetic acid, 1% were used as catholyte and anolyte, respectively. By applying a protein concentration of only 0.08 mg/ml prepared from *D. radiodurans*, approximately 210 putative protein peaks were detected in a pH 3–10 gradient (Pharmalyte, 1% v/v). Besides the high resolution obtained with such a low protein concentration, the risk for protein precipitation was decreased. The addition of 8 M urea and 2% Igepal CA630 (detergent) to the same sample decreased the separation performance. In most cases, HPC-coated capillaries provided a greater resolution than those obtainable with PVA- and LPA-coated capillaries. The authors estimated the resolution (based on model proteins and baseline resolution, i.e. Rs 1.5) to be Δp*I* 0.013. Also cell lysate proteins from *E. coli* and *S. cerevisiae* were investigated by CIEF using these capillary coatings.

References pp. 130–133

Tran et al. [49] developed a one-step CIEF method to resolve the glycoforms of the recombinant human immunodeficiency virus (HIV) envelope glycoprotein 160s-MN/LAI. Earlier attempts using CZE at acidic or alkaline pH for this purpose have not been successful. One of the main problems encountered with this protein is its poor solubility at a pH close to its p*I* and its tendency to aggregate. Various parameters such as capillary coatings, additives, ampholyte pH ranges and concentration, and field strengths were studied in an attempt to optimize this analysis. CIEF was performed using the Beckman P/ACE 5500 instrument. Anolyte and catholyte were 100 mM H_3PO_4 and 20 mM NaOH, respectively. Focusing was performed in the shorter 7 cm section of the capillary (27 cm total length (20 cm effective), 50 μm i.d.), between the detection window and anolyte, by using a reversed polarity configuration. A poly(vinyl alcohol) (PVA) and polyacrylamide (PAA) coated capillary were compared. Both capillaries gave similar resolutions but the PVA capillary showed better migration time reproducibility, for both intra-day (6 runs) and long-term use (60 runs). Optimized separation conditions were found using 5% carrier ampholytes with pH ranges 3.5–5, 5–8, and 3–10 (mixed in a 71:12:17 v/v/v ratio), 0.085 M CAPS, 1% TEMED, 6% saccharose, 0.1% HPMC, and a field strength of 425 V/cm. Results demonstrated different acidic glycoforms of the protein between pH 4 and 5. Moreover, 4 M urea was found to be a better solubilizing agent from the above mentioned but since it denatured the protein its p*I* became different from the native form and was therefore not employed, also the capillary coatings were deteriorated by urea. It was only possible to analyze only a very narrow concentration interval of the protein (0.2–0.9 μg/μl). Furthermore, the migration times of the standard proteins were found to be affected by the presence of the glycoprotein.

In a more recent report, Tran et al. [50] characterized a highly *O*-glycosylated casein-macropeptide (CGMP), Mw ~7000 (the soluble fraction of κ-casein in bovine milk when cleaved by chymosin) by CIEF, high-performance anion-exchange chromatography with pulsed amperometric detection (HPAEC-PAD), and normal-phase chromatography with fluorescence detection. CGMP consists of a number of glycosylated and phosphorylated forms. Five glycosylation sites have so far been identified. The peptides/proteins are very acidic (p*I* values < 4). A two-step CIEF method was employed using the Beckman P/ACE 5500 CE system and a polyacrylamide coated capillary. A 2% mixture of wide (pH 3–10) and narrow (pH 3.5–5) range carrier ampholytes (25/75 v/v mix) was found optimal for the separation of at least 14 glycoforms. Focusing was performed in the short segment of the capillary, similar to that mentioned above; mobilization was by pressure with maintained voltage. Migration time RSD was less than 0.6%. Peak area, however, was less reproducible (partly because of unresolved peaks hampering accurate peak integration).

Jochheim et al. [51] reported the CIEF separation of different glycoforms of a recombinant human tumor necrosis factor receptor (p75) Fc fusion protein (rhu TNFR:Fc), named Enbrel. Enbrel is approved for the treatment of rheumatoid arthritis. It includes several glycosylation sites partially terminated with negatively charged sialic acid residues. CIEF was performed both with an imaging detection system and a conventional instrument (two-step method). The imaging system used was the recently commercialized CIEF instrument iCE280 (see above, Ref. [14]). With imaging detection, the sample (500 μg/ml) was mixed with 4% Servalyte pH 2–11 and 0.35% MC and injected into the 50 mm × 100 μm i.d. fluorocarbon coated capillary. Anolyte and catholyte were 100 mM phosphoric acid and 40 mM sodium hydroxide, respectively. Focusing was at 3 kV for 4 min (i.e. total analysis time). The two-step CIEF method was performed with the BioFocus 3000 instrument using an eCAP neutral coated capillary. The sample (350 μg/ml) here was mixed with 2% Ampholines, pH range 3.5–9.5, 2.5% TEMED and 0.2% MC. Anolyte and catholyte were 20 mM H_3PO_4 and 40 mM NaOH, respectively. Focusing was at 15 kV for 4 min followed by chemical mobilization using a 'cathodic mobilizer' (Bio-Rad), total analysis time ~30 min. Adddition of TEMED resulted in a better migration time reproducibility and sharper peaks. Results showed several major and minor peaks in respective electropherograms. The peak profiles, however, looked slightly different. The major TNFR:Fc peak focused around p*I* 4.8 with p*I* RSD $< 1.5\%$, as determined with synthetic p*I* markers. In an effort to increase the resolution the authors mixed narrow range Ampholines, pH 3.5–5.5, with the wide range ampholytes (pH 3.5–9.5). Interestingly, this resulted in less reproducibility. Also a comparison with slab gel IEF was made and comparable peak profiles were achieved with an enhanced resolution of some isoforms in the CIEF profile (CIEF employing the BioFocus instrument).

Hui et al. [52] employed CIEF to profile the composition of cellulases in complex fermentation samples of secreted proteins from various strains of the filamentous fungus *Trichoderma reesei*. The enzyme cellobiohydrolase I (CBH I), a major component in these extracts, was further purified by ion-exchange chromatography and characterized by high-performance anion-exchange chromatography with pulsed amperometric detection (HPAEC-PAD), capillary liquid chromatography-electrospray ionization mass spectrometry (cLC-ESI MS), and tandem mass spectrometry (MS-MS). The main cellulases from *T. reesei* are acidic glycoproteins (p*I* < 5.5), and shows extensive N- and/or O-linked glycosylation. Separation was carried out using the Beckman P/ACE 5000 instrument and an eCAP neutral coated capillary (20 cm effective length). A crude cellulase sample from various strains was injected into the capillary together with 3% blended carrier ampholytes (Beckman ampholytes pH 3–10 and Servalyte pH 3–7 (3:1 v/v mixture)) and an internal standard. The narrow range ampholytes were added to enhance the resolution in the acidic pH region. Anolyte and catholyte were 10 mM phosphoric acid and 20 mM sodium hydroxide, respectively. Focusing was at 13.5 kV for 10 min followed by chemical (cathodic) mobilization replacing the catholyte with 1% acetic acid in 50% methanol. Results demonstrated that 5–6 different cellulases were detected; however, corresponding glycoforms were not resolved. The attached carbohydrates were preferably of GlcNAc and mannose type lacking charged residues except an unusual carbohydrate assigned as a phosphorylated disaccharide. Detection limits were in the low μg/ml range.

Hagmann et al. [53] reported the characterization of a $F(ab')_2$ fragment of murine monoclonal antibody using CIEF, ESI MS, HPAEC-PAD, and LC-MS peptide mapping (see above, Ref. [52] for acronyms used). The antibody was first cleaved by pepsin leading to the $F(ab')_2$ and Fc' fragments. CIEF was performed both under non-denaturing (two-step method) and denaturing (one-step method) conditions using the Beckman P/ACE 5500 instrument. For non-denaturing conditions a μ-Sil DB-1 coated capillary (J&W Scientific) was used. The sample (0.3–0.5 mg/ml) mixed with Pharmalyte pH 3–10, MC, and TEMED, was injected into the capillary. MC was necessary to reduce the residual EOF existing in the DB-1 coated capillary. After focusing, chemical mobilization was performed by salt addition (30 mM sodium chloride) to the catholyte (20 mM sodium hydroxide).

Results showed several peaks in the electropherogram corresponding to $F(ab')_2$ isoforms. Digestion with neuraminidase removed charged sialic acid groups and simplified the interpretation of the electropherogram and the mass spectrum. By performing CIEF under denaturing and reducing conditions the peak pattern was further simplified. An eCAP neutral coated capillary (Beckman) was here used with enough residual EOF. The capillary was filled with reduced and denatured (DTT/urea) sample, carrier ampholytes, and TEMED. By this sample treatment, the heavy and the light chains were separated into two peaks each. Further characterization was performed by the above-mentioned techniques.

Ru et al. [54] reported the analysis of apolipoprotein (ApoH). ApoH is a plasma glycoprotein partially associated with lipoproteins. Analysis was performed using the Beckman P/ACE MDQ instrument and a neutral coated capillary. Anolyte and catholyte were 100 mM H_3PO_4 in 4% HPMC and 100 mM NaOH, respectively. ApoH (10 μg/ml), purified from the serum of a subject, was mixed with Pharmalyte pH 3–10, focused and mobilized by pressure (0.5 psi). Several peaks were seen in the electropherogram. According to the authors, split peaks indicated ApoH from a heterozygote gene while similar non-split peaks expressed a homozygote gene.

Manabe et al. [55] examined CIEF conditions to separate human cerebrospinal fluid (CSF) proteins in the absence of denaturing agents. A comparison was made to already established CIEF procedures for the analysis of human plasma/serum proteins. Since the protein content in CSF is only about 0.5% that of plasma while the level of salt content is similar a desalting procedure by microdialysis was established. The authors constructed a dialysis device able to handle 20 μl volumes. The desalted sample together with 1% ampholines pH 3–10 and 0.2% TEMED was injected into a polyacrylamide coated capillary. With TEMED a better resolution was achieved for basic proteins. Chemical (cathodic) mobilization was performed; analysis time ~60 min. About 70 putative proteins were detected. Similar electrophoretic conditions, with minor modifications, could be used as already established with CIEF of plasma/serum proteins; comparative studies are therefore possible.

Hiraoka et al. [56] analyzed, by CIEF, glycoforms of the β-trace protein (β-TP) in the CSF of patients with various neurological disorders. β-TP has been identified as a lipocalin-type prostaglandin D synthase and is the most abundant component among human CSF proteins produced within the central nervous system. A one-step CIEF method was employed using the Beckman P/ACE 2000 system and an eCAP neutral coated capillary, 50 μm i.d. × 27 cm (20 cm effective length). Anolyte and catholyte were 10 mM H_3PO_4 and 20 mM NaOH, respectively. Fractionated and desalted CSF proteins were mixed with Pharmalyte pH 3–10, TEMED, HPMC, and standard proteins as p*I* markers, and focused at 4.67 kV in the short segment of the capillary (similar to that mentioned above, see Ref. [49]). Four peaks with p*I* values between pH 4 and 6 were identified as isoforms of β-TP. Some of the isoforms were expressed at various ratios in association with diseases as determined by peak area ratios.

Lupi et al. [57] developed a method for the analysis of alpha$_1$-antitrypsin (α_1AT) protein isoforms in human serum. α_1AT is the major plasma inhibitor of a variety of serine proteases and its deficiency in humans is characterized by the development of emphysema (destruction of alveoli). At least 20 different α_1AT alleles have been identified in association with this disease. CIEF analysis was performed with the Beckman P/ACE 2200 system and a fused-silica capillary dynamically coated with polyethylene oxide (PEO). A crude sample from serum or a similarly enriched/partially purified sample was mixed with 5% carrier ampholytes of various pH ranges. Focusing was performed for 2 min at 900 V/cm and mobilization by pressure maintaining the voltage. The anolyte consisted of 91 mM phosphoric acid in 0.75% PEO and catholyte 20 mM sodium hydroxide. Electroosmosis was not observed with this coating and by simple rinsing, the capillary with phosphoric acid and PEO in water between runs and the coating was re-established. An acidic pH gradient, Pharmalyte pH 4.2–4.9, was most efficient with regard to resolution of the different isoforms but gave a low migration time reproducibility. Highest run-to-run reproducibility was obtained in a pH 3.5–5 gradient (Sigma). Phenotype identification from various subjects was possible from the obtained protein profile.

Lasdun et al. [58] examined the validity of a two-step CIEF method for quantitative measurement of product-related impurities found in production lots of a protein drug substance, according to guidelines published by the International Conference on Harmonization (ICH). The protein drug used in this study was myelopoietin (MPO; belonging to a family of engineered dual interleukin-3 (IL-3) and granulocyte colony-stimulating factor (G-CSF) receptor agonists) containing a monodeamidated degradation product and an aggregated form of MPO as impurities. The impurities were isolated and quantitated by cation-exchange chromatography and spiked at various concentrations into the drug sample before analysis by CIEF. Triplicate analyses were performed for each spike level. The method specificity, accuracy, linearity, precision (including repeatability, intermediate precision and reproducibility), and limit of quantitation (LOQ) were established from the measurement of the impurity peak area percentage for spiked samples. CIEF analysis was performed with the Beckman P/ACE MDQ system and two different brands of coated capillaries (eCAP neutral coated capillary and Zeroflow capillary from Microsolv). The sample was mixed with several carrier ampholytes of various pH ranges (p*I* 3–10, 3.5–10, 5–8, 5–7, and 6–8). Focusing was at 25 kV for 20 min and chemical mobilization (cathodic mobilizing solution from Bio-Rad) applying the same voltage. The results demonstrated the importance of including p*I* markers since absolute migration time was observed to shift as much as 3 min between days. With p*I* markers included, consistent relative migration times between different capillary lots from the same vendor and from different vendors were obtained. Large RSD values in peak area and peak area percent were, however, obtained upon changes in instruments and capillary brand. Despite this, the authors proposed the method to satisfactorily meet the various validation criteria outlined.

6.6.4. Hemoglobin

Jenkins et al. [59] discussed the potential for routine implementation of CIEF in the clinical laboratory for the analysis of hemoglobin variants. For CIEF to be accepted in the clinical laboratory it must be reproducible and cost effective. Various brands of p*I* markers were tested here together with clinical hemoglobin samples and hemoglobin with known p*I* values. A one-step CIEF method was carried out using the BioFocus 2000 instrument and a polyAAEE (acryloylaminoethoxyethanol) coated capillary (BioRad). A rinse procedure between each run was performed using 10 mM phosphoric acid, 1% Brij detergent, water, and ampholyte mixture. Anolyte was 100 mM phosphoric acid in 0.4% MC while the catholyte consisted of 20 mM NaCl in 20 mM NaOH + 0.4% MC. The capillary did show reproducible results for over 100 runs. Isoelectric point markers from Pharmacia gave the most accurate p*I* values in this experimental set-up (pH 3–10 gradient; less accuracy was obtained with more narrow pH gradients). The consumables and labor costs per five samples, including HbA_2 quantitation, were compared with traditional electrophoresis and anion-exchange chromatography and found to be ~4 times less expensive. Also, in comparison with slab-gel IEF, CIEF was better (faster analyses, real-time detection and lower background).

Hempe and Craver [60] analyzed hemolysates containing hemoglobin (Hb) variants to evaluate the diagnostic sensitivity and specificity using CIEF. Analysis of several thousand patient samples by CIEF have shown that Hb C, S, D, and G can be confidently identified by CIEF in one single run based on automatic post-analytical p*I* calculation. In contrast, Hb E, C-Harlem, and O-Arab have similar p*I*s and are not separated by CIEF alone; other techniques may be necessary here. The analyses were performed using the Beckman P/ACE 2200 CE system (capillary and conditions not mentioned). p*I* was estimated using a control hemolysate as an external standard. Linear regression analysis was used to calculate the p*I* of Hb variants in unknown samples based on normalized migration times using HbA as a reference peak. Resolution values of 0.017 p*I* units were achieved. In comparison with HPLC, CIEF is more cost effective. CIEF can also simultaneously identify and quantify both major (e.g. Hb A, S, E, and C) and minor (e.g. Hb A_2 and F) hemoglobins, while HPLC needs two separate programs for this analysis. In relation to various other CE methods used for analysis of Hb variants, the authors propose the assay presented herein to offer better sensitivity, specificity, precision, and throughput.

Sugano et al. [61] reported a two-step CIEF method to randomly analyze both hemoglobins (Hb) and globin chains. According to the authors' information from CIEF, separations of globin chains are lacking. CIEF was performed using the Beckman P/ACE 5510 instrument and an eCAP-CIEF 3–10 kit (Beckman). Red blood cells were washed, lysed, and centrifuged while the globin chains were precipitated from the hemolysates. Samples were mixed with internal standards and carrier ampholytes (pH 3–10). Focusing was at 13.5 kV for 2 min followed by pressure mobilization. Results indicated that UV detection at 280 nm was the most suitable for analysis of both hemoglobin and globin chains. Identification of various hemoglobins as well as globin chains was based on a mobilization time ratio using carbonic anhydrase p*I* 6.6 as an internal standard. However, to distinguish a mutated globin chain (if any) from a non-mutated, the ratio of peak heights was used as a parameter. This was possible since the peak height ratio between normal α- and β-globin chains was almost constant (2.5 ± 0.1, a value which was obtained from four healthy subjects) so any other ratio indicated a mutated globin chain.

6.6.5. Microorganisms

Two reports recently addressed the analysis of bacteria [62], and yeast cells [63], by CIEF. Currently the characterization of microorganisms by traditional techniques suffers, among other things, from long analysis times. It is proposed that CE techniques like CIEF potentially can alleviate this drawback as well as add its inherent advantage of high resolution and peak capacity. Since microorganisms/cells often are more fragile and sticky towards surfaces than proteins, sample preparation procedures, capillary conditioning and an inert capillary inner surface are very important for successful analyses.

Armstrong et al. [62] separated three different bacteria of similar size (rodlike with diameter ~1 μm) at around 20 min analysis time obtaining high symmetry peaks and high peak capacity. A Beckman P/ACE 5000 CE system was used with UV detection at 280 nm. The freeze-dried or cultured bacteria was dissolved in 0.5% v/v Bio-Lyte pH 3–10 ampholytes and injected into a methylcellulose coated capillary (40 cm effective length × 50 μm i.d.). Before each analysis, the capillary was rinsed with water + ampholyte. The capillary was partly filled with sample and ampholyte solution and partly with ampholyte alone. Focusing was at 20 kV for 5 min followed by pressure mobilization. Anolyte and catholyte were 20 mM phosphoric acid and 20 mM sodium hydroxide, respectively. According to the authors, the nature and concentration of ampholyte exerted a strong influence on the separation. Furthermore, meticulous sample preparation and exact separation conditions were needed to ensure reproducibility. In the same report, they also applied capillary gel electrophoresis (employing poly(ethylene oxide) as a sieving matrix) for the separation of other microbes; a similar separation performance was obtained as with CIEF at a shorter analysis time (< 10 min).

Shen et al. [63] reported the separation and identification of yeast cells by CIEF. The authors proposed using the p*I* value to establish an identification index that would potentially be more attractive than electrophoretic mobility (e.g. using CZE) since it is essentially independent of experimental measurement conditions. Yeast cells (4 μm diameter) were cultured to various cell densities (early log, mid log, and stationary growth phase) before being mixed with 1% v/v Pharmalyte pH 3–10 and injected into a HPMC-coated capillary (54 cm effective length × 100 μm i.d.). Focusing was at 20 kV and mobilization by gravity. Anolyte and catholyte were 1% acetic acid and 1% ammonium hydroxide, respectively. The CIEF separation of a mixture of yeast cells cultured to various cell densities is shown in Fig. 6.10. Analysis time was ~50 min. Differences of up to 1.2 pH units were found for the p*I* of the yeast cells at different culture stages. Very narrow peaks (~2 s basewidth) were observed with peak asymmetry of ~1 and an estimated peak capacity of > 4000. It was found that the cell concentration was a critical factor affecting the reproducibility, only between 0.6 and 2.5 cells/μl could be injected, below or above this range no peaks were detected (UV detection at 280 nm). p*I* reproducibility was ±0.1 as determined by standard protein p*I* markers.

6.7. CIEF-MASS SPECTROMETRY

Currently there is a trend of combining capillary separation techniques (LC, CE, CIEF) with mass spectrometry (MS), especially electrospray ionization (ESI) MS. One reason for this is the current era of proteomics. Here, it is of interest to analyze complex protein fractions containing hundreds to thousands of different proteins at various ratios expressed from, e.g. a cell line or microorganism. By mass spectrometry, very accurate molecular

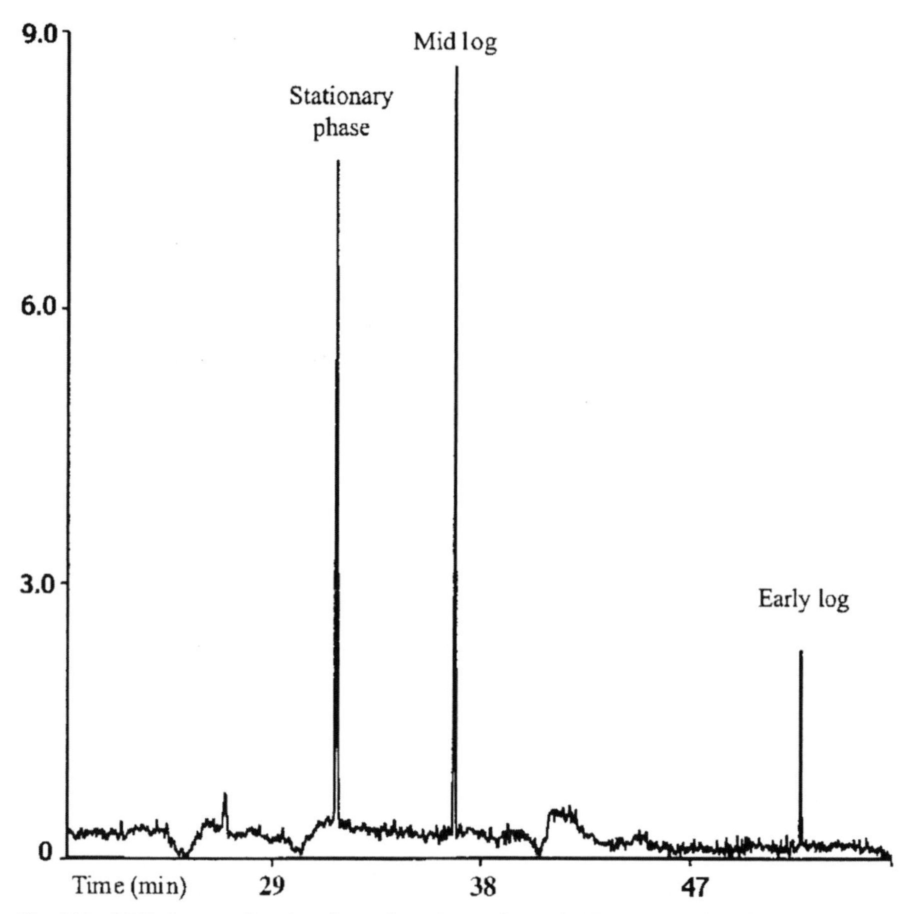

Fig. 6.10. CIEF of yeast cells cultured to various phases of growth. The concentration of cells suitable for CIEF was around 0.4–1 cell/μl. Differences of up to 1.2 p*I* units were found for the various culture stages. From Ref. [63], with permission.

masses of proteins can be gained, especially peptides, including their amino acid sequence. By analyzing the molecular masses in biological databases, identification of protein(s) can be achieved. Since mass spectrometry itself has a limited peak capacity upstream sample fractionation methods are used in order to reduce the sample complexity (in analogy with two-dimensional gel electrophoresis). CIEF is one such technique that also provides the added benefit of isoelectric point (p*I*) values which are important search parameters in databases for protein identification.

CIEF deals with similar issues as that of the interfacing of CE in general with ESI MS. For a more in-depth description of various configurations, technical aspects and arrangements, see Ref. [64]. Matters of importance are, briefly, as follows: (1) interface configuration. Three kinds are employed: (i) liquid junction; (ii) coaxial sheat flow; and (iii) sheatless interfaces. A coaxial sheat-flow interface is most commonly used in CIEF-MS due to its relatively easy arrangement and robustness; however, one drawback with this configuration is sample dilution and zone broadening. (2) Buffer/solvent compatibility. (i) The use of volatile buffers and organic solvents are important for efficient analyte ionization, a stable spray, and also for avoiding fouling of the MS apparatus. Since CIEF includes carrier ampholytes that are not MS compatible, they are often kept at a lower concentration than optimal with regard to resolution and/or diluted in the sheat flow before entrance into the mass spectrometer. (ii) If a two-step CIEF method is used, employing chemical mobilization, the formation of moving ionic boundaries can significantly affect migration time and separation [65]. Acetic acid is often used as a chemical mobilizer and besides being MS compatible has been shown to retain a high-resolution during mobilization [66]. Another aspect of CIEF MS is the lack of automation; however, recently this seems to have been solved [67,68].

In 1995, Tang et al. [69] demonstrated for the first time the on-line combination of CIEF with ESI using a triple quadrupole MS. The influence of carrier ampholyte concentration on separation performance and ESI detection sensitivity of some standard proteins was studied. A laboratory made set-up was used and separation performed with a polyacrylamide coated capillary. The results indicated that a 0.5% v/v solution of carrier ampholytes (Pharmalytes pH 3–10) gave the best overall performance. A higher concentration showed a better CIEF resolution between the proteins but at the expense of a decreased MS sensitivity (a reduction in protein peak intensity as well as a decrease in protein net charge). A lower ampholyte concentration ($<0.5\%$) showed a significant zone broadening. Chemical (cathodic) mobilization towards the MS inlet was initiated by replacing the catholyte (20 mM NaOH) with sheat-liquid consisting of 1% acetic acid in 50% methanol. Detection limits were in the 10 μg/ml range.

Using a similar CIEF MS set-up Yang et al. [70] separated glycoforms of transferrin. Di-, tri-, and tetrasialo transferrins were resolved by CIEF. Additional transferrin variants within each of the CIEF-separated di-, tri-, and tetrasialic-acid glycosylated transferrins were distinguished by the mass spectrometer. To speed up protein mobilization and minimize the effects of the moving ionic boundary inside the capillary, a gravity flow was superimposed on the chemical/cathodic mobilization by simply raising the inlet reservoir 8 cm above the electrospray needle. The induced hydrodynamic flow partly counter-balanced the migration of acetate ions into the capillary.

For other CIEF-MS reports by Lee et al. published before 1999, see Refs. [71–76].

Lyubarskaya et al. [77] coupled CIEF on-line to ESI ion-trap MS for the selection and structural identification of strong non-covalently bound ligands (tyrosine-phosphorylated peptides) to a specific protein domain (*src* homology 2 domain). Ion trap MS is a scanning instrument with MS^n capabilities and was useful here for the complete structural characterization (MS^3). Complex formation between ligands and protein was performed off-line in solution and mixed with 1–2% carrier ampholytes pH 3.5–10 and standard protein p*I* markers before injection into a PVA coated capillary. CIEF was performed using the Beckman P/ACE 5510 system. After focusing the complexes were mobilized by pressure and dissociated in the mass spectrometer whereby the free ligands were identified. The sheat liquid, at the catholyte side (1% acetic acid in 75% methanol), together with a high skimmer voltage, facilitated the dissociation of the complex. It was possible to detect protein concentration as low as 1.2 μM mixed with only 5 μM of peptide ligand. According to the authors the presence of 1% carrier ampholytes was not seen to interfere with the analysis, affecting neither sensitivity nor specific affinity interactions. Moreover, the ampholytes assisted solubilization of the hydrophobic protein–ligand complexes.

One potential drawback when interfacing high-speed high-resolution separation techniques like CE and CIEF to quadrupole mass spectrometers are their relatively low scanning rate compared with the peakwidths of the analytes (typically a few seconds at half peak height). If too limited a number of data points are collected over an analyte peak, sensitivity as well as separation performance will be affected. Lee et al. [78] reported the interfacing of CIEF to an ESI linear time-of-flight (TOF) MS. The TOF analyzer acquired spectra at a 5000 Hz frequency from which one averaged spectra per second was obtained. CIEF MS was performed using a laboratory made set-up and a 30 cm long polyacrylamide coated capillary. Standard proteins (0.5–1 mg/ml) were mixed with 0.5% Pharmalyte (pH 3–10 or 5–8). Focusing took place at 15 kV and chemical (cathodic) mobilization by exchanging the catholyte (20 mM sodium hydroxide) with sheat liquid consisting of 1% acetic acid in 80% methanol. Besides chemical mobilization, hydrodynamic mobilization by gravity was also applied in order to minimize the formation of a moving acetate ion boundary from the sheat liquid, as mentioned above (see Ref. [70]). A non-linear pH gradient was achieved during mobilization, which resulted in a greater separation among basic proteins than that of acidic proteins. With the current set-up, however, only a modest resolution (around 400) was achieved. The authors proposed using a reflector TOF design to enhance the resolution.

Fourier transform ion cyclotron resonance (FTICR) MS offers unsurpassed mass resolution and mass accuracy, as well as high sensitivity and the potential to perform multistage MS^n for the analyses of, e.g. peptides and proteins. In 1996, Severs et al. [79] demonstrated the first on-line interfacing of CIEF with FTICR as a mass detector. In this initial application the technique was used to analyze a red blood cell lysate. A laboratory made CIEF system was used. A polyacrylamide coated capillary was filled with 1% carrier ampholyte pH 5–8 and sample (2% red blood cell lysate). Anolyte and catholyte were 20 mM acetic acid and 20 mM ammonium hydroxide, respectively. After focusing, cathodic mobilization was achieved by a combination of chemical (acetic acid in methanol constituting the sheat liquid) and hydrodynamic (positive pressure) mobilization. Hemoglobin and carbonic anhydrase from the cell lysate were separated by CIEF and identified by FTICR.

In a number of more recent publications, the group of Smith et al. [80–88] has extended and developed the use of on-line CIEF-FTICR MS for the study of more complex protein samples, e.g. for proteome studies.

Jensen et al. [80] analyzed cell lysates from *E. coli*, cultured on both isotope depleted cell media and normal media, by CIEF-FTICR MS. A laboratory made set-up was used. The sample (~300 ng of protein extract) was mixed with 0.5% Pharmalyte pH 3–10 and injected into a polyacrylamide coated capillary. After focusing, a combination of cathodic and gravity mobilization was performed in a sheat-flow configuration (similar to that mentioned above, see Ref. [70]). Results from the experiments revealed ~900 putative proteins in one single analysis. The protein molecular mass was plotted against scan number resembling a 2D-gel image. While isotope depletion significantly extended achievable mass accuracy, dynamic range and detection limit, the scan rate (a duty cycle of 3 s) was somewhat of a limitation and several proteins could go undetected here during the time between scans. Intact proteins were used for identification but also on-line protein sequencing of intact proteins by MS-MS was shown, where specific mass tags were created for unambiguous protein identification. With the present instrumentation, however, this on-line sequencing is a slow process (~ 10 s) and not practically implemented with the present set-up. While CIEF is a highly resolving technique several proteins can still be accumulated in one band, as was clearly demonstrated by the authors in this report where more than 10 proteins sometimes were detected in the mass spectrum from one single CIEF peak. This is illustrated in Fig. 6.11A and B. Part A shows high-resolution FTICR MS spectra for charge states representing different molecular masses present in a single MS scan. Part B shows a mass spectrum containing peaks corresponding to at least 10 putative protein masses. The mass spectra were taken from the indicated peaks in the reconstructed IEF-FTICR electropherogram (bottom of Fig. 6.11) for a cell lysate grown in isotopically depleted media.

Zhang et al. [81] developed a stepwise mobilization method to improve CIEF resolution of complex protein mixtures and simplify on-line MS-MS for protein sequencing. Carrier ampholytes, due to their content of positively charged oligoamines, have been shown to adsorb to the surface of uncoated fused-silica capillaries and there suppress, eliminate,

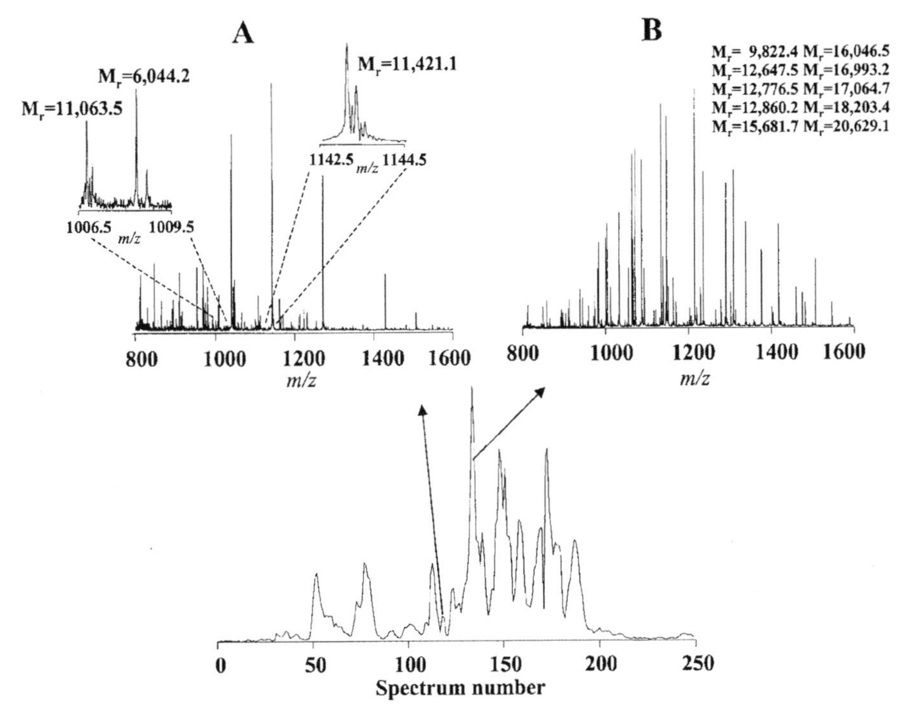

Fig. 6.11. CIEF-FTICR mass spectrometry. (A) Shows high-resolution FTICR MS spectra for charge states representing different molecular masses present in a single MS scan. (B) Shows a mass spectrum containing peaks corresponding to at least 10 putative protein masses. The mass spectra were taken from the indicated peaks in the CIEF-FTICR electropherogram (bottom of the figure) for a cell lysate grown in isotopically depleted media. From Ref. [80], with permission.

and sometimes reverse the electroosmotic flow (i.e. induce anodic flow). This phenomenon was used in this report to establish the pH gradient as well as to control the EOF. Mass spectrometry was performed using either an ion-trap or FTICR MS. A laboratory made CIEF system was used. An uncoated fused-silica capillary was filled with either standard proteins or *E. coli* lysate (protein concentration ~300 ng/ml) containing 0.5% Pharmalyte pH 3–10. Catholyte and anolyte were 60 mM ammonium acetate (pH 9.3) and 0.2 M acetic acid, respectively. Meanwhile focusing took place, mobilization of proteins towards the anode was achieved by controlling the inlet pressure (at cathode) and applied voltage, leaving the remaining protein zones focused inside the capillary. Protein zones were hereby stepwise mobilized into the mass spectrometer by changing the mobilization conditions. A five times higher resolution was obtained for the standard proteins by this approach compared with constant pressure mobilization conditions. The authors propose that the ability to increase the time interval between adjacent peaks may allow for efficient on-line coupling of CIEF with MS-MS experiments. Besides the increased resolution individual peaks can be sampled for extended times allowing either MS^n studies or more spectra to be collected (increasing the signal-to-noise ratio). The authors also modified the ESI sheat flow interface to facilitate automation of on-line CIEF-ESI-MS. The configuration described allowed for hydrodynamic mobilization of the separated proteins without interruption of applied voltages (over the capillary and electrospray needle, respectively) as well as avoiding the need for repositioning of the capillary 'tip' between focusing and mobilization.

Veenstra et al. [82] developed a CIEF-FTICR MS method for identifying intact proteins from genomic databases using a combination of partial amino acid content and accurate molecular mass measurements. Knowledge of the number of one (or several) amino acid residue(s) in a protein, besides the molecular mass, puts additional constraints on the database search, thus increasing the possibility of accurate identification. Protein (~0.25 mg/ml) was extracted from *E. coli* grown in minimal medium with and without isotopically labeled leucine (Leu-D_{10}), mixed with 0.5% Pharmalyte pH 3–10, and injected into a polyacrylamide coated capillary. Phosphoric acid 20 mM and sodium hydroxide 20 mM were used as anolyte and catholyte, respectively. After focusing, chemical (cathodic) mobilization was achieved by replacing the catholyte with sheat liquid (1% acetic acid in 50% methanol) concurrently with gravity mobilization. The mass difference between any same protein species from respective medium with or without isotopic labeling was used to determine the number of Leu present for detected proteins. In this report, the authors identified three proteins based on the number of Leu and molecular mass.

Jensen et al. [83] described recent progress in the development of on-line CIEF-FTICR mass spectrometry for proteome characterization. Studies have revealed 400–1000 proteins in the mass range of 2–100 kDa from a total injection of ~300 ng protein in one single analysis from cell lysates of both *E. coli* and *D. radiodurans*. The authors also demonstrated the use of isotope labeling of cell growth media to improve mass accuracy measurements and provide a means for quantitative proteome-wide measurements of protein expression. Protein expression profiling was performed by mixing isotopically depleted cell media (stress induced by Cd^{2+}) and normal (non-perturbed) media before protein extraction and CIEF-FTICR MS analysis. The peak intensity ratio of the two isotopically different and resolvable versions of each protein accurately reflected any change in protein abundance. In this report several such proteins were identified and relatively quantified. According to the authors, the features of FTICR MS, coupled with the advantages of CIEF separations, provides the basis for high throughput studies of proteomes with greater speed, resolution, and sensitivity than feasible with methods based on alternative technologies.

Gao et al. [84] developed a preparative CIEF method for separation and fractionation of proteins extracted from yeast cells grown in natural isotopic abundance and ^{15}N-enriched media, to measure changes in protein abundance. Protein fractions collected from the CIEF separation were digested with trypsin and the peptides further separated and analyzed by capillary LC online to FTICR MS. CIEF was performed in a large-bore format (200 μm i.d.) to increase the sample load. Because of this a flow restriction (20 μm i.d.) had to be fabricated in one end of the capillary to avoid siphoning, and IEF was performed in a constant current mode to reduce Joule heating. A laboratory made CIEF system was used. Protein (100 μg/ml) was extracted from the yeast cells (an equal mixture from respective media) mixed with 1% Pharmalyte pH 3–10 and injected into a HPC coated capillary. After focusing, more than 50 fractions were, by robotics, automatically collected into a 384 well plate by gravity mobilization (< 1 μl aliquots). A sheat liquid (1% ammonium hydroxide) was delivered (flow rate 2 μl/min) over the effluent during fractionation. The results demonstrated that from the analysis of one single CIEF fraction, 95% of the peptide mass measurements, from a complete LC run, matched predicted peptides from the yeast genome database within 5 ppm. Quantitative proteomic measurements were shown by the use of isotope labeling strategies, creating a comparative display between similar proteins expressed in respective media, one such protein was displayed and identified.

Isotopic labeling of *E. coli* cells (perturbed and non-perturbed) for high-throughput proteome precision measurements of protein expression using CIEF-FTICR MS has previously been reported by the same group, see Ref. [85].

A recent report by Smith et al. [86] describes the status of their efforts towards the development of a high-throughput proteomics capability.

Martinović et al. [87] described on-line CIEF-FTICR MS for the analysis of intact or dissociated non-covalent protein complexes. Model proteins consisting of tetramers and dimers were separated under native CIEF conditions and the complexes characterized in the mass spectrometer. A laboratory made CIEF system was used. The sample was mixed with 1% Pharmalyte pH 3–10 and injected into a polyacrylamide coated capillary. After focusing, gravity mobilization was applied concurrently with cathodic mobilization. The sheat liquid (cathodic mobilizer) together with the ESI interface parameters determined whether the protein complexes were to remain intact or dissociated in the mass spectrometer. The complexes dissociated using 1% acetic acid (50% methanol/49% water), pH 2.6, as sheat liquid, but were kept intact using 10 mM ammonium acetate, pH 6.8 (10–20% methanol). A procedure called 'in-trap' ion cleanup was also described. This ion cleanup removed extensive adduction to the multiprotein complexes whereby a higher mass accuracy and better sensitivity could be achieved. According to the authors, two successive experiments permitted a fast and efficient characterization of intact complex stoichiometry, the individual complex subunits and the possible presence of metal or other adducted species.

Martinović et al. [88] used infusion ESI-FTICR and on-line CIEF-FTICR MS to separate and characterize mixtures of αα, $\beta_1\beta_1$ and $\beta_3\beta_3$ alcohol dehydrogenase (ADH) isoenzymes

References pp. 130–133

with very narrow isoelectric points and similar molecular masses. For the CIEF-FTICR MS experiment, the sample was mixed with 0.5% v/v each of Pharmalyte pH 3–10 and 8–10.5 and separated in a 50 cm × 50 μm i.d. capillary (coated with linear polyacrylamide). A 2D display illustrated resolution of seven different ADH species, out of nine possible isomers, resolved in the p*I* 8.26–8.67 range. The species were detected in the mass spectra as denatured monomers. Mass spectrometry infusion experimental analysis of native ADHs further revealed that each non-covalent ADH complex contained two monomeric protein units and four zinc atoms. This was performed under non-denaturing conditions with the sample contained in ammonium acetate solution (neutral pH) and applying gentle ESI source conditions for the preservation of complexes, as described above, see Ref. [87].

6.8. MINIATURIZED CIEF AND chip-IEF

Microfabricated/microfluidic platforms (chips) for bioanalytical applications are receiving growing interest due to their potential for faster separations/analyses; multiplex sample processing/increased throughput; small sample and reagent consumption, and waste production; as well as lower production cost and instrument size. There is a strong drive to integrate several features on these miniaturized platforms in accordance with the 'lab-on-a-chip' or 'micro-total analysis' (μTAS) concept. Most microfluidic applications use electrokinetic force to manipulate and drive liquid flow in the microchannels while other means for fluid displacement like pressure-driven flow and centrifugal force also exist. Most modes of capillary electrophoresis, including CIEF, have been implemented on such platforms. Also some reports regarding the coupling of chip-IEF to mass spectrometry have recently appeared. For recent reviews regarding microfluidic systems, see Ref. [89] ('capillary electrophoresis on microchip'), Ref. [90] ('recent developments in electrokinetically driven analysis on microfabricated devices'), Ref. [91] ('current developments in electrophoretic and chromatographic separation methods on microfabricated devices'), Ref. [92] ('analytical microdevices for mass spectrometry'), Ref. [93] ('chip-MS: coupling the large with the small') and Ref. [94] ('recent developments in detection methods for microfabricated analytical devices'). In this part, covering the years 1999–2001, miniaturized CIEF as well as chip-IEF are reviewed.

6.8.1. Imaging detection

Mao and Pawliszyn [95] demonstrated chip-IEF with absorption imaging detection (for detection principle, see above, Refs. [7,8]). Microchannels were fabricated on a quartz wafer by chemical etching and covered with a quartz plate. The channels were coated with linear polyacrylamide. Myoglobin was mixed with 4% Pharmalytes pH 3–10 and injected into a 40 mm long × 100 μm wide × 10 μm deep separation channel. Vials containing anolyte (10 mM phosphoric acid) and catholyte (20 mM sodium hydroxide) were connected to the chip through hollow fibers. Focusing was performed at 3 kV. Results showed that myoglobin A and B (p*I* 7.2 and 6.8) were highly resolved after 10 min. The detection limit was 30 μg/ml, employing UV detection at 280 nm (64 scans were averaged per electropherogram to increase the signal-to-noise ratio). According to the authors this low concentration was not possible to achieve in a non-chip format despite a detection optical pathlength of only 10 μm; a better optical alignment was possible with the chip and less spike noise appeared as compared with their standard CIEF set-up with imaging detection.

Wu et al. [96] demonstrated CIEF in a very short length fused-silica capillary using a commercial imaging instrument (see Ref. [14]) with UV detection at 280 nm or a laboratory made set-up using a light-emitting diode (LED) at 430–470 nm as a light source. While not performed in a chip format the authors propose the simplification of integrating LEDs in a future chip format. The capillary dimensions were 1.2 cm × 200 μm i.d. and coated with linear polyacrylamide. Myoglobins (p*I* 6.8 and 7.2), and p*I* markers, were individually mixed with 3% Pharmalyte pH 6–8 (and pH 3–10) and 0.35% methylcellulose, injected, and focused in 160 s at 500 V (80 s with pH gradient 3–10). Peak width at half peak height was about 0.07 pH units (0.13 units with pH gradient 3–10). With a LED as a light source one challenge is to match the wavelength of emitting light from the LED with the analyte absorption maxima to gain efficient detection sensitivity. In this example, because of this mismatch, the authors needed a high concentration of 1.2 mg/ml of myoglobin to visualize a peak once focused. Several kind of LEDs (red, yellow, and blue) have recently been developed and there is a trend towards developing shorter wavelength emissions, e.g. 280 nm, thus making LEDs universally applicable for CIEF of most proteins in a chip format.

A similar miniaturized CIEF system with imaging detection, aimed at being a portable hand-held analysis system, was reported by Herr et al. [97] using an array of blue LEDs as excitation sources directed towards a 8 mm long × 75 μm i.d. bare fused-silica separation channel (methylcellulose was included in sample to suppress EOF). A CCD camera was used as a fluorescence detector. Focusing of two Cy5 labeled peptides (500 nM each) in Biolyte pH 3–10 was achieved after only 46 s (500 V/cm). The resolution, however, was moderate at 0.65 pH units.

In a more recent report by Herr et al. [98] means were explored to extract a purified, concentrated sample zone from a chip-IEF separation employing a one-step focusing/mobilization method. Their future aim would be to couple chip-IEF to other micro separation techniques as part of a multidimensional analysis/separation system. Fluorescence imaging was accomplished using an epi-fluorescent microscope, a mercury lamp as an excitation source, and a CCD camera for detection. The chip was fabricated from polymethylmethacrylate (PMMA) by hot embossing and contained a 6 cm long separation channel (200 μm wide and 40 μm deep). No coating of the channel was performed since the EOF was low enough for an efficient protein focusing. Green fluorescent protein (GFP), 190 nM was used as the sample. Focusing took place at 500 V/cm; meanwhile protein zones were transported by EOF towards a T-junction (an intersection between two orthogonal microfluidic channels) where electrokinetic extraction of the sample took place in the orthogonal direction. The protein migration velocity in the orthogonal direction was over a magnitude higher due to a higher field strength applied. Therefore, the extracted protein (or mixture of proteins in one zone) potentially can be further analyzed/resolved in a second dimension before a new sample fraction from the IEF separation channel is introduced. This is illustrated in Fig. 6.12a–d where the focused sample zone was transported towards—and orthogonally extracted at—the intersection.

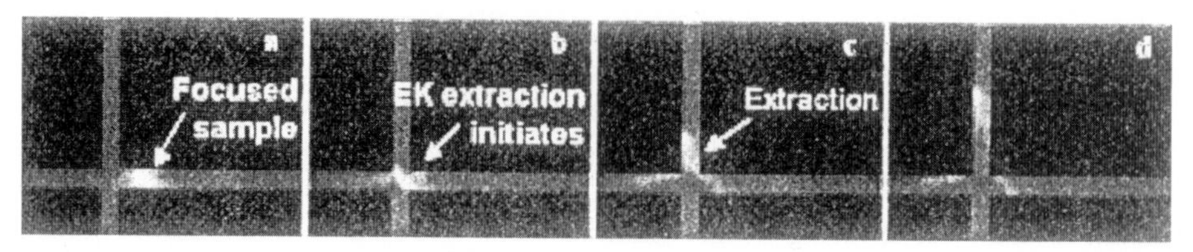

Fig. 6.12. Chip-IEF. Green fluorescent protein was focused in a microchannel and simultaneously mobilized by EOF towards the intersection where it was electrokinetically extracted into the orthogonal channel (by a higher applied potential). The system demonstrated aims at coupling various microseparation techniques as part of a multidimensional separation system. From Ref. [98], with permission.

6.8.2. Mobilization strategies

VanderNoot et al. [99] proposed that one potential barrier to the miniaturization of IEF has been the incorporation of mobilization strategies that are compatible with micro-scale devices. In this report they demonstrated the feasibility of using a capillary electrokinetic pump coupled online to CIEF (non-chip) for the mobilization of focused protein zones past a single-point detector. A schematic design of the set-up is shown in Fig. 6.13. The electrokinetic pump consisted of a positively charged porous polymer monolith polymerized in-situ in a short capillary piece. A reversed EOF was generated with direction towards the IEF separation capillary. Using a 'floating' power supply two different voltages could be applied over the monolith and IEF capillary, respectively. Therefore, while focusing occurred in the IEF capillary, analyte zones were slowly transported past the detector by the electroosmotically generated hydrodynamic flow (electrokinetically generated pressure mobilization) from the pump. Evaluation was performed with an ampholyte/p*I* marker mixture. Detection was by UV absorbance at 280 nm. Focusing was at 15 kV for 5 min while mobilization at different voltages were demonstrated. Results showed the slowest mobilization rate (−200 V/cm) afforded the best resolution and linearity of pH gradient; the run was completed in ~50 min. (Though not mentioned in this report it must be assumed that the IEF capillary was coated with a following reduced- or eliminated-EOF, reviewers remark.)

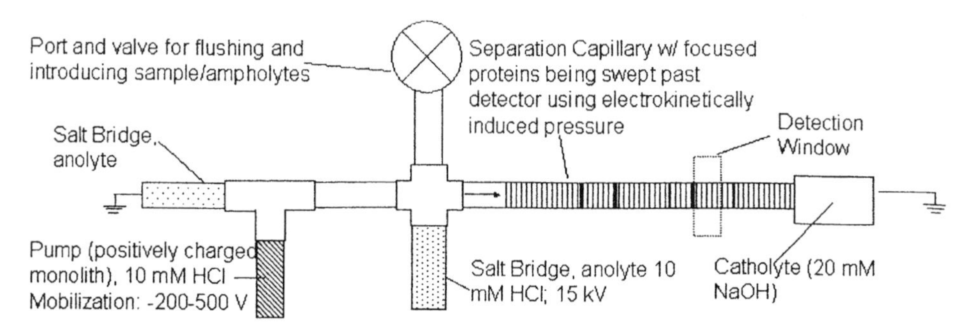

Fig. 6.13. CIEF using an electrokinetically driven pump (creating reversed EOF flow) to mobilize focused sample zones past a detector by pressure, see text for details. Two different potentials were simultaneously applied over the separation capillary and pump, respectively. From Ref. [99], with permission.

Hofmann et al. [100] evaluated, by 'standard' CIEF, different mobilization methods: chemical, hydrodynamic, and EOF, in an effort to transfer the most suitable method to chip-IEF. 'Standard' CIEF was performed with the Beckman P/ACE 5510 system employing an eCAP neutral coated capillary for the two-step method and an uncoated fused-silica capillary for the EOF one-step method. Results showed that both chemical and hydrodynamic mobilization gave better separation efficiency and reproducibility than the EOF method. However, the authors proposed EOF mobilization to be most suitable for miniaturization because of the high speed. The chip-IEF channels were fabricated on glass by chemical wet etching (isotropic etching) and a glass cover plate was thermally bonded onto its top. A sample consisting of Cy5-labeled (fluorophore) peptides was mixed with 40% glycerol and 1.85% Pharmalytes pH 3–10 and injected into a 7 cm long (200 μm wide and 10 μm deep) separation channel. Focusing was performed at 1000 V/cm. At a run time of less than 5 min six peptides were baseline separated with an expected peak capacity of 30–40. On-line laser-induced fluorescence (LIF) detection was used at half channel length (excitation wavelength at 632 nm) for best sensitivity and resolution.

6.8.3. Miscellaneous applications

Rossier et al. [101] proposed the use of chip-IEF. The device was fabricated from 100 μm thick PET sheets and microchannels created by UV excimer laser photoablation. A PET film was used to cover the channels. The separation channel had the dimensions 2 cm × 100 μm wide × 50 μm deep wherein a polyacrylamide gel was polymerized containing two polymerized ampholytes (pK_a 4.6 and 6.2, pH set to 5.4). A mixture of cytochrome C and β-lactoglobulin was dissolved in water and placed in the anode reservoir and voltage applied. The immobilized ampholytes formed an isoelectric trap. Only the more basic protein cyt C migrated through the gel to the cathode while the more acidic lactoglobulin never entered the gel ($pI < 5.4$). Fractions (i.e. cyt C) were collected at the cathode and analyzed/detected off-line by capillary electrophoresis.

Xu et al. [102] demonstrated chip-IEF on a microfluidic device made of poly(methyl methacrylate) (PMMA). Microchannels were imprinted using a silicon template and high pressure at room temperature. A poly(dimethylsiloxane) (PDMS) film was used to cover and seal the channels. Chip IEF was performed in a non-coated 3 cm long × 91 μm width × 32 μm depth channel. Anolyte and catholyte were 20 mM phosphoric acid and 20 mM sodium hydroxide, respectively. A sample consisting of 10 μg/ml GFP was mixed with 2% Pharmalyte pH 3–10 and focused at 500 V/cm. Focusing was studied on-line with fluorescence microscopy (excitation wavelength 488 nm). A final sample zonewidth of 0.2 mm was obtained. PMMA is optically transparent at this wavelength and also shows a low background fluorescence compared with many other plastic materials. No EOF was detected, probably because of ampholyte adsorption as proposed earlier (see above, Ref. [81]).

6.8.4. Mass spectrometry

Wen et al. [103] described a microfabricated (chip) device for IEF incorporating an electrospray ionization emitter 'tip' for on-line chip-IEF electrospray ionization mass spectrometry (ESI MS). The tip was an integrated part of the device. The chip was made of polycarbonate and contained a serpentine separation channel 16 cm long × 50 μm deep × 30 μm wide, machined by excimer laser. A PET sheet was thermally bonded onto the chip closing the microchannels. The separation channel was non-coated and EOF was found to be 25–30% less (between pH 4 and 10) that of bare fused-silica capillary. The sample consisting of myoglobin (p*I* 7.2, 50 μg/ml) and carbonic anhydrase (p*I* 5.9, 50 μg/ml) was mixed with 1% Pharmalyte pH 3–10 and injected into the chip. Anolyte and catholyte were 10% acetic acid and 0.3% ammonium hydroxide, respectively. Porous membranes were used to isolate the electrolytes (and electrodes) from the separation channel allowing the electrical connections to be made without fluid flow as well as preventing air bubbles formed at the electrodes from entering the channel. Focusing was performed for 10 min with −8 kV applied at the cathode, whereas the anode was grounded. Mobilization was towards the anode by hydrostatic pressure; meanwhile the voltage at the cathode was changed to −6 and +2 kV applied on the electrospray emitter (anode side) to spray the analytes into an ion-trap mass spectrometer. The spray was found to be more stable with sheat gas and sheat liquids. Figure 6.14 shows a photograph (left side) of part of the chip including the electrospray emitter, the right side shows the assembly of the chip. Figure 6.15 shows a reconstructed electropherogram of three proteins separated by chip-IEF as detected by the mass spectrometer. The resolution obtained was typically lower than CIEF-ESI MS using coated fused-silica capillaries and the authors proposed several improvements in future designs.

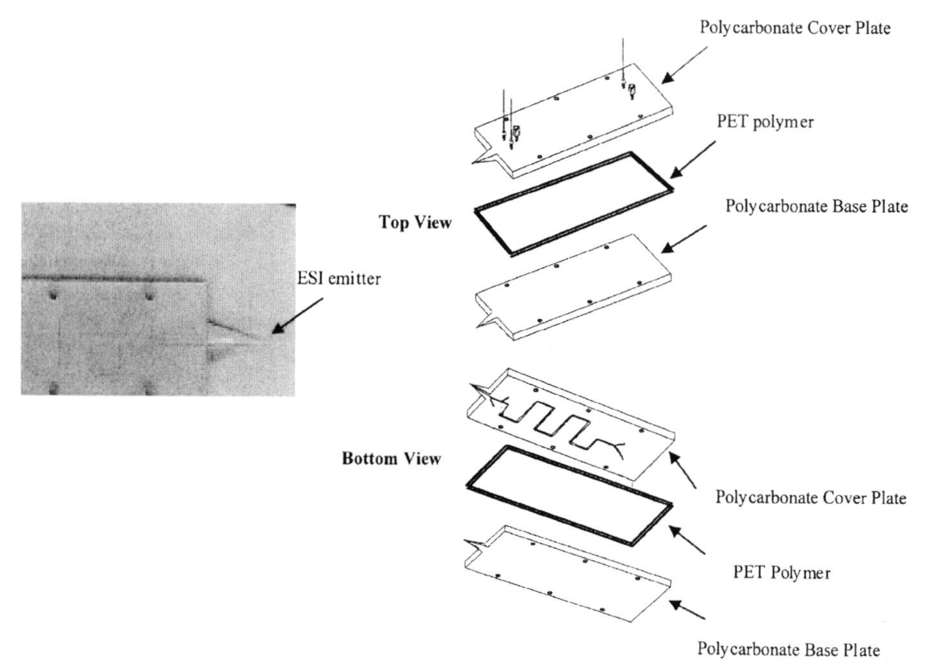

Fig. 6.14. (Left) Photograph of a microfabricated IEF chip including part of the separation channel and the electrospray emitter. (Right) Assembly of the various parts. From Ref. [103], with permission.

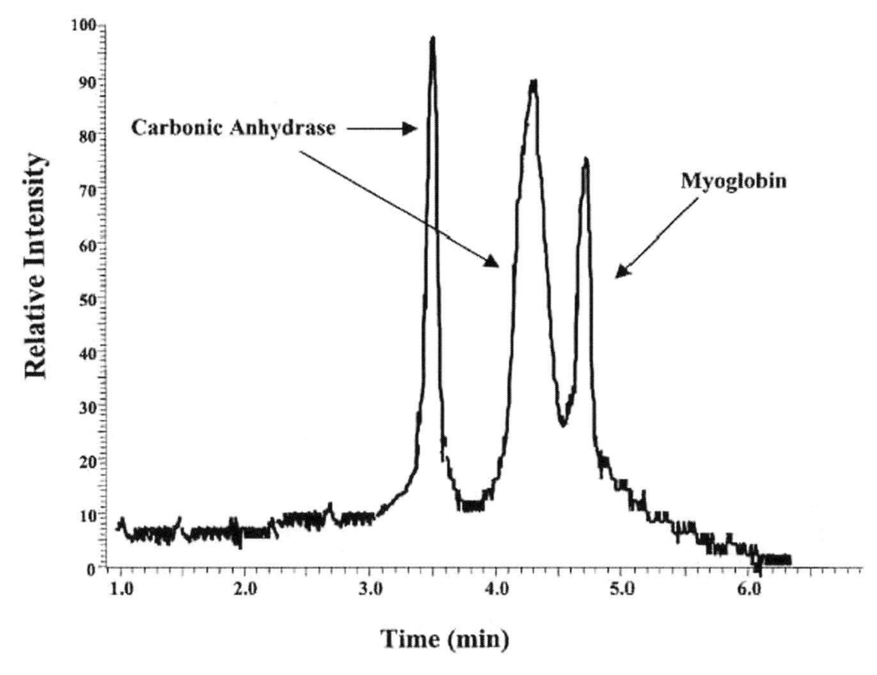

Fig. 6.15. Reconstructed electropherogram showing protein separation and detection by chip-IEF-ESI mass spectrometry. 50 μg/ml each of myoglobin (p*I* 7.2) and carbonic anhydrase (p*I* 5.9 and 6.8) were injected. From Ref. [103], with permission.

Free flow electrophoresis (FFE) is mostly used for preparative purposes and can be performed in various electrophoretic modes [104]. Chartogne et al. [105] reported interfacing 'standard' CIEF to electrospray ionization mass spectrometry (ESI MS) via a FFE chip. The purpose was to reduce the concentration of ampholytes in the sample by passing the CIEF resolved proteins over the FFE chip. Many ampholytes have mobilities different from proteins and will therefore take other trajectory paths in FFE. Ampholytes will therefore be removed from the proteins whereby the purified proteins are collected via a transfer capillary at the exit of the FFE device and transported by hydrodynamic flow into the electrospray source. An enhanced sensitivity and spray stability was therefore expected by ESI MS. A schematic representation of the instrumental set-up is shown in Fig. 6.16. A 90 cm long × 50 μm i.d. fused-silica capillary (2 in Fig. 6.16) was interfaced to the FFE chip via a finger-tight inlet. The capillary (including transfer capillary) was coated with siloxanediol-polyacrylamide while no coating was applied on the polycarbonate made FFE chip. UV detection at 280 nm of focused proteins was performed near the FFE inlet (UV 1). A syringe pump was connected via tubing to a second inlet on the FFE chip to create a hydrodynamic flow sweeping the CIEF separated proteins through the chip to its exit side (3 in Fig. 6.16). An electric field was imposed on the transverse flow in the chip. Proteins traveling through the chip were collected by a transfer capillary (rightmost in the figure). This capillary also acted as part of a coaxial sheat flow system interfaced to the inlet of the MS for electrospray of proteins. The transfer capillary was one of three capillaries, where

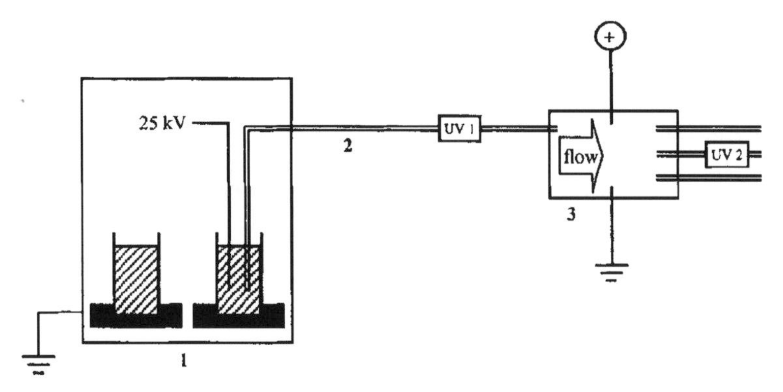

Fig. 6.16. Instrumental set-up for CIEF interfaced to chip-FFF (and further to ESI mass spectrometry). Two UV detectors were used for optimization of sample transfer from the IEF capillary, by the FFE device, and into the mass spectrometer. From Ref. [105], with permission.

the other two were wastes. UV detection (UV 2) was performed on-line with the transfer capillary. The CIEF capillary was filled with 1 mg/ml each of myoglobin (p*I* 7.2), carbonic anhydrase (p*I* 6.6) and β-lactoglobulin (p*I* 5.15), mixed with 5% Pharmalytes pH 5–8. Focusing/mobilization was at 25 kV and applying 45 mbar. FFE voltage was 56 V and carrier flow rate 2.75 μl/min. A CIEF-FFE separation observed at UV1 and UV2 is shown in Fig. 6.17a and b, respectively. Interestingly, acetic acid (1%) was used as both anolyte and catholyte. The catholyte constituted the carrier flow through the FFE chip. This arrangement, however, did not create a stationary pH gradient condition; instead the carrier ampholytes, while focusing, also migrated towards the cathode (i.e. the FFE chip). While resembling a one-step EOF driven CIEF method this focusing/mobilization process is different and contained capillary zone electrophoresis superimposed upon IEF. Therefore,

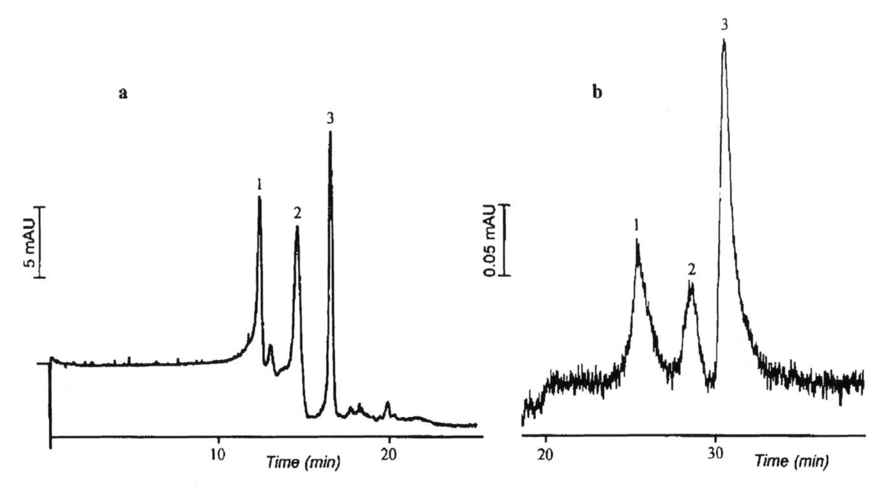

Fig. 6.17. (a) CIEF-FFE separation of myoglobin (p*I* 7.2), carbonic anhydrase (p*I* 6.6) and β-lactoglobulin (p*I* 5.15) as observed at UV1 and (b) UV2, see Fig. 6.16. From Ref. [105], with permission.

migration/mobilization times were related to electrophoretic mobility besides the isoelectric point. Moreover, mobilization was partly due to an over-pressure applied at the inlet of the IEF capillary to compensate for the pressure-induced backflow into the same capillary by the pressure generated in the FFE chip (induced by the carrier flow). While the authors demonstrated the feasibility of CIEF-FFE/ESI MS, future experiments will work on improving the FFE device to reduce peak broadening and sample dilution.

6.8.5. Natural pH gradients

In a series of reports Yager et al. [106–109] investigated the use of IEF in microfluidic devices, i.e. chip-IEF. Key features of the chip-IEF device are the absence of electrolyte reservoirs and anolyte and catholyte, integration of the electrodes with the walls of the fluidic separation channel, and an electric field perpendicular to the fluid flow direction (similar to that mentioned above, see Ref. [105]). No ampholytes are used; a natural pH gradient is generated across the channel by electrolysis of water generating H^+ at the anode and OH^- at the cathode (see also Ref. [15]). The voltage applied is ~2–2.5 V. This low absolute voltage reduces or eliminates the generation of gas bubbles that otherwise could interfere with the uniform flow in the channel. A schematic representation of the microfluidic channel (electrochemical flow cell) containing gold electrodes are shown in Fig. 6.18 (see Ref. [107]). An optical detection axis (not shown in the figure) is positioned in the middle and perpendicular to the microfluidic channel. Visualization is typically performed by optical microscopy and images recorded by a CCD camera.

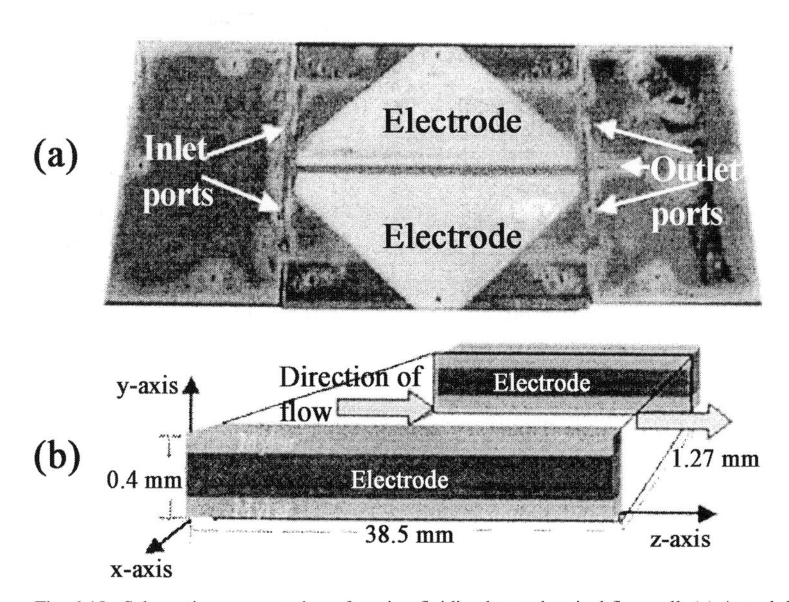

Fig. 6.18. Schematic representation of a microfluidic electrochemical flow cell. (a) Actual device, consisting of two gold electrodes sandwiched between layers of Mylar and held together with pressure-sensitive adhesive; (b) schematic of the main sample channel. From Ref. [107], with permission.

Macounová et al. [106] described the manufacturing of microfluidic channels in polyester (Mylar) material formed by two parallel gold electrodes (40 mm long with a distance between the electrodes of 2.54 mm). They also quantified, by optical imaging, the developing natural pH gradient formed under non-flow (static) conditions using acid–base indicators, and run proteins in the microchannel by IEF. Results showed that the pH profile generated in the channel was sensitive to the initial pH, and in comparison with a non-buffered electrolyte, a buffered electrolyte (0.1 mM histidine) was necessary to create a stable pH gradient. The time needed for stable pH gradient development ranged from 8 min at 2 V to 4 min at 2.3 V. Hemoglobin and fluorescently labeled serum albumin were seen focused into a single and narrowed band.

In a later report, Cabrera et al. [107] extended this study to the production of natural pH gradient formation in a similar microfluidic device under flow (non-static) conditions. The channel dimensions were 40 mm long × 0.4 mm thick × 1.25 mm between electrodes. In conjunction a theoretical model was developed describing the various phenomena occurring in the system such as electrolysis at the electrodes, buffering effects of weak acids and bases, and mass transport caused by diffusion and electrophoresis. The model prediction was shown to agree well with experimental results of IEF of a protein (fluorescently labeled serum albumin) in the chip device.

Macounová et al. [108] further developed the concept by concentration and separation of a binary mixture of proteins using a novel three-inlet configuration of the microfluidic device (the proteins were introduced into the channel through the middle stream while the two streams closer to the electrodes contained buffer only). The properties of pH gradient formed were studied by variation of parameters such as initial pH, ionic strength and composition of buffers, flow rate, and applied voltage. It was demonstrated that by changing the electrode material made of gold for palladium, higher initial voltages could be used enabling more rapid focusing and narrower focused bands. This was due to the 'non-gassing' character of palladium. It was also shown that a shallower pH gradient formation was produced in multiple buffers as compared with single buffer solutions. An example of IEF separation was demonstrated for a mixture of fluorescently labeled serum albumin and lectin.

Cabrera et al. [109] demonstrated the concentration of bacteria (*E. herbicola*) in a flowing stream using a similar microfluidic device. A positively charged detergent (CTAB) had to be added to suppress EOF while the excess detergent was removed before sample injection to keep the bacteria viable during analysis. The position and narrowness of the focused zone could be adjusted by altering the IEF buffer composition.

6.9. REFERENCES

1. J.P. Landers (Ed.), Handbook of Capillary Electrophoresis, CRC Press, London, 1997
2. T. Wehr, R. Rodriguez-Diaz and M. Zhu, Chromatographia Suppl., 53 (2001) S45
3. P. Glukhovskiy and G. Vigh, Anal. Chem., 71 (1999) 3814
4. A.M. Rizzi and L. Kremser, Electrophoresis, 20 (1999) 3410
5. Q. Mao, J. Pawliszyn and W. Thormann, Anal. Chem., 72 (2000) 5493
6. Z. Huang and C.F. Ivory, Anal. Chem., 71 (1999) 1628
7. Q. Mao and J. Pawliszyn, J. Biochem. Biophys. Meth., 39 (1999) 93

8. J. Wu, C. Tragas, A. Watson and J. Pawliszyn, Anal. Chim. Acta, 383 (1999) 67
9. J.-L. Liao and R. Zhang, J. Chromatogr. A, 684 (1994) 143
10. R. Zhang and S. Hjerten, Anal. Chem., 69 (1997) 1585
11. J. Wu, C. Tragas, A. Watson and J. Pawliszyn, Anal. Chim. Acta, 383 (1999) 67
12. N.J. Clarke, A.J. Tomlinson and S. Naylor, Am. Soc. Mass Spectrom., 8 (1997) 743
13. N.J. Clarke, A.J. Tomlinson, G. Schomburg and S. Naylor, Anal. Chem., 69 (1997) 2786
14. Q. Mao and J. Pawliszyn, J. Chromatogr. B, 729 (1999) 355
15. T. Huang, X.-Z. Wu and J. Pawliszyn, Anal. Chem., 72 (2000) 4758
16. T. Huang and J. Pawliszyn, Analyst, 125 (2000) 1231
17. X. Fang, M. Adams and J. Pawliszyn, Analyst, 124 (1999) 335
18. X.-Z. Wu, T. Huang, W.M. Mullet, J.M. Yeung and J. Pawliszyn, J. Microcol. Sep., 13 (2001) 322
19. C. Tragas and J. Pawliszyn, Electrophoresis, 21 (2000) 227
20. D.P. Richards, C. Stathakis, R. Polakowski, H. Ahmadzadeh and N.J. Dovichi, J. Chromatogr. A, 853 (1999) 21
21. T.J. Nelson, J. Chromatogr. A, 623 (1992) 357
22. C. Gelfi, M. Curcio, P.G. Righetti, R. Sebastiano, A. Citterio, A. Ahmadzadeh and N.J. Dovichi, Electrophoresis, 19 (1998) 1677
23. S. Hjertén and M.-D. Zhu, J. Chromatogr., 346 (1985) 265
24. M. Hashimoto, K. Tsukagoshi, R. Nakajima and K. Kondo, J. Chromatogr. A, 852 (1999) 597
25. M. Hashimoto, K. Tsukagoshi, R. Nakajima and K. Kondo, Anal. Sci., 15 (1999) 1281
26. G.F. Verbeck and S.C. Beale, J. Microcol. Sep., 11 (1999) 708
27. R.A. Strong, H. Liu, I.S. Krull, B.-Y. Cho and S.A. Cohen, J. Liq. Chrom. Rel. Technol., 23 (2000) 1775
28. M. Minarik, F. Foret and B.L. Karger, Electrophoresis, 21 (2000) 247
29. M. Horka, T. Willimann, M. Blum, P. Nording, Z. Friedl and K. Slais, J. Chromatogr. A, 916 (2001) 65
30. K. Shimura, W. Zhi, H. Matsumoto and K. Kasai, Anal. Chem., 72 (2000) 4747
31. K. Shimura, Z. Wang, H. Matsumoto and K. Kasai, Electrophoresis, 21 (2000) 603
32. Y. Shen, F. Xiang, T.D. Veenstra, E.N. Fung and R.D. Smith, Anal. Chem., 71 (1999) 5348
33. Y. Shen and R.D. Smith, J. Microcol. Sep., 12 (2000) 135
34. A. Palm, M. Zaragoza-Sundqvist, G. Marko-Varga, submitted to J. Sep. Sci.
35. K. Srinivasan, C. Pohl and N. Avdalovic, Anal. Chem., 69 (1997) 2798
36. Y. Shen, S.J. Berger and R.D. Smith, J. Chromatogr. A, 914 (2001) 257
37. T. Manabe, Electrophoresis, 20 (1999) 3116
38. J.R. Mazzeo, J.A. Martineau and I.S. Krull, Anal. Biochem., 208 (1993) 323
39. K.A. Cruickshank, J. Olvera and U.R. Muller, J. Chromatogr. A, 817 (1998) 41
40. K. Shimura, H. Matsumoto and K. Kasai, Electrophoresis, 19 (1998) 2296
41. Y. Shen, S.J. Berger, G.A. Anderson and R.D. Smith, Anal. Chem., 72 (2000) 2154
42. M. Taverna, N.T. Tran, T. Merry, E. Horvath and D. Ferrier, Electrophoresis, 19 (1998) 2572
43. M.A. Strege and A.L. Lagu, Electrophoresis, 18 (1997) 2343
44. I.S. Krull, X. Liu, J. Dai, C. Gendreau and G. Li, J. Pharm. Biomed. Anal., 16 (1997) 377
45. X. Liu, Z. Sosic and I.S. Krull, J. Chromatogr. A, 735 (1996) 165
46. S. Tang, D.P. Nesta, L.R. Maneri and K.R. Anumula, J. Biopharm. Biomed. Anal., 19 (1999) 569
47. L.C. Santora, I.S. Krull and K. Grant, Anal. Biochem., 275 (1999) 98
48. A. Cifuentes, M.V. Moreno-Arribas, M. de Frutos and J.C. Diez-Masa, J. Chromatogr. A, 830 (1999) 453
49. N.T. Tran, M. Taverna, M. Chevalier and D. Ferrier, J. Chromatogr. A, 866 (2000) 121
50. N.T. Tran, Y. Daali, S. Cherkaoui, M. Taverna, J.R. Neeser and J.-L. Veuthey, J. Chromatogr. A, 929 (2001) 151
51. C. Jochheim, S. Novick, A. Balland, J. Mahan-Boyce, W.-C. Wang, A. Goetze and W. Gombotz, Chromatographia Suppl., 53 (2001) S59
52. J.P.M. Hui, P. Lanthier, T.C. White, S.G. McHugh, M. Yaguchi, R. Roy and P. Thibault, J. Chromatogr. B, 752 (2001) 349
53. M.-L. Hagmann, C. Kionka, M. Schreiner and C. Schwer, J. Chromatogr. A, 816 (1998) 49
54. Q.-H. Ru, G.-A. Luo, S.-F. Sui and S.-X. Wang, J. Cap. Elec. Microchip Tech., 5/6 (1999) 177
55. T. Manabe, H. Miyamoto, K. Inoue, M. Nakatsu and M. Arai, Electrophoresis, 20 (1999) 3677

56. A. Hiraoka, K. Seiki, H. Oda, N. Eguchi, Y. Urade, I. Tominaga and K. Baba, Electrophoresis, 22 (2001) 3433
57. A. Lupi, S. Viglio, M. Luisetti, M. Gorrini, P. Coni, G. Faa, G. Cetta and P. Ladarola, Electrophoresis, 21 (2000) 3318
58. A.M. Lasdun, R.R. Kurumbail, N.K. Leimgruber and A.S. Rathore, J. Chromatogr. A, 917 (2001) 147
59. M.A. Jenkins and S. Ratnaike, Clin. Chim. Acta, 289 (1999) 121
60. J.M. Hempe and R.D. Craver, Electrophoresis, 21 (2000) 743
61. M. Sugano, H. Hidaka, K. Yamauchi, T. Nakabayashi, Y. Higuchi, K. Fujita, N. Okumura, Y. Ushiyama, M. Tozuka and T. Katsuyama, Electrophoresis, 21 (2000) 3016
62. D.W. Armstrong, G. Schulte, J.M. Schneiderheinze and D.J. Westenberg, Anal. Chem., 71 (1999) 5465
63. Y. Shen, S.J. Berger and R.D. Smith, Anal. Chem., 72 (2000) 4603
64. A. von Brocke, G. Nicholson and E. Bayer, Electrophoresis, 22 (2001) 1251
65. F. Foret, T.J. Thompson, P. Vouros, B. Karger, P. Gebauer and P. Boček, Anal. Chem., 66 (1994) 4450
66. T. Manabe, H. Miyamoto and A. Iwasaki, Electrophoresis, 18 (1997) 92
67. D.P. Kirby, J.M. Thorne, W.K. Gotzinger and B. Karger, Anal. Chem., 68 (1996) 4451
68. C.-X. Zhang, F. Xiang, L. Paša-Tolić, G.A. Anderson, T.D. Veenstra and R.D. Smith, Anal. Chem., 72 (2000) 1462
69. Q. Tang, A.K. Harrata and C.S. Lee, Anal. Chem., 67 (1995) 3515
70. L. Yang, Q. Tang, A.K. Harrata and C.S. Lee, Anal. Biochem., 243 (1996) 140
71. Q. Tang, A.K. Harrata and C.S. Lee, Anal. Chem., 68 (1996) 2482
72. Q. Tang, A.K. Harrata and C.S. Lee, Anal. Chem., 69 (1997) 3177
73. P.K. Jensen, A.K. Harrata and C.S. Lee, Anal. Chem., 70 (1998) 2044
74. J. Wei, L. Yang, A.K. Harrata and C.S. Lee, Anal. Chem., 70 (1998) 2044
75. L. Yang, C.S. Lee, S.A. Hofstadler, L. Paša-Tolić and R.D. Smith, Anal. Chem., 70 (1998) 3235
76. L. Yang, C.S. Lee, S.A. Hofstadler and R.D. Smith, Anal. Chem., 70 (1998) 4945
77. Y.V. Lyubarskaya, S.A. Carr, D. Dunnington, W.P. Pritchett, S.M. Fisher, E.R. Appelbaum, C.S. Jones and B.L. Karger, Anal. Chem., 70 (1998) 4761
78. J. Wei, C.S. Lee, I.M. Lazar and M.L. Lee, J. Microcol. Sep., 11 (1999) 193
79. J.C. Severs, S.A. Hofstadler, Z. Zhao, R.T. Senh and R.D. Smith, Electrophoresis, 17 (1996) 1808
80. P.K. Jensen, L. Paša-Tolić, G.A. Anderson, J.A. Horner, M.S. Lipton, J.E. Bruce and R.D. Smith, Anal. Chem., 71 (1999) 2076
81. C.-X. Zhang, F. Xiang, L. Paša-Tolić, G.A. Anderson, T.D. Veenstra and R.D. Smith, Anal. Chem., 72 (2000) 1462
82. T.D. Veenstra, S. Martinović, G.A. Anderson, L. Paša-Tolić and R.D. Smith, Am. Soc. Mass Spectrom., 11 (2000) 78
83. P.K. Jensen, L. Paša-Tolić, K.K. Peden, S. Martinovic, M.S. Lipton, G.A. Anderson, N. Tolić, K.-K. Wong and R.D. Smith, Electrophoresis, 21 (2000) 1372
84. H. Gao, Y. Shen, T.D. Veenstra, R. Harkewicz, G.A. Anderson, J.E. Bruce, L. Paša-Tolić and R.D. Smith, J. Microcol. Sep., 12 (2000) 383
85. L. Paša-Tolić, P.K. Jensen, G.A. Anderson, M.S. Lipton, K.K. Peden, S. Martinović, N. Tolić, J.E. Bruce and R.D. Smith, J. Am. Chem. Soc., 121 (1999) 7949
86. R.D. Smith, L. Paša-Tolić, M.S. Lipton, P.K. Jensen, G.A. Anderson, Y. Shen, T.P. Conrads, H.R. Udseth, R. Harkewicz, M.E. Belov, C. Masselon and T.D. Veenstra, Electrophoresis, 22 (2001) 1652
87. S. Martinović, S.J. Berger, L. Paša-Tolić and R.D. Smith, Anal. Chem., 72 (2000) 5356
88. S. Martinović, L. Paša-Tolić, C. Masselon, P.K. Jensen, C.L. Stone and R.D. Smith, Electrophoresis, 21 (2000) 2368
89. V. Dolnik, S. Liu and S. Jovanovich, Electrophoresis, 21 (2000) 41
90. G.J.M. Bruin, Electrophoresis, 21 (2000) 3931
91. J.P. Kutter, Trends Anal. Chem., 19 (2000) 352
92. R.D. Oleschuk and D.J. Harrison, Trends Anal. Chem., 19 (2000) 379
93. A.J. de Mello, Lab on a Chip, 1 (2001) 7N
94. M.A. Schwarz and P.C. Hauser, Lab on a Chip, 1 (2001) 1
95. Q. Mao and J. Pawliszyn, Analyst, 124 (1999) 637

96. X.-Z. Wu, N.S.-K. Sze and J. Pawliszyn, Electrophoresis, 22 (2001) 3968
97. A.E. Herr, J.I. Molho, J.G. Santiago, T.W. Kenny, D.A. Borkholder, G.J. Kintz, P. Belgrader and M.A. Nothrup, in: A. van den Berg et al. (Eds.), Micro Total Analysis System, Kluwer Academic Publishers, Dordrecht, 2000, p. 367
98. A.E. Herr, J.I. Molho, R. Bharadwaj, J.C. Mikkelsen, J.G. Santiago, T.W. Kenny, D.A. Borkholder and M.A. Northrup, in: J.M. Ramsey and A. van den Berg (Eds.), Micro Total Analysis Systems, Kluwer Academic Publishers, Dordrecht, 2001, p. 51
99. V.A. VanderNoot, G. Hux, J. Schoeniger and T. Shepodd, in: J.M. Ramsey and A. van den Berg (Eds.), Micro Total Analysis Systems, Kluwer Academic Publishers, Dordrecht, 2001, p. 127
100. O. Hofmann, D. Che, K.A. Cruickshank and U.R. Muller, Anal. Chem., 71 (1999) 678
101. J.S. Rossier, A. Schwarz, F. Reymond, R. Ferrigno, F. Bianchi and H.H. Girault, Electrophoresis, 20 (1999) 727
102. J. Xu, L. Locascio, M. Gaitan and C.S. Lee, Anal. Chem., 72 (2000) 1930
103. J. Wen, Y. Lin, F. Xiang, D.W. Matson, H.R. Udseth and R.D. Smith, Electrophoresis, 21 (2000) 191
104. L. Krivankova and P. Boček, Electrophoresis, 19 (1998) 1064
105. A. Chartogne, U.R. Tjaden and J. van der Greef, Rapid Commun. Mass Spectrom., 14 (2000) 1269
106. K. Macounová, C.R. Cabrera, M.R. Holl and P. Yager, Anal. Chem., 72 (2000) 3745
107. C.R. Cabrera, B. Finlayson and P. Yager, Anal. Chem., 73 (2001) 658
108. K. Macounová, C.R. Cabrera and P. Yager, Anal. Chem., 73 (2001) 1627
109. C.R. Cabrera and P. Yager, Electrophoresis, 22 (2001) 355

G.A. Marko-Varga and P.L. Oroszlan (Editors)
Emerging Technologies in Protein and Genomic Material Analysis
Journal of Chromatography Library, Vol. 68

CHAPTER 7

Bio-affinity Extraction for the Analysis of Cytokine and Proteomic Samples

GYÖRGY MARKO-VARGA

Department of Analytical Chemistry, Lund University, Box 124, 221 00 Lund, Sweden

7.1. INTRODUCTION TO CYTOKINE ANALYSIS

Cytokine is a general term for a large group of molecules involved in signaling between cells during immune response. A major challenge in biology today is to understand the mechanisms of integration of the many components that make up the cell, or organism, as the initial step to understanding the function of the whole.

Cytokines elicit their responses by binding to specific high affinity cell-surface receptors on target cells and initiating a series of intracellular signal transduction pathways. The receptors of several cytokines and growth factors are homologous within their extracellular domains. Cytokines are involved in many phases of the inflammation process. Inflammation diseases may develop following acute inflammation and may last for weeks or months, and in some instances as a life-time disease [1–3].

During this phase of inflammation, cytokine interactions result in monocyte chemotaxis to the site of inflammation where macrophage activating factors (MAF), such as IFN-gamma, monocyte chemoattractant protein-1 (MCP-1), and other molecules then activate the macrophages, while cytokines such as GM-CSF [4] and IFN-gamma, retain them at the inflammatory site. Several cytokines play key roles in mediating acute inflammatory reactions; these mediators contribute to the inflammatory process by chronically elaborating low level cytokine expressions in macrophages [3–5]. Cytokines such as interleukin (IL)-4–7, IL-10, and transforming growth factor-β (TGF-β) are key regulators in these events [6–10].

IL-8 and other low molecular weight chemokines, e.g. MCP-1 and RANTES, belong to a chemotactic cytokine family and are responsible for the chemotactic migration and activation of neutrophils and other cell types such as monocytes, lymphocytes, basophils, and eosinophils at sites of inflammation [6–14].

Selective nano-extraction techniques can be applied to retrieve trace levels of these cytokines using biocompatible 'restricted access' (RA) supports. The principle of the RA

E-mail address: gyorgy.marko-varga@analykem.lu.se.

support is based upon the possession of two chemically different surface properties of the porous silica particles. The spherical shape of these beads is well defined and the size ranges between 25 and 40 μm. The outer surface of the particles is highly biocompatible, it possesses diol modification and is hydrophilic, while the pore surface chemistry is made up of a hydrophobic dispersion phase with C-18 functionality [15,16].

The outer surface of these packings possesses a defined diffusion barrier and a non-adsorptive outer particle surface towards macromolecular matrix proteins larger than 15 kDa [17,18]. Low molecular weight proteins <15 kDa will diffuse into these freely accessible affinity centers, i.e. the pores (approximately 60 Å), and be bound to the hydrophobic C-18 groups; this part of the binding mechanism forms the physical diffusion barrier [15]. High selectivity can be reached by exclusion of high molecular weight proteins >30 kDa, while simultaneous trace enrichment is gained from the small molecular weight peptides and proteins.

In the present chapter, an interface between nano-capillary extraction of IL-8 and IL-10 and MALDI-TOF MS at femtomole levels is described. The selectivity of the RA silica support allows efficient and simultaneous sample clean-up and trace enrichment of cell supernatant samples.

7.2. CAPILLARY MICRO EXTRACTION LINKED TO MALDI-TOF MS FOR THE ANALYSIS OF CYTOKINES AT FEMTOMOLE LEVELS

Determining the levels of cytokines in cellular samples, a challenge that causes the signaling communication, is highly complex and requires especially dedicated means of technology developments [15]. The time course whereby interleukins are present in biological samples may vary to a large extent. Kinetic profiling is one important line of experimental investigations undertaken to determine at what timing window the cytokine at hand has its operational activity. This is clearly correlated in such studies by making both qualitative and quantitative time course measurements.

The biocompatible surface of the RA extraction material ensures high recoveries and good selectivity in the analysis of biological material such as cell and plasma samples [16–18]. The Mw restriction of the RA-support was found not to be absolute at 15 kDa, with a smaller part of the pores being larger, which results in a binding capacity for somewhat larger proteins as well, and this makes IL-10 also an excellent solute to extract. Reaching the molecular range of 34 kDa, as for Lysosyme, makes the extraction less efficient, but still it is possible to 'fish out' the protein. Proteins of this size will yield lower extraction efficiencies compared with smaller sized proteins. The nano-capillaries are prepared by packing the RA support prepared from a slurry into the 200 μm ID capillary using an adapter that transfers the stationary phase. The technique used here is similar to that of Gobum et al. [19] and other groups [20–22].

Dried droplet experiments, produced using α-cyano-hydroxycinnamic acid (CHCA) and sinapinic acid, were conducted. Both matrices were purchased from Hewlett–Packard. The ready-made matrices were diluted with Millipore water five times prior to use. Stock solutions of the proteins were made by dissolving them in phosphate buffer (10 mM), and working solutions were made by diluting the stock solution in Millipore water and 2.5% formic acid. The water was purified using a Millipore apparatus (Bedford, MA, USA).

A RA support (kindly provided by Prof. Karl-Siegfried Boos, Clinical Hospital, Gross Hadern, Germany) with a particle size of 25–40 μm and pore size of approximately 60 Å was packed in capillary columns (2 mm × 0.2 μm i.d. and 10 mm × 0.2 μm i.d.). A sample containing IL-8 or IL-10 was loaded on the RA media and later eluted with acetonitrile (50:50 v/v) containing 10 mg/ml sinapinic acid. The capillaries used were purchased microloader tips (Eppendorf, Hamburg, Germany, cat. no.: 5242956003).

Fig. 7.1 shows the two types of extraction devices used. The high capacity column (10 mm × 0.2 μm i.d.) is used in applications where typically up to 5 pmol of both IL-8 and IL-10 could be applied without major losses. The upper column in Fig. 7.1 shows the low capacity column (2 mm × 0.2 μm i.d.), where trace levels down to 900 attomole (amol) of IL-8 could be determined. The actual support amount that is packed therein is indicated by the arrow in the figure. Each time one handles low levels of proteins, losses due to adsorption are a major concern. By utilizing capillaries with minute amounts of RA

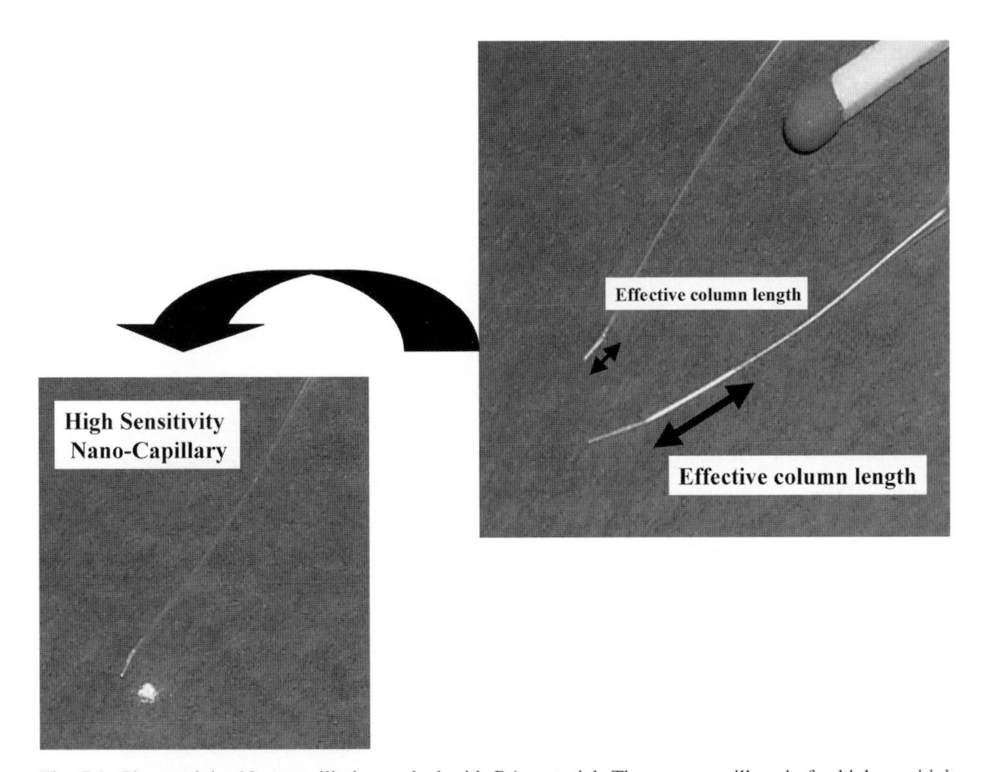

Fig. 7.1. Picture right: Nano-capillaries packed with RA material. The upper capillary is for high sensitivity purposes with a 2 mm packed bed size and the lower capillary is a high capacity capillary with 10 mm length. Picture left: The high sensitivity nano-capillary with the actual amount of packing material needed to pack it.

material resulted in decreased losses and better delectability of the cytokines. The starting material needed to manufacture these types of nano-capillary is shown in Fig. 7.1 (the left-hand side picture).

A pre-concentration procedure was developed for cytokine sample clean-up. It consisted of a solvatization step where the affinity material was activated using an acetonitrile/formic acid mixture (50:50 v/v, for more detail see experimental). Next, the capillary was washed with 200 μl of FA (2.5%). A sample containing IL-8 or IL-10 was loaded onto the column, followed by a washing step using 200 μl of FA (2.5%) to remove loosely bound matrix components. The cytokines were eluted from the nano-capillaries using a solution of MALDI matrix diluted in acetonitrile (50:50 v/v).

Typically 80 nl elution fractions were spotted onto the stainless steel MALDI target plate. Measurements were performed using a Voyager DE-PRO MALDI-TOF instrument. A method using a 25 kV accelerating voltage and a 150 ns delayed time in the linear mode was used.

Elution of cytokines from RA capillaries using the MALDI matrices immediately initiates the MALDI-crystallization and dries down in less than 60 s, depending on the size of the droplet. Fig. 7.2 shows the configuration of the experimental setting, as well as the discrete MALDI spots that are generated upon desorption from the nano-capillaries. Equally sized uniform spots are formed that can be deposited with fraction densities ranging up to 300 spots on a normal sized target plate. Typically around 30–40 samples can be positioned onto the target plate surface.

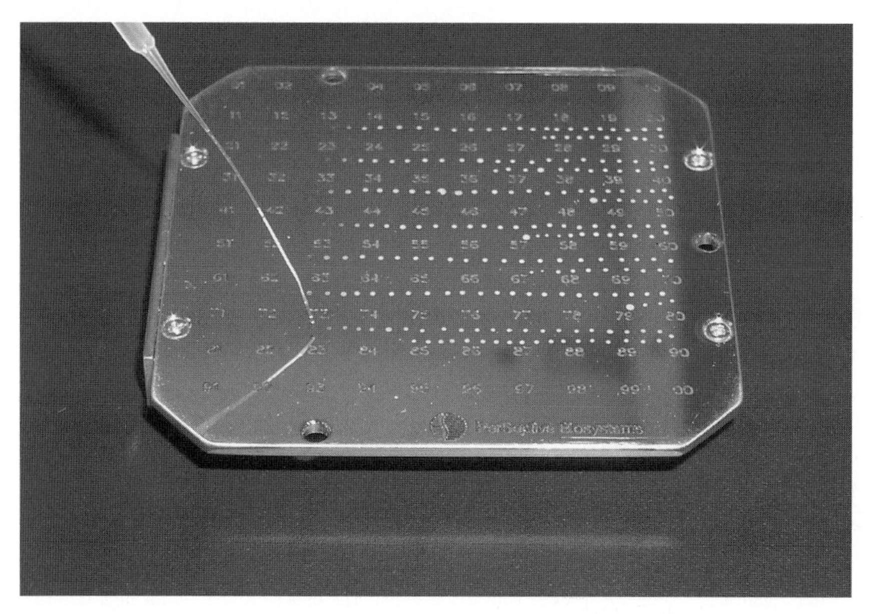

Fig. 7.2. Picture of the nano-capillary extraction using the RA material where cell sample extracts are spotted onto the MALDI target plate.

7.3. APPLICATION TO THE ANALYSIS OF IL-8 AND IL-10 IN CELL SAMPLES

The cells were lysed by sonication after cultivation and harvested by centrifugation (1000*g*). The supernatant was collected and diluted by a factor of two with PBS followed by cytokine addition. The cytokine containing sample was then loaded onto the capillary columns packed with RA material. The three nano-capillaries were then desorbed using the three MALDI matrices, CHCA, SA and FA, respectively, and sampled onto the MALDI plate. The process line from start to end is illustrated in Fig. 7.3. The multi-disposable capillary methodology was highly efficient and allowed a screening type of optimization choosing the most appropriate matrix for MALDI analysis. This can be further extended to larger sets of matrices for unknown immunoregulating proteins, which makes this selection procedure advantageous.

The cytokines are removed well using the CHCA matrix. It is possible to generate both singly charged and doubly charged ions from IL-8 and IL-10 by simply adding the matrix to the eluting solvent that is used in the desorption step of the micro-extraction procedure. Sinapinic acid, most commonly used as the matrix for protein identifications in MALDI, did not generate doubly charged cytokine ions at all.

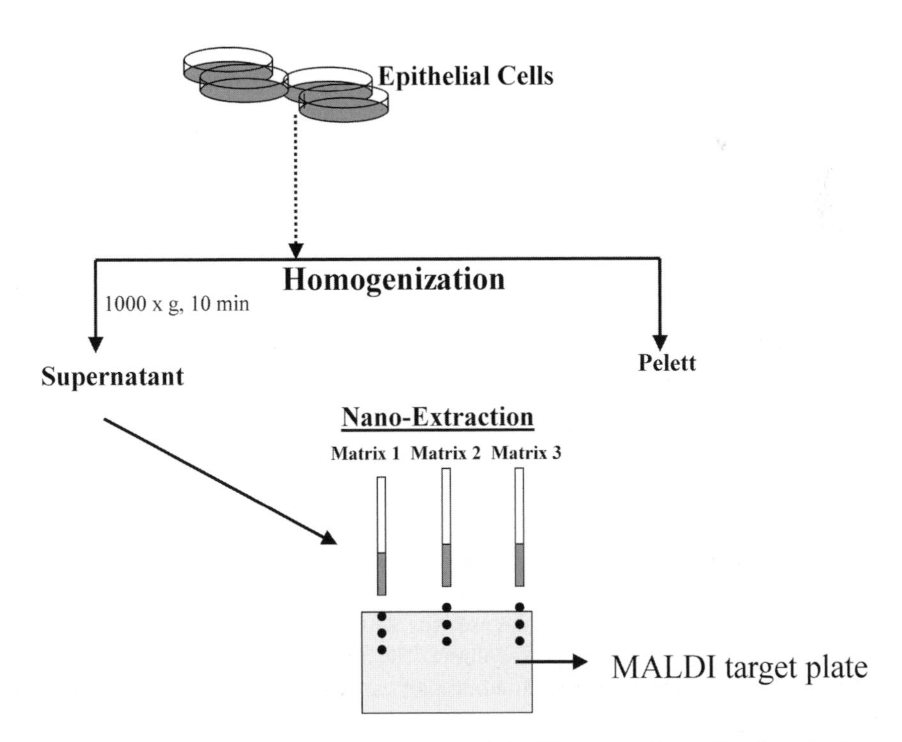

Fig. 7.3. Illustration of all steps included in the sample handling process for cytokine determinations.

7.3.1. Analysis of IL-10

The samples generated from the cell incubations were taken directly from the cell wells and frozen at − 80 °C or analyzed directly. Prior to extraction all samples were acidified in order to stabilize these samples and processed afterwards. A good MALDI spectrum can be obtained from IL-10 (400 fmol/μl) present in epithelial cell samples as shown in Fig. 7.4A using CHCA as the MALDI matrix. The singly charged peak is seen at 18692.4 and the doubly charged at 9346.2. Some insulin is also clearly identified in the sample, which is to be expected since insulin is used during cell cultivation as an additive to the cell medium. Figure 7.4B shows the same sample when sinapinic acid is used as the MALDI matrix. A very weak signal is generated that can be identified upon magnification of the spectrum, which is shown in Fig. 7.4C. The choice of matrix is clearly of great importance when analyzing IL-10. The sensitivities that it was possible to reach with IL-10 were also found to be clearly lower than those obtained with other cytokines, the reason for this is not really clear.

7.3.2. Analysis of IL-8

In order to evaluate the behavior of IL-8 in an immunological response, cell lines are commonly stimulated with various activators, e.g. drugs, viruses or cytokines. Consequently, it is important to be able to follow, in our case, the expression levels of IL-8 when an activator triggers the immunological reactions. This results in a fundamental understanding of the cytokine and chemokine action and regulation mechanisms. The mammalian cell samples were analyzed using the selective nano-capillary extraction technique. Human epithelial cells were lysed by sonication after cultivation, centrifuged (1000*g*), and the resulting supernatants were used as the reference cell samples. The epithelial cell samples above were spiked with recombinant human IL-8 in the concentration range 1–500 pM and recombinant human IL-10, 50–1000 nM.

Direct injection of untreated epithelial cell supernatants were made onto the nano-capillaries. The micro-extraction step effectively removed the interfering compounds. The extracted cell sample included only low mass ions <800 mass units. Successful micro-fractionation of IL-8 was achieved from the cell sample containing IL-8 as shown in Fig. 7.5A where the signal from the IL-8 ion was detected at a mass to charge ratio of 8391.2. The IL-8 levels in the MS-spectra corresponds to 150 fmol/μl. The extraction was achieved by the RA material from a sample volume of 5 μl that was adsorbed on a 2 mm × 0.2 μm ID capillary column. Desorption was carried out into a volume of approximately 80 nl that formed the crystal on the MALDI target plate from where the MS-spectrum was generated. Sharp and narrow peaks with good resolution (greater than 500) are accomplished. Next, Fig. 7.5B shows a MS-spectrum where the dried droplet technique was used, the IL-8 level in this case was 150 fmol/μl in 10 mg/ml of human serum albumin using the larger sized nano-column. The albumin amount is a typical level found in human blood plasma. In these dried droplet experiments the Teflon target plate from Applied Biosystems was used (400 positions) with 3 × 1 μl depositions. A poor signal is generated and in four out of five preparations there were clear problems to

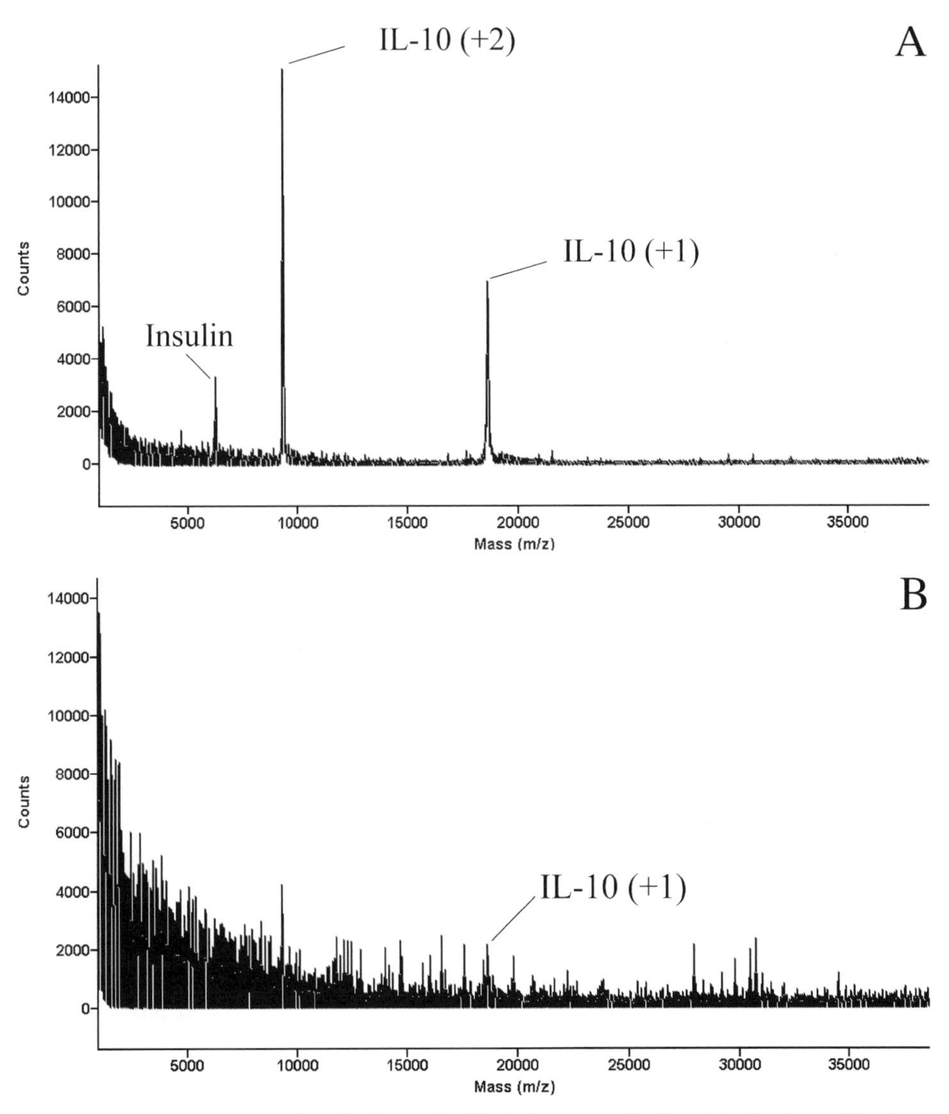

Fig. 7.4. (A) MALDI-spectrum of IL-10 (400 fmol) with singly charged IL-10 and doubly charged IL-10, and insulin present in the epithelial cell supernatant sample using CHCA as the MALDI matrix. (B) MALDI spectrum of IL-10 using sinapinic acid as the matrix. (C) Shows a magnification of the correct mass of IL-10 (Mw: 18692.4). Experimental conditions; operated in the linear mode using an accelerating voltage of 25 000 V and a delayed extraction time of 150 ns.

generate any signal at all. By applying the nano-extraction a selective sample clean up was achieved where the resulting MS-spectrum was found to generate high signals for both the mono and the doubly charged peak (see Fig. 7.5C), thereby circumventing the ion suppression that high levels of human serum albumin generates. Extraction was made on the same sample volume as with the dried droplet experiment (3 μl). Using the larger sized

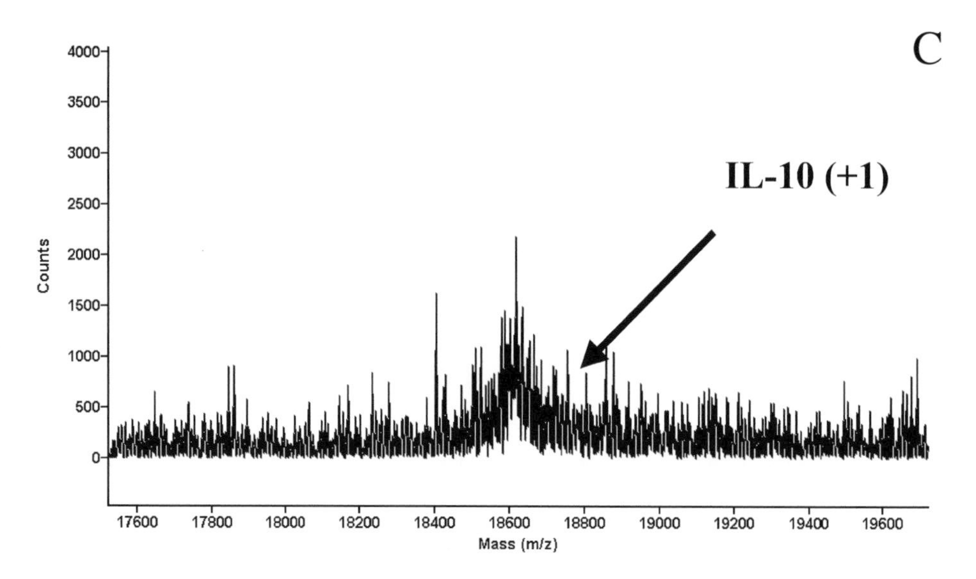

Fig. 7.4 (*continued*)

column, loading capacity was not a limitation, as sample volumes up to 200 μl with a concentration of 350 fmol/μl of IL-8 could be applied without showing any breakthrough. This was determined by analyzing the eluate of the sample coming out from the capillary.

Highly sensitive determinations can be achieved by using the small-sized capillaries (shown in Fig. 7.1). Fig. 7.5D shows the analysis of 900 amol IL-8 present in the cell sample; this spectrum is generated from a sample volume of 2 μl. Only the singly charged peak can be seen at these levels due to the increase in enrichment factor obtained, since the desorption of the sample could be spotted on the MALDI target plate in droplets of around 20–30 nl. It was also found that the absolute volumes using these small capillaries are limited to around 5 μl. An important experimental parameter to take into account is the binding kinetics. One needs to elute at a slow flow rate generating a weak pressure so that the protein has time to diffuse into the pores of the support material.

The recovery of this methodology for IL-8 quantitations in epithelial cell supernatants was investigated. Three concentration levels were chosen: 50, 100, and 200 fmol/μl. Sample volumes of 5 and 10 μl were used for all the three concentration levels. In these samples, cytochrome *C* (300 fmol/μl) was used as the internal standard. The percentage recovery was determined by using the ratios of internal standards as follows; IL-8 sample/IL-8 standard × 100. Triplicate micro-extractions were performed on each concentration and volume of the samples are shown in Table 7.1 including the relative standard deviations (RSD). There is a linear relationship between the three concentration levels used in the recovery study. It resulted in a correlation coefficient of 0.992 where sample intensity was measured of the $[M + H^{+}]^{+}$. The recovery data were also generated from sample intensity points, based on the average of 256 spectra obtained at three different positions. Typically five positions where chosen within each spot to generate

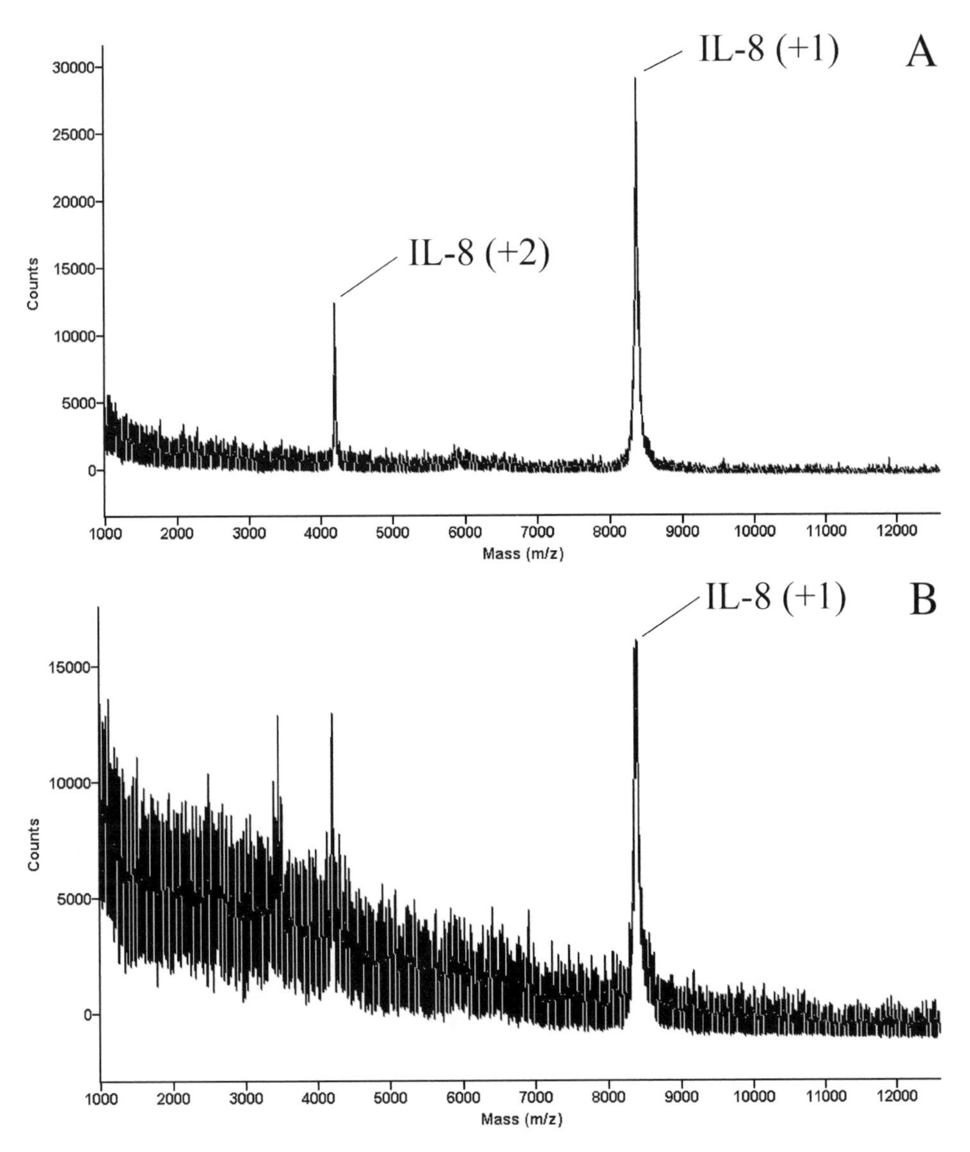

Fig. 7.5. (A) MALDI-spectrum of IL-8 (100 fmol) with singly charged IL-8 and doubly charged IL-8, in the epithelial cell supernatant sample using CHCA as the MALDI matrix. (B) MALDI spectrum of IL-8, 150 fmol/μl in 10 mg/ml of human serum albumin using the larger sized nano-column. Teflon target plate was used with 400 positions with 3×1 μl depositions. (C) Same as (A), 30 fmol IL-8 present in the epithelial sample. (D) 900 fmol/μl of IL-8.

the spectrum. In some cases one and two out of five positions resulted in weak signals that were not included in the average signal intensity calculations.

Recoveries obtained ranged between 54 and 104% for all samples. There is a clear tendency of the data shown in Table 7.1. The lowest recoveries are obtained from samples

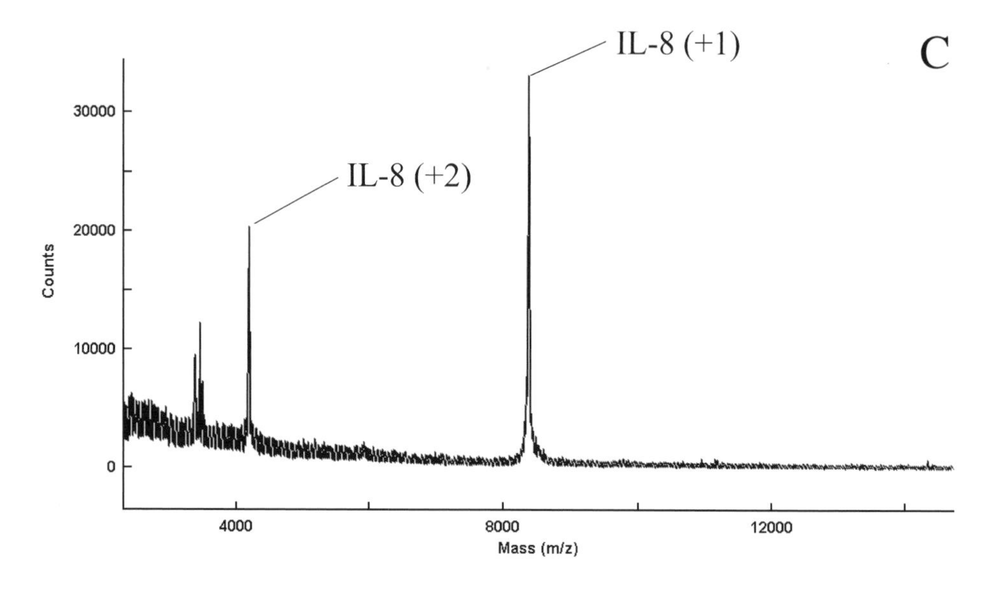

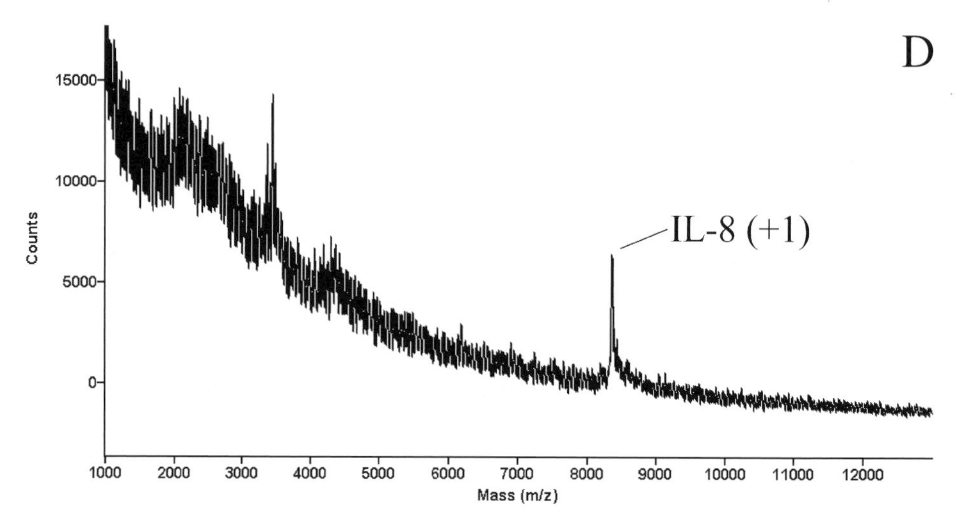

Fig. 7.5 (*continued*)

TABLE 7.1

RECOVERIES OF IL-8 FOR THE OBTAINED MOLECULAR MASS $[M + H^+]^+$ IN EPITHELIAL CELL SUPERNATANTS BASED UPON SIGNAL INTENSITY ($n = 3$) INCLUDING THE RELATIVE STANDARD DEVIATIONS (%)

Sample volume (μl)	50 fmol/μl	100 fmol/μl	200 fmol/μl
5	68 (57%)	86 (44%)	89 (38%)
10	74 (53%)	88 (48%)	104 (40%)

with the lowest levels, as would be expected at 50 fmol/μl, resulting in 68% recovery using 5 μl as the sample volume. By using a 10 μl sample volume of the same concentration, the recovery increased to 74%. At the middle concentration (100 fmol/μl), there were no major differences found. The recoveries were 86 and 88% using these two sample volumes. The highest concentration (200 fmol/μl) resulted in recoveries closest to 100, 89% using 5 μl, and 104% with 10 μl, respectively.

Recovery experiments were performed as follows; the two sample volumes (5 and 10 μl) were loaded onto the capillaries according to the methodology described in the experimental section. Next, the desorption was performed by using 250 nl of the saturated CHCA matrix dissolved in a mixture of 5% formic acid and 50% methanol and deposited as a single spot on the MALDI target plate. This was, however, not the standard procedure used throughout the work described above. Rather, small volume droplets were used there, where droplets were eluted onto the target surface with the aim of reaching highest possible enrichment factors. It was found that smaller volumes of less than 200 nl could not generate reproducible data in terms of resulting signal intensities. It is not trivial experimentally to make reproducible spotting manually to obtain the same amount of proteins in a series of spots. There is always a variation in the amount of cytokine that is seen in a train of matrix spots. Another issue is the memory effects of the capillary. By a second repetitive desorption, the cytokines were found to be bound onto the sorbent. Even at 75% organic modifier mixtures, it was not possible to generate a kinetically favorable elution of the proteins from the column, e.g. using 100 nl. The memory effects were still left on the following spectrum from the second pass elution. Mainly qualitative data are obtained at high sensitivities, as shown in Figs. 7.4 and 7.5. One possibility to generate improved quantitative data would be to use an on-line coupling of the capillary inlet with a consistent flow, typically in the 500 nl/min region. This would solve the irreproducibility of the elution and sample spot deposition. In the case of high sensitivities, the smaller sized nano-capillaries should be used as shown in Fig. 7.1. It was found that the highest sensitivity using the micro-extraction technique for both IL-8 and IL-10, when the desorption step is performed in a way that the eluting droplets are as small as possible, can be made < 50 nl. This is nicely shown experimentally with the capture of a nano-capillary desorption; the picture image is frozen and the actual droplets are eluted as discrete segments. These droplets then form the micro-crystals by fast evaporation of the eluting solvent.

Typically, enrichment factors of 200 can be achieved with this technique if, e.g. 10 μl sample volume is loaded onto the capillary column and the desorption is made in a 50 nl matrix volume, and all of the interleukin is in that spot. In situations where larger sample volumes are available, the enrichment factor can be increased. Usually these levels are above the real sensitivities that the instrument is able to reach. Considering IL-8 and IL-10, the delectabilities are set by the differences in structure characteristics. In the case of IL-8 the lowest level that can be handled is around 1 fmol/μl, determined by immunoassay techniques [23,24]. The reason is unspecific adsorption from exposed surfaces in pipettes, vial tube walls, and probably the silica surface of the RA support material itself. For IL-10, we have delectabilities at a higher level. This is not limited by losses due to surface chemistry phenomena. In a previous work we have shown that IL-10 is a highly stable cytokine that we can quantify at 1 fmol levels by flow immunochemical techniques [25]. The reason for the restriction in sensitivity is probably due to its poor MALDI properties.

7.4. CYTOKINE ANALYSIS CONCLUSIONS

Cytokines are key modulators of inflammation. They participate in acute and chronic inflammation in a complex network of interactions. Several cytokines exhibit some redundancy in function and share overlapping properties, as well as subunits of their cell surface receptors. This selective nano-capillary extraction technique for the analysis of IL-8 and IL-10 in cell samples can be of high importance for cytokine screening with optimized protocols presented above. Important features of the methodology such as handling of the samples at femtomole levels of cytokine presence with respect to adsorption, washing, desorption and regeneration of the nano-capillaries is of utmost importance. Identification and confirmation of IL-8 and IL-10 in epithelial cell samples was carried out using MALDI-TOF MS reaching high selectivity and high sensitivity, down to low femtomole levels. In order to reach high sensitivities, one order of magnitude improvement is certainly on the lower limit of what should be reached, and would facilitate quantification at the lower end of IL-8 levels, i.e. at the beginning of extra cellular transport of IL-8 in the epithelial cell samples. One ongoing research line we have within our research team to meet these demands is the use of a silicon microchip approach [26]. A new RA material has been synthesized and characterized having a polymer backbone [27]. High selectivity has been proven towards albumin when applied to human plasma samples. This polymeric material also has a pH operational range of pH 1–14, which allows conditioning using both acidic and alkaline conditions, circumventing the cross-contamination between samples.

For high speed analysis, further improvements can be achieved by using a more quantitative 'seed-layer' sample preparation technique developed for MALDI-MS [28] reaching very good sample homogeneity. Transfer to a robotic instrumentation would enhance the throughput whereby a fully automated and unattended operation would save both money and the laborious work. An additional gain in precision, accuracy and sensitivity would be feasible.

The above summary will ultimately be a key point in order to obtain better understanding of the pathways regulated by cytokines and this will allow the identification and/or development of agents for improved modulation of the inflammatory response for the treatment of autoimmune, infectious, and neoplastic diseases. A disposable nano-scale capillary extraction technique is described utilizing a silica based extraction support with molecular weight RA properties. The purification step was carried out using the C-18 fictionalized RA chromatography. The technique is fast, simple and freely allows simultaneous concentration and removal of high molecular weight >30 kDa sample components. Additional benefits are effective elimination of salts and buffers present in the sample that normally result in unwanted adduct formations in the MS-spectrum at femtomole levels of mapping interleukin-8 (IL-8) and interleukin-10 (IL-10) with the RA support interfaced to MALDI-TOF MS.

Applications to cell samples prove that a strict discrimination towards other matrix proteins present in the samples could be achieved, and cytokines could be extracted resulting in clean mass spectrometry spectra. The additional gain that is obtained by a high degree of purification of the sample using MALDI is rather special. The limitations in protein determination arise when many additional high abundant proteins and endogenous sample components are present in the MS-spectrum. Ion suppression effects occur that will mask the ionization process as well as the time of flight of analytes. In our experience, purified samples at a high degree can influence the detectability of proteins up to an order of magnitude and sometimes even more when the interleukin levels are low in samples. Typical recovery data analyzing epithelial cell supernatants reaching between 68 and 104% at 50–200 fmol/μl levels is also a window narrow enough to perform relevant cell based assays in pharmacological experiments.

The high sensitivity of the nano-capillary extraction methodology using the restriction of molecular mass of solutes is also due to the limited access to adsorptive surfaces that the proteins have to bind unspecifically and be wasted. This is illustrated by the mapping of interleukins in epithelial cell lines where IL-8 could be detected at 900 amol levels and IL-10 at 400 fmol levels, respectively.

7.5. SAMPLE PREPARATION FOR PROTEOMICS STUDIES

In protein expression profiling, the main interest and focus is set on finding the identities of regulated proteins in biological systems. The specific disease or biological system, or model, regardless of sample type, e.g. cells, tissues or organs, or specific compartments in the cells are all dependent on proper sample handling and sample preparation. The success in these pre-sample handling steps will have a direct influence on the final expression data results generated in the study.

The most efficient high-resolution separation technique available today for protein separations is still two-dimensional gel electrophoresis, which is why RA affinity protocols compatible with the conditions required by gel electrophoresis separations were developed [29]. TGF-β belongs to the TGF-β super family, which regulates the growth and differentiation of various types of cells. A novel family of signaling proteins (SMAD proteins) mediates the intracellular signaling of the TGF-β superfamily. When TGF-β binds to its type II receptor in concert with the type I serine-threonine kinase receptors, the R-Smads (Smad 2,3) becomes phosphorylated. The R-Smads then associate with the Co-Smad (Smad 4) and this heteromeric Smad complex translocates to the nucleus where it binds to DNA and induces transcription of downstream target genes. In this way, several new proteins were revealed and some were regulated as a consequence of TGF-β stimulation.

7.5.1. Sample preparation of cell lysates

In this application, the total protein levels were high, typically between 100 and 500 μg, which is why larger sized sample preparation columns were used, as described earlier

[30–32]. In this case, support materials with particle sizes of 25–40 μm and a pore size of approximately 60 Å were used, packed in a column (LiChroCART© 25-4, Merck, Darmstadt, Germany). The scheme of starting from the cell samples to the final gel spot excision and identification by mass spectrometry is presented in Fig. 7.6.

The following sample preparation protocol for cell lysate samples was developed:

(a) column was washed with 5 ml 75% acetonitrile (ACN), 5% trifluoroacetic acid (TFA) in water and then with 4 ml 1% aqueous TFA,
(b) sample loading, 0.1–1 ml, was gently passed through the column, after which the flow through the fraction was collected,
(c) washing step included 3 ml 1% aqueous TFA,
(d) elution of desorbed material was made by 2 ml 60% ACN, 0.1% TFA in water,
(e) cleaning and conditioning of the column was made by washing with 1 ml 75% ACN, 5% TFA in water followed by 4 ml 1% aqueous TFA.

7.5.2. 1D-gel analysis utilizing restricted access affinity purification

With the defined pore size of the RA stationary phase, proteins with a molecular mass region of choice can be enriched whereby discrimination is achieved towards larger sized proteins. This is nicely illustrated in Fig. 7.7 where the two first lines show the non-retained sample fraction passing through the column followed by the washing step. Lines two and five represent higher molecular weight proteins that are restricted entry to the pores of the beads, thereby passing through without retention. The molecular cut-off of the support

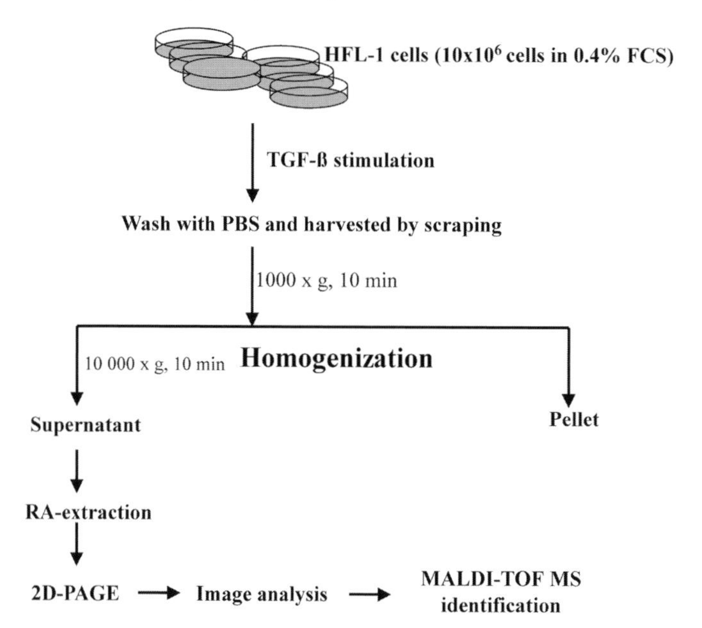

Fig. 7.6. Working process summary including the solid phase fraction procedure from the cultured and stimulated fibroblasts to an enrichment and sample clean up.

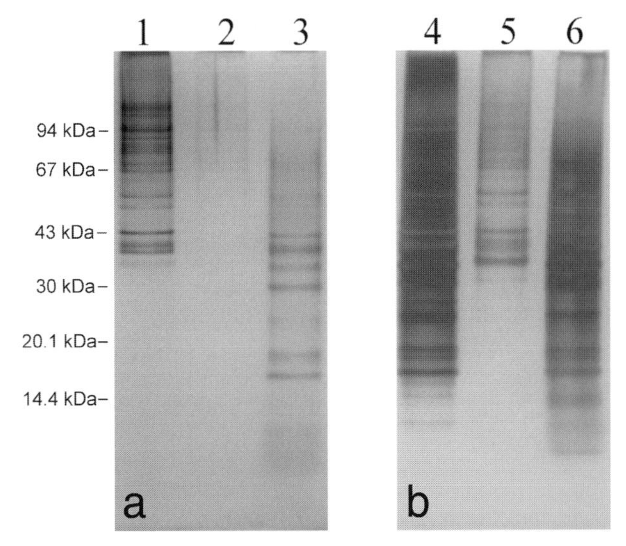

Fig. 7.7. 1D-PAGE (4–12% NuPAGE with MES as running buffer) of fractions from the solid phase extraction, flow through (lanes 1,4), wash fraction (lanes 2,5) and eluted fraction (lanes 3,6). To the column (a) 100 μg and (b) 200 μg protein of TGF-β stimulated HFL-1 cells were loaded.

material is not a definite value but rather a molecular mass window with a lower end at around 30 kDa. Line three shows the enrichment obtained by the bio-affinity principle where more than 10 bands could be amplified in absolute protein levels at the <30 kDa molecular weight region.

Line six shows a doubling of the total amount loaded onto the column, i.e. 200 μg. This is somewhat of a high level, as can be seen from the stained fractions where the non-retained fraction also contains a lot of <30 kDa fractions. This is an example of low capacity of the column for these types of samples.

7.5.3. 2D-gel analysis utilizing restricted access affinity purification

Fig. 7.8 shows reference gel images generated from HFL-1 cells [29,32]. These images are represented by silver, post-gel staining using Sypro Ruby, a fluorescent dye, and ^{35}S-methionine labelled cells. Similar sensitivities are obtained by silver and fluorescence staining of HFL-1 as shown in Fig. 7.8. More than 1000 spots can be detected as high quality image spots. ^{35}S-methionine metabolic labeling increases the possibility to visualize newly synthesized proteins, whereby 3500–4000 spots can be made visual, as shown in Fig. 7.8. It should be clarified, however, that the metabolic labeling images are not directly comparable with the silver and fluorescent stained gels because of the differences between the two experimental approaches. The stainings in Fig. 7.8 are post gel image read-outs, while the labeling is an isotope incorporation that displays the ^{35}S-methionine, which is used as the amino acid upon protein synthesis.

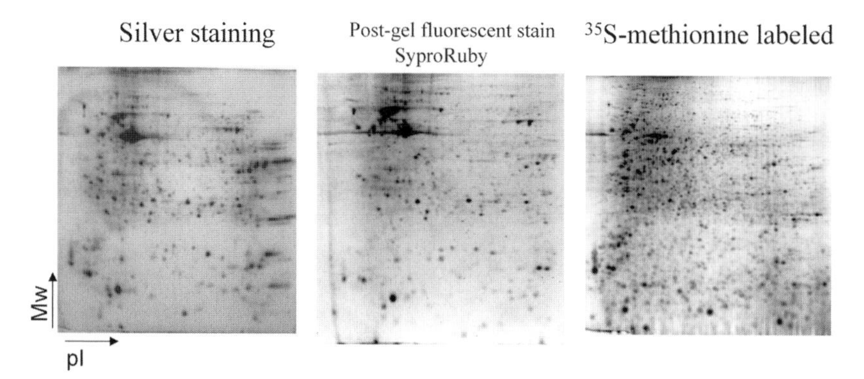

Fig. 7.8. 2D-PAGE of HFL-1 cells stimulated with 10 ng TGF-β/ml for 20 h. Isoelectric focusing in the first dimension was carried out on non-linear pH 3–10; 18 cm immobilized pH gradients strips, second dimension, 14% Duracryl gels were used.

From these experiments it was clear that the expression profiles could be increased at <40 kDa by making HFL-1 enrichment utilizing the RA extraction technique. These differences are clearly seen in Fig. 7.9 where the non-retained fraction is absolutely blank, at the highest sensitivity that the image analysis software could analyze. A large subset of protein could be identified at the lower molecular mass region that otherwise is not readily accessible by normal 2D-gel separations where no bio-affinity enrichment is performed on the samples.

7.6. PROTEOMICS ANALYSIS CONCLUSIONS

A solid phase extraction methodology involving C-18 functionalized RA affinity chromatography has been developed.

A methodology was established in this work whereby highly complex proteomic samples with more than 4000 proteins present could be low molecular weight amplified. The application of this extraction principle made it possible to study enriched protein fraction with low masses from HFL-1 cells and their changes in protein expression as a consequence of TGF-β stimulation. It has to be noted that the washing steps using strong eluents did not result in any memory effects. A pre-requisite for the stable operation of repeated use of the column is that the washing procedures are carefully made after each and every sample purification cycle. It is our experience that carryover effects can be expected in some instances from complex samples, as described previously by our group upon plasma sample analysis [26]. This unwanted negative selectivity effect can, however, be circumvented almost entirely by the addition of a lower degree of organic modifier in the processing eluents. It was verified that the pre-extraction increased the overall protein expression of the HFL-1 cells revealed on both 1D- and 2D-PAGE, especially in the neutral p*I*-region. The method described here will be useful in proteomics work for enrichment and identification of proteins in the lower mass region from biological samples such as solubilized cells of many different types.

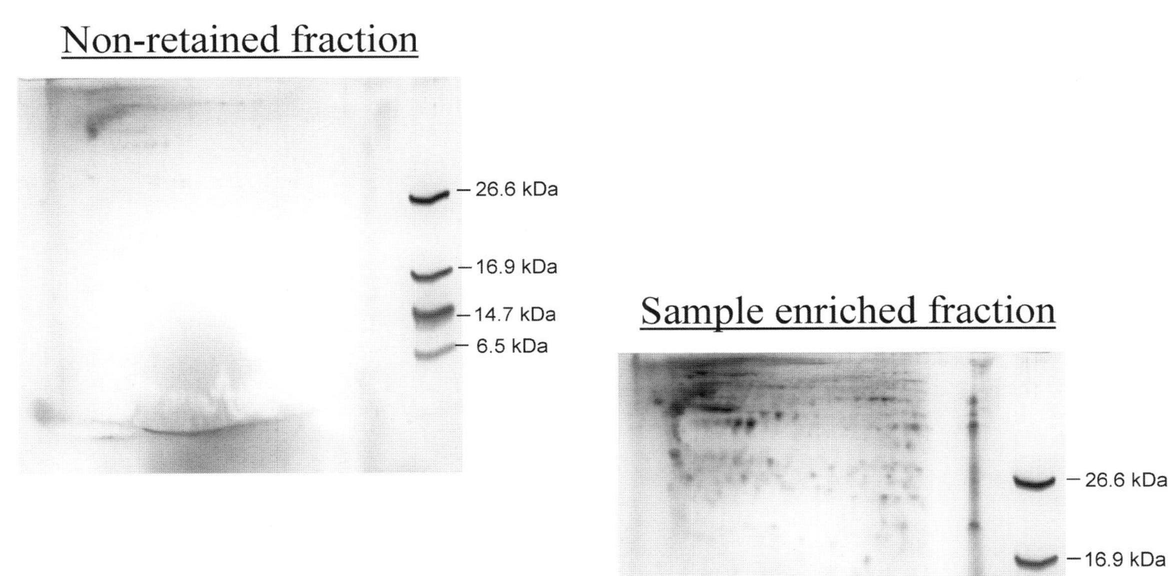

Fig. 7.9. Resulting 2D-PAGE of fraction from sample preparation using the extraction column on human cell samples. Isoelectric focusing in the first dimension was carried out on non-linear pH 3–10; 7 cm immobilized pH gradients strips. For the second dimension, 18% Duracryl gels were used.

Specific regulations were found for many spots in the low mass range identified by protein sequencing and peptide mass fingerprinting using mass spectrometry. The difference in expression profiles is not strikingly obvious by just looking at the gel images. However, by software spot-imaging, we find that weaker spots appear on the gel where enrichment of the cell supernatants were made, while these are not detectable on the untreated silver stained gel.

7.7. ACKNOWLEDGEMENTS

Prof. Karl-Siegfried Boos, Clinichum, University of Munich, Germany is greatly acknowledged for the free samples of synthesized 'restricted access' support.

7.8. REFERENCES

1. K. Koike, M. Ogawa, J.N. Ihle, T. Miyake, T. Shimizu, A. Miyajima, T. Yokota and K.-I. Arai, J. Cell Physiol., 131 (1987) 458–464
2. J.S. Warren, Crit. Rev. Clin. Lab. Sci., 28 (1990) 37–59
3. J.I. Gallin, I.M. Goldstein and R. Snyderman (Eds.), Inflammation. Basic Principles and Clinical Correlates, 2nd ed. Raven Press, New York, 1992
4. C.A. Dinarello, Int. J. Tissue React., 14 (1992) 65–75
5. C.A. Dinarello, in: A. Thomson (Ed.), The Cytokine Handbook, Academic Press, San Diego, 1994
6. M.Y. Stoeckle and K.A. Barker, New Biologist, 2 (1990) 313–323
7. R.M. Strieter, T.J. Standiford, G.B. Huffnagle, L.M. Colletti, N.W. Lukacs and S.L. Kunke, J. Immunol., 156 (1996) 3583–3586
8. M.D. Miller and M.S. Krangel, Crit. Rev. Immunol., 12 (1992) 17–46
9. M. Baggiolini and I. Clark-Lewis, FEBS Lett., 307 (1992) 97–101
10. M. Seitz, B. Dewald, M. Ceska, N. Gerber and M. Baggiolini, Rheumatol. Int., 12 (1992) 159–164
11. Y.R. Mahida, M. Ceska, F. Effenberger, L. Kurlak, I. Lindley and C.J. Hawkey, Clin. Sci., 82 (1992) 273–275
12. T.R. Mosmann and K.W. Moore, Immunol. Today, 12 (1991) A49–A53
13. W.F. Chen and A. Zlotnik, J. Immunol., 147 (1991) 528–534
14. R.M. de Waal, H. Yssel, M.G. Roncarolo, H. Spits and J.E. de Vries, Curr. Opin. Immunol., 4 (1992) 314–320
15. R. Callard, A.J.T. George and J. Stark, Immunity, 11 (1999) 507–513
16. K.S. Boos and A. Rudolphi, LC-GC, 15 (1997) 602–611
17. K.S. Boos, A. Walfort, D. Lubda and F. Eisenbein, German Patent DE41, 30 475 A1 (1991)
18. A. Rudolphi, S. Vielhauer, K.S. Boos, D. Seidel, I.-M. Bäthke and H. Berger, J. Pharm. Biomed. Anal., 13 (1995) 615–619
19. J. Gobum, E. Nordhoff, E. Mirgorodskaya, R. Ekman and P. Roepstorff, J. Mass. Spectrom., 34 (1999) 105–116
20. H. Zang, P.E. Andren and R.M. Caprioli, J. Mass Spectrom., 30 (1995) 1768–1771
21. M. Wilm and M. Mann, Anal. Chem., 68 (1996) 1–8
22. H. Erdjument-Bromage, M. Lui, L. Lacomis, A. Grewal, R.S. Annan, D.E. McNulty, S.A. Carr and P. Tempst, J. Chromatogr. A, 826 (1998) 167–181
23. E. Burestedt, S. Kjellström, J. Emnéus and G. Marko-Varga, Anal. Biochem., 279 (2000) 46–54
24. S. Kjellström, J. Emnéus and G. Marko-Varga, Int. J. Bio-chromatogr, 246 (2000) 119–130
25. S. Ekström, P. Önnerfjord, J. Nilsson, T. Laurell and G. Marko-Varga, Anal. Chem., 72 (2000) 286–293

26. K.-H. Grimm, K.-S. Boos, C. Appel, K.K. Unger, L. Heintz, P. Önnerfjord, L.-E. Edholm and G. Marko-Varga, Chromatographia, 52 (2000) 703–709
27. P. Önnerfjord, S. Ekström, J. Bergquist, J. Nilsson, T. Laurell and G. Marko-Varga, Rapid Commun. Mass Spectrom., 13 (1999) 315–322
28. C. Bratt, C. Lindberg and G. Marko-Varga, J. Chromatogr. A, 909 (2001) 279–288
29. A. Oosterkamp, H. Irth, G. Marko-Varga, L. Heintz, S. Kjellström and U. Alkner, J. Lig. Assay, 20 (1997) 40–45
30. P. Önnerfjord, S. Eremin, J. Emnéus and G. Marko-Varga, J. Chromatogr. A, 800 (1998) 219–226
31. J. Malmström, G. Westergren Thorsson and G. Marko-Varga, Electrophoresis, 22 (2001) 177–186
32. G. Westergren-Thorson, J. Malmström and G. Marko-Varga, J. Pharm. Biomed. Anal., 24 (2001) 815–824

signal transduction pathways [6]. The activity of various hormones and neuropeptides (neuromodulators) functions in this manner. Simultaneously, there are many biologically active peptides that do not match the above-mentioned characteristics of classical peptide bioregulators. Proteins are degraded after completion of their function by specific proteolytic enzymes generating fragments that can be defined as a tissue specific pool of peptides [7].

In the past, characteristics predominantly possessed by biologically active peptides (e.g. C-terminal amidation or C-terminal basic amino acid residues) have been successfully used for isolation of putative novel neuropeptides [8–11]. The sequences of these peptides have, in most cases, been resolved by extensive purification steps followed by Edman degradation. Evidently, this strategy has led to the discovery of numerous biologically important peptides [6,12] but its low sensitivity and requirement of high sample amount is a considerable drawback. Recently, MS has been used to make advances in the area of direct analysis of biological molecules because of its high sensitivity and the possibility to combine it with aqueous-based buffer systems. Capillary nanoscale liquid chromatographic (nanoLC) separation coupled on-line with electrospray ionization (ESI) quadrupole time-of-flight (Q-TOF) MS is uniquely advantageous for the study of complex native neuropeptide mixtures. Additional analysis by tandem MS (MS/MS) using collision-induced dissociation (CID) fragmentation reveals the amino acid sequences and thus the identity of the selected peptides. However, experiments using MS for neuropeptide profiling in tissues have previously been hampered by the complex and time-consuming purification steps [13,14].

In the present chapter, we describe a novel peptidomic approach for studying a large number of neuropeptides in the same analysis. We have employed the methodology in an investigation of the endogenous neuropeptide content of brain tissue samples from rats and mice. Using this combination of sample preparation protocol and nanoLC/ESI Q-TOF MS, we have been able to detect more than 800 endogenous peptides from only 1 mg brain tissue. Several of these neuropeptides were found to be novel, whereas others have been previously described. Moreover, unknown post-translational modifications of some of these peptides have also been identified. Thus, this methodological approach makes it possible to extensively investigate the mammalian brain peptidome.

8.2. PEPTIDOMICS SAMPLE PREPARATION

In our peptidomics studies we use focused microwave irradiation (4.5–5 kW for 1.4 s) utilizing a small animal microwave for the sacrificing of rats and mice. This procedure is important to instantly denaturate enzymes to prevent post-mortem degradation of the proteins. The brain region of interest is thereafter rapidly dissected out and stored at −80 °C. The brain tissue is suspended in cold extraction solution (0.25% acetic acid) and homogenized by microtip sonication to a concentration of 0.2 mg tissue/μl. The suspension is centrifuged and the protein- and peptide-containing supernatant was transferred to a centrifugal filter device with a nominal molecular weight limit of 10 kDa, and centrifuged at 14 000*g* for 45 min at 4 °C. Using this strategy we can obtain both a peptide-rich sample fraction from the brain and the proteins that are larger than 10 kDa,

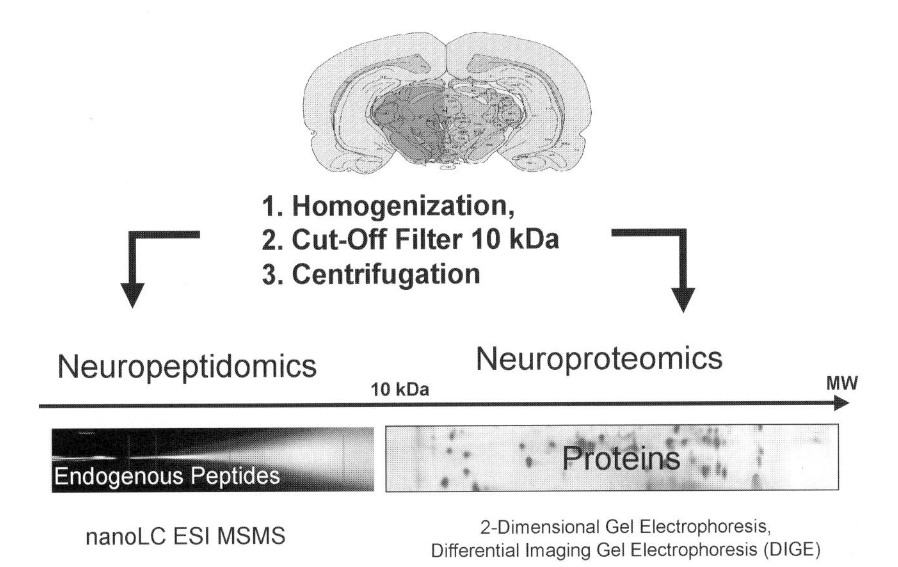

Fig. 8.1. Schematic workflow of the sample preparation of the brain. The brain samples are divided into a peptide fraction for the peptidomic analysis and a protein fraction for the proteomic analysis. The endogenous peptides are analyzed using nanoLC ESI Q-TOF MSMS. The proteins are separated using 2-DE or DIGE, digested by trypsin, and subsequently identified using MS.

which are retained on the centrifugal filter (Fig. 8.1). Finally, the peptide filtrate is immediately frozen and stored at −80 °C until analysis.

8.3. MASS SPECTROMETRY OF BRAIN TISSUE PEPTIDE EXTRACTS

We analyze the peptide extracts using nanoliter flow per minute liquid chromatography MS. Five microliters of peptide filtrate (equivalent to 1.0 mg hypothalamic brain tissue) is injected onto a fused silica capillary column (75 μm i.d., 15 cm length), packed with 3 μm diameter reversed phase C18 particles (NAN75-15-03-C18PM, LC Packings, Amsterdam, the Netherlands). The particle bound sample is desalted by an isocratic flow of buffer A (0.25% acetic acid) for 35 min and eluted during a 60 min gradient from buffer A to B (35% acetonitrile in 0.25% acetic acid), delivered using an Ultimate LC system (LC Packings). The eluate is directly infused into the ESI Q-TOF mass spectrometer (Q-Tof, Micromass Ltd, Manchester, United Kingdom) at a flow rate of 120 nl/min for analysis.

An 'in-house' constructed spray emitter is manually drawn from a 75 μm i.d. fused-silica capillary to obtain a tapered spray tip with an i.d. of approximately 5 μm and an o.d. of 10 μm. The electrospray potential of 1.9 kV is applied to a zero dead volume metal union placed 5 cm from the emitter tip [15–17].

MS data collected during the 60-min chromatographic peptide separation are then exported as a text file and the mass information was prepared using a proprietary software tool currently being developed (Amersham Biosciences, Uppsala, Sweden) to visualize the elution profile both as a virtual two-dimensional (2D) gel and as a 3D graph. This program visualizes the 60-min elution and separation as a virtual 2D gel and facilitates the interpretation of the MS analysis.

8.4. TANDEM MASS SPECTROMETRY OF ENDOGENOUS NEUROPEPTIDES

Sequence information of the peptides is obtained from the peptide precursor ions by an automatic switching function of the MS software from MS to MS/MS mode. The mass spectrometer automatically selects the precursor peptide ions for fragmentation during the nanoLC separations. The switching depends on the intensity of the ions. The collected collision-induced dissociation fragmentation spectra are integrated into a single spectrum twice every second in the m/z-ratio range of 40–1200 Da. These spectra are deconvoluted and interpreted by the BioLynx (MassLynx 3.4) software tools and/or manually. The proposed peptide sequences are compared with the non-redundant database of the National Center for Biotechnology Information (NCBI) to establish the peptide identities using the Basic Local Alignment Search Tool (BLAST) to 'search for short nearly exact matches' (http://www.ncbi.nlm.nih.gov/BLAST).

8.5. SIMULTANEOUS DETECTION AND IDENTIFICATION OF A LARGE NUMBER OF ENDOGENOUS NEUROPEPTIDES

More than 800 peptide ions producing distinct MS peaks can be detected during the 60-min nanoLC/ESI Q-TOF MS analysis of hypothalamic brain tissue from rats and mice (Fig. 8.2). The MS data are processed using a proprietary computer program and displayed as 2D or 3D virtual peptide maps, which improved the visualization for qualitative and semi-quantitative interpretation (Fig. 8.3). These peptide profile maps consisted of both novel and known neuropeptides that are sequenced and identified using MS/MS analysis (Fig. 8.4).

The Q-TOF mass spectrometer is set to an automatic function-switching mode (MS-MS/MS-MS) during analysis for sequence identification of the endogenous hypothalamic peptides. The MS software selected approximately 10% of the precursor ions from each sample injection for fragmentation using CID during the 60-min gradient. The generated MS/MS spectra are interpreted manually and/or by software from the N-terminal (b-ions) and C-terminal (y-ions) producing the amino acid sequences (Fig. 8.4). Matching of the sequences from the MS/MS spectra using protein databases identified neuropeptides with masses ranging up to 4960 Da (Table 8.1).

8.6. IDENTIFICATION OF NOVEL NEUROPEPTIDES

In our peptidomic approach, we have been able to identify several neuropeptides that have not previously been described in the literature. These originated from known neuropeptide-containing precursors and one novel neuropeptide originated from stathmin, a protein not known to be a peptide precursor (Table 8.1). The majority of the identified

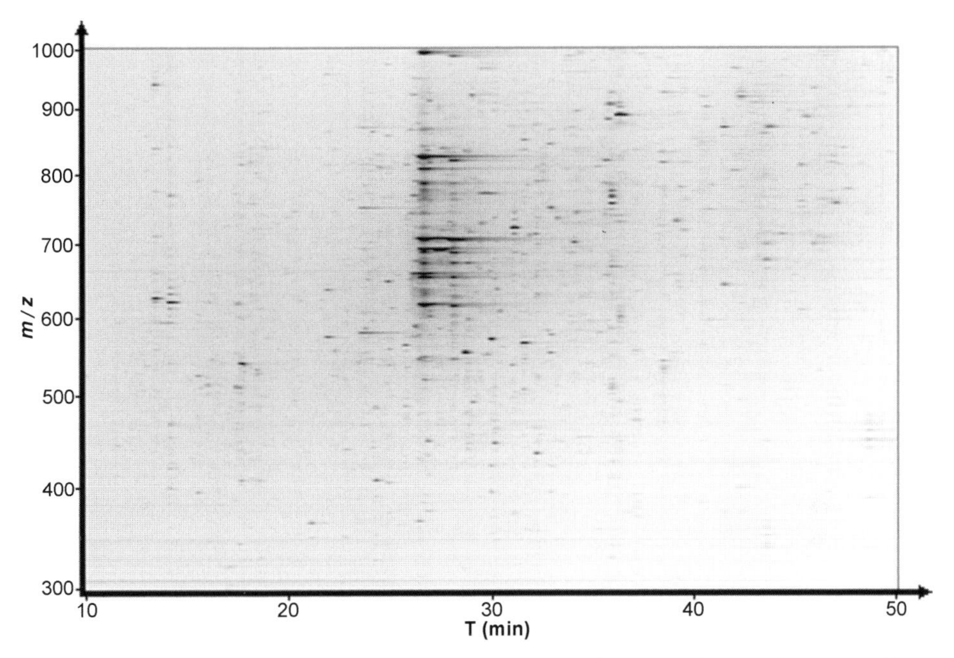

Fig. 8.2. Virtual 2D graph of the neuropeptide content from 1 mg of rat hypothalamus. The neuropeptide map displays peptides in the m/z range 300–1000 in a 60-min gradient elution from the nanoLC separation. Spot intensity is represented by color changes, with black being the most intense reading and white the lowest.

novel peptides are proteolytically cleaved at pairs of the basic amino acids lysine and arginine or less frequently, at single basic residues. However, two neuropeptides originating from neurosecretory protein VGF and the protein stathmin were shown to be processed at non-basic sites.

The proteolytic cleavage sites of the novel peptides found in the pro-enkephalin A, chromogranin A, neuroendocrine protein VGF and pro-opiomelanocortin precursors were located at the same positions in both the rat and mouse. The pro-opiomelanocortin precursor derived joining peptide was post-translationally amidated and contained multiple inter-species amino acid substitutions (Fig. 8.4a and b). In addition, two forms of the peptide I (1–11 and 1–12), originating from the pro-enkephalin A precursor were also identified (Fig. 8.4c).

8.7. IDENTIFICATION OF 'CLASSICAL' NEUROPEPTIDES

Many of the classical neuropeptides, such as neurotensin, substance P, neurokinin A, corticotrophin-like intermediate lobe peptide (CLIP) and beta-endorphin have been identified both in rat and mouse brain tissue (Table 8.1). The two most abundant MS peaks during the 60-min elution have been sequenced and identified as the 4.96 and 4.93 kDa thymosin beta-4 and thymosin beta-10 peptides, respectively. From the pro-SAAS

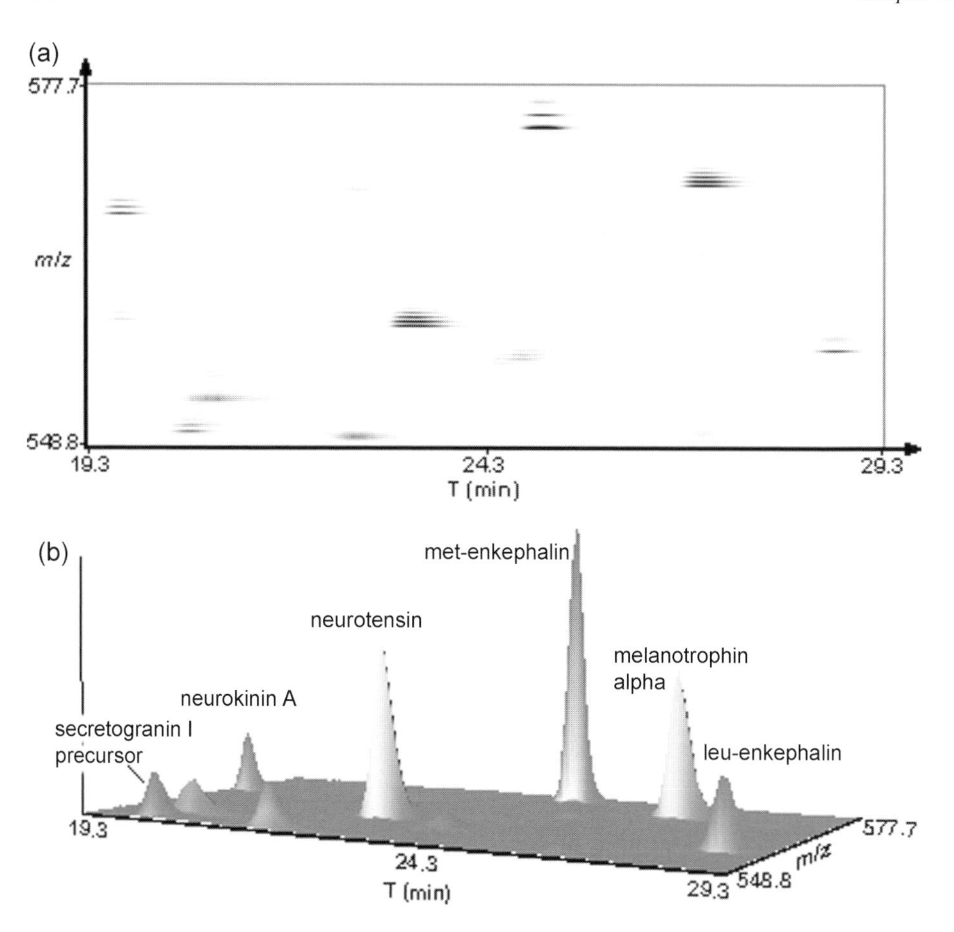

Fig. 8.3. Selected region of a neuropeptide map of the rat hypothalamus displayed as 2D and 3D graphs. Both graphs present the nanoLC elution profile for 19.2–29.6 min and the m/z range 548.7–577.8. (a) Relative spot intensity in the 2D graph is represented by color changes, black being the most intense reading and white the lowest. (b) 3D graph showing the relative intensities of identified neuropeptides.

precursor big PEN and little SAAS were identified. Moreover, relatively high levels of the enkephalin variants leucine-enkephalin (leu-enk), methionine-enkephalin (met-enk), met-enkRF, met-enkRGL/met-enkRSL (rat/mouse), as well as the 4.44 kDa lipotropin gamma peptide from the pro-opiomelanocortin precursor were sequenced and identified (Fig. 8.4d).

8.8. IDENTIFICATION OF POST-TRANSLATIONAL MODIFICATIONS OF NEUROPEPTIDES

A novel peptide from the cocaine- and amphetamine-regulated transcript (CART) protein was identified in the rat hypothalamus carrying a phosphorylation at Ser48.

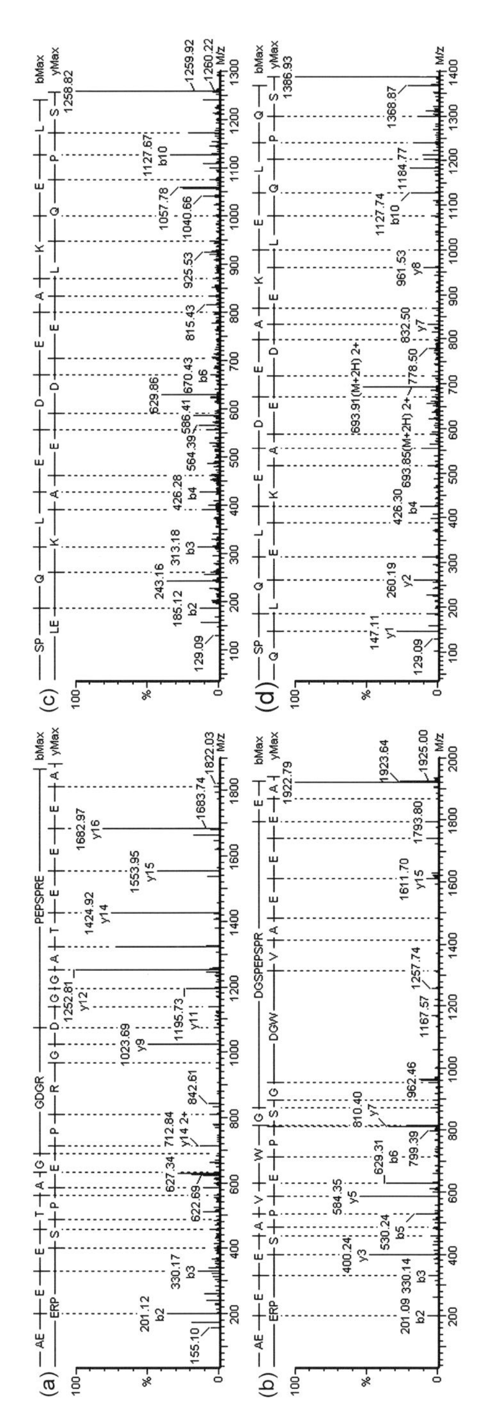

Fig. 8.4. Tandem MS analysis of selected neuropeptides. (a, b) The sequences of the C-terminally amidated joining peptide derived from pro-opiomelanocortin precursor in rat and mouse, respectively. (c) The complete b- and y-ion series of peptide I 1–12, which is derived from the pro-enkephalin A precursor. (d) The sequence identification of the neuropeptide lipotropin gamma. The MS/MS fragment ion labels used are based on the Roepstorff nomenclature [25].

TABLE 8.1

NOVEL AND KNOWN PEPTIDES IDENTIFIED FROM HYPOTHALAMIC TISSUE OF RAT AND MOUSE USING CAPILLARY NANOFLOW LIQUID CHROMATOGRAPHY AND ELECTROSPRAY QUADRUPOLE TIME-OF-FLIGHT TANDEM MS

Novel Peptides			
Precursor[a]	Acc. no.: rat/mouse[c]	Sequence[d]	Species[e]
Chromogranin A precursor	P10354/P26339	395AYGFRDPGPQL405/392AYGFRDPG-PQL402	Rat/mouse
Cocaine-amphetamine-regulated transcript protein	P49192	82IPIYE86	Rat
Cocaine-amphetamine-regulated transcript protein	P49192	37ALDIYSAVDDASHEKELPR55 Ser48 phosphorylation	Rat
Neurosecretory VGF protein	P20156/NA	491PPEPVPPPRAAPAPTHV507	Rat/mouse
Neurosecretory VGF protein	P20156/ND	Pyroglutamic acid-180QETAAAETETR-THTLTRVNLESPGPERVW209	Rat/mouse
Pituitary adenylate cyclase activating polypeptide	P13589	111GMGENLAAAAVDDRAPLT128	Rat
Pro-enkephalin A precursor	P04094/P22005	198SPQLEDEAKELQ209	Rat/mouse
Pro-enkephalin A precursor	P04094	198SPQLEDEAKEL208	Rat
Pro-enkephalin A precursor	P04094	264GGFMRF269	Rat
Pro-enkephalin A precursor	P04094/P22005	219VGRPEWWMDYQ229	Rat/mouse
Pro-enkephalin A, B, pro-opiomelanocortin	P04094, P06300, P01194	YGGF multivalent	Rat
Pro-enkephalin B precursor	GI:204040	133SSEMAGDEDRGQDGDQVGHE-DLY155	Rat
Pro-hormone convertase 2	P28841	94IKMALQQEGFD104	Rat/mouse
Pro-MCH precursor	P14200	131EIGDEENSAKFPIG144	Rat
Pro-opiomelanocortin	P01193	Acetylation-205YGGFMTSEKSQTPL-VTL221	Mouse
Secretogranin I precursor	O35314	585SFAKAPHLDL594	Rat
Secretogranin II precursor	Q03517	300ESKDQLSEDASKVITYL316	Mouse
Stathmin	P13668	Acetylation-2ASSDIQVKELEKRAS-GQAF20	Rat
Vasopressin-neurophysin 2-copeptin	P01186	151VQLAGTQESVDSAKP-RVY168	Rat

TABLE 8.1 (*CONTINUED*)

Known Peptides				
Precursor[a]	Peptide[b]	Acc. no.: rat/mouse[c]	Sequence[d]	Species[e]
Cerebellin	Cerebellin	P23436	1SGSAKVAFSAIRSTNH16	Rat
Chromogranin A precursor	WE-14	P26339	358WSRMDQLAKELTAE371	Mouse
Neurotensin/neuromedin N	Neurotensin	P20068/Q9D3P9	Pyroglutamic acid-150QLYEN-KPRRPYIL162	Rat/mouse
Pro-enk A precursor, POMC	Met-enk	P04094, P01194/P22005, P01193	YGGFM, multivalent	Rat/mouse
Pro-enk A, pro-enk B	Leu-enk	P04094, P06300/P22005, P06300	YGGFL, multivalent	Rat/mouse
Pro-enkephalin A precursor	Met-enk RGL	P04094/P22005	188YGGFMRGL195/188YGG-FMRSL195	Rat/mouse
Pro-enkephalin A precursor	Met-enk RF	P04094	263YGGFMRF269	Rat/mouse
Pro-gonadoliberin 1	Pro-gonadoliberin I	P07490	Pyroglutamic acid-24QHWSY-GLRPG33-amidation	Rat
Pro-MCH precursor	Neuropeptide E-I	P14200	131EIGDEENSAKFPI143-amidation	Rat
Pro-opiomelanocortin	Melanotropin alpha	P01194/P01193	124SYSMEHFRWGKPV136-amidation Ser124, Ser126 acetylation	Rat/mouse
Pro-opiomelanocortin	CLIP	P01193	141RPVKVYPNVAENESAEAFP-LEF162	Mouse
Pro-opiomelanocortin	Lipotropin gamma	P01193	165ELEGERPLGLEQVLESDAEKD-DGPYRVEHFRWSNPPKD202	Mouse
Pro-opiomelanocortin	Beta-endorphin	P01193	Acetylation-205YGGFMTSEKSQTP-LVTLFKNAIIKNAH231	Mouse
Pro-opiomelanocortin	Joining peptide	P01194/P01193	103AEEETAGGDGRPEPSPRE120-amidation/103AEEEAVWGDGS-PEPSPRE120-amidation	Rat/mouse
Pro-SAAS	Big PEN	Q9QXU9/Q9QXV0	245LENSSPQAPARRLLPP260/245LEN-PSPQAPARRLLPP260	Rat/mouse
Pro-SAAS	Little SAAS	Q9QXU9/Q9QXV0	42SLSAASAPLAETSTPLRL59/-42SLSAASAPLVETSTPLRL59	Rat/mouse
Pro-somatostatin	Somatostatin-28 (1-12)	P01167	89SANSNPAMAPRE100	Rat
Pro-tachykinin 1	Neurokinin A	P06767	98HKTDSFVGLM107-amidation	Rat
Pro-tachykinin 1	Substance P	P06767/P41539	58RPKPQQFFGLM68-amidation	Rat/mouse

(*continued on next page*)

TABLE 8.1 (*CONTINUED*)

Known Peptides				
Precursor[a]	Peptide[b]	Acc. no.: rat/mouse[c]	Sequence[d]	Species[e]
Pro-tachykinin 1	C-term flanking peptide	P06767	111ALNSVAYERSAMQNYE126	Rat
Thymosin beta-4	Thymosin beta-4	P01253/P20065	Acetylation-2SDKPDMAEIEKFD-KSKLKKTE-TQEKNPLPSKETIEQEKQAGES44	Rat/mouse
Thymosin beta-10	Thymosin beta-10	P13472/GI:13384628	Acetylation-2ADKPDMGEIASFDK-AKLKKTETQEKNTLPTKETIE-QEKRSEIS44	Rat/mouse

[a]Peptide precursor name.
[b]The name of the peptide.
[c]Precursor accession number in SWISS-PROT or NCBI.
[d]Amino acid sequence of the identified peptide and localization in the precursor.
[e]Species in which the peptide has been identified.

The neuropeptide CLIP was sequenced and identified in mouse with and without a phosphate group at Ser154. Melanotropin alpha, which carries a C-terminal amidation in rat and mouse, was found to contain two additional modification sites and states. Ser124 and Ser136 were acetylated in both species and in the mouse a variant lacking the Ser136 acetylation was also identified. A neuropeptide E-I variant originating from the pro-MCH precursor protein was identified lacking its C-terminal isoleucine amide. Two forms of the relatively abundant thymosin beta-4 were discovered of which the less abundant form was oxidized at the Met7 position. One of the novel peptides from the neuroendocrine protein VGF had a pyroglutamic acid modification of the Gln180 residue.

8.9. EFFECT OF MICROWAVE IRRADIATION OF BRAIN TISSUE

In order to avoid protein degradation and to obtain an unaltered concentration of the neuropeptides, the rats and mice are sacrificed by focused microwave irradiation as mentioned above (see Fig. 8.5). Such microwave irradiation raises the brain temperature to 90 °C within 1.4 s and thereby rapidly prevents enzymatic degradation [18–20]. Post-mortem activity of proteases has been shown to play an important role with regard to the level of peptide concentrations in the brain [4,19,21], as well as for detecting post-translational modifications of proteins and peptides [22,23]. In our previous experiments using traditional (non-microwave irradiation) sacrificing methods, peptide fragments from hemoglobin were a major peptide source detected at high levels in the brain. Interestingly, none of these hemoglobin fragment peptides have been detected in microwaved brain tissue. Furthermore, in our previous investigation of hypothalamic tissue from decapitated rats we have detected a C-terminally degraded form of thymosin beta-4, which was not detected in the present study. These findings suggest that substantial protein/peptide degradation occurs in hypothalamic tissue within minutes post-mortem [4]. The microwave irradiation procedure of the brain tissue therefore has several advantages; it

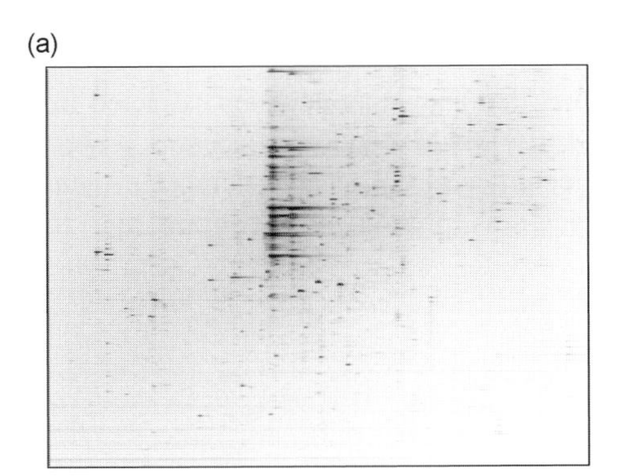

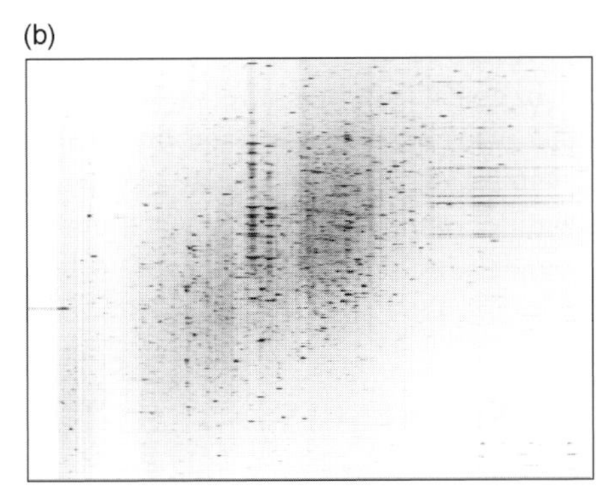

Fig. 8.5. Virtual 2D graphs showing the difference in peptide expression between a microwaved brain and decapitated brain. (a) Shows the peptides from a microwaved brain and displays approximately 800 endogenous neuropeptides. (b) Shows that after 3 min post-mortem of a decapitated brain the degradation of the proteins have increased the number of peptides to about 1500 peptides (protein fragments).

enables the relatively low-abundant neuropeptides to remain intact, minimizes degradation of proteins by proteolysis and conserves the post-translational modifications of the neuropeptides.

8.10. CONCLUSIONS

In this chapter, we have described a methodological approach to simultaneously detect and identify a large number of endogenous neuropeptides in complex tissue extracts from the brain (hypothalamus) of rats and mice. There are several major findings using our peptidomic approach. (a) Many of the identified peptides represent previously uncharacterized and novel processed fragments of protein precursors. (b) Novel peptides were discovered from proteins previously not known to be neuropeptide precursors. (c) Novel post-translational modifications of neuropeptides were identified and located at specific amino acids in both novel and previously characterized neuropeptides.

In future studies, our goal is to examine the precise biological function of the novel neuropeptides that we have discovered. Moreover, we will study changes in peptide patterns ('differential display' peptidomics) and levels following various treatments and disorders. For these purposes, it will be important to develop approaches to quantify the levels of neuropeptides in a reproducible manner. Traditionally, peptides in nervous tissue have been studied by immunoassays. The limitations of this technique include cross-immunoreactivity with structurally similar peptides, thus preventing the unequivocal identification of a specific neuropeptide, and the restriction in the number of peptides that can be analyzed simultaneously. Another common method includes measurement of mRNA expression levels. These methods have the disadvantage of not displaying changes related to post-translational events, e.g. proteolytic cleavage and amino acid modifications, which are common features of neuropeptides. In addition, the latter method may be misleading since mRNA levels have not always been shown to be directly correlated to protein levels [24]. Therefore, a method requiring low amounts of brain tissue that is capable of endogenous neuropeptide detection and quantification is pertinent in neuroscience research.

In conclusion, we have shown in this chapter that analyzing peptides extracted from microwaved tissue using on-line nanoLC/ESI Q-TOF MS and MS/MS is a powerful combination for simultaneous detection and identification of a large number of neuropeptides and their post-translational modifications present in the brain and thus complements standard proteomic methods.

8.11. REFERENCES

1. S. Kennedy, Toxicol. Lett., 120 (2001) 379
2. P. Schulz-Knappe, H.D. Zucht, G. Heine, M. Jurgens, R. Hess and M. Schrader, 4 (2001) 207
3. P. Verhaert, S. Uttenweiler-Joseph, M. de Vries, A. Loboda, W. Ens and K.G. Standing, Proteomics, 1 (2001) 118
4. K. Skold, M. Svensson, A. Kaplan, L. Bjorkesten, J Astrom and P.E. Andren, 2 (2002) 447

5. A. Karelin, E. Blishchenko and V. Ivanov, FEBS Lett., 428 (1998) 7
6. T. Hokfelt, C. Broberger, Z. Xu, V. Sergeyev, R. Ubink and M. Diez, Neuropharmacology, 39 (2000) 1337
7. V.T. Ivanov, A.A. Karelin, M.M. Philippova, I.V. Nazimov and V.Z. Pletnev, Biopolymers, 43 (1997) 171
8. W.E. Schmidt, J.M. Conlon, V. Mutt, M. Carlquist, B. Gallwitz and W. Creutzfeldt, Eur. J. Biochem., 162 (1987) 467
9. K. Tatemoto and V. Mutt, Proc. Natl Acad. Sci. USA, 75 (1978) 4115
10. K. Tatemoto and V. Mutt, Nature, 285 (1980) 417
11. F.Y. Che, L. Yan, H. Li, N. Mzhavia, L.A. Devi and L.D. Fricker, Proc. Natl Acad. Sci. USA, 98 (2001) 9971
12. V. Mutt, Eur. J. Clin. Invest., 20 (1990) 2
13. J.R. Slemmon, T.M. Wengenack and D.G. Flood, Biopolymers, 43 (1997) 157
14. A.A. Karelin, M.M. Philippova, O.N. Yatskin, O.A. Kalinina, I.V. Nazimov, E.Y. Blishchenko and V.T. Ivanov, J. Pept. Sci., 6 (2000) 345
15. P.E. Andren, M.R. Emmett and R.M. Caprioli, Am. Soc. Mass Spectrom., 5 (1994) 867
16. P.E. Andren and R.M. Caprioli, J. Mass Spectrom., 30 (1995) 817
17. M. Emmett and R.M. Caprioli, J. Am. Soc. Mass Spectrom., 5 (1994) 605
18. E. Theodorsson, C. Stenfors and A.A. Mathe, Peptides, 11 (1990) 1191
19. A.A. Mathe, C. Stenfors, E. Brodin and E. Theodorsson, Life Sci., 46 (1990) 287
20. I. Nylander, C. Stenfors, K. Tan-No, A.A. Mathe and L. Terenius, Neuropeptides, 31 (1997) 357
21. X. Zhu and D.M. Desiderio, J. Chromatogr., 616 (1993) 175
22. M.Z. Hossain, L.J. Murphy, E.L. Hertzberg and J.I. Nagy, J. Neurochem., 62 (1994) 2394
23. R.M. Lewis, I. Levari, B. Ihrig and M.J. Zigmond, J. Neurochem., 55 (1990) 1071
24. S.P. Gygi, Y. Rochon, B.R. Franza and R. Aebersold, Mol. Cell Biol., 19 (1999) 1720
25. P. Roepstorff and J. Fohlman, Biomed. Mass Spectrom., 11 (1984) 601

G.A. Marko-Varga and P.L. Oroszlan (Editors)
Emerging Technologies in Protein and Genomic Material Analysis
Journal of Chromatography Library, Vol. 68

CHAPTER 9

The Beauty of Silicon Micromachined Microstructures Interfaced to MALDI-TOF Mass Spectrometry

THOMAS LAURELL,[a,*] JOHAN NILSSON[a] and GYÖRGY MARKO-VARGA[b]

[a]*Department of Electrical Measurements, Lund Institute of Technology, Lund University, P.O. Box 118, S-221 00 Lund, Sweden*
[b]*Department of Analytical Chemistry, Lund University, P.O. Box 124, SE-221 00 Lund, Sweden*

9.1. BACKGROUND TO MINIATURIZATION AND MICROSTRUCTURE DEVELOPMENTS WITHIN THE PROTEOMICS RESEARCH AREA

It is obvious that miniaturization utilizing microstructures and microfluidic developments will be an important way forward in the quest for proteome expression analysis. It has already been proven that liquid phase separation techniques will be an important complement to 2D-gel electrophoresis (2D-GE). In order to keep analyte losses at a minimum, multidimensional capillary systems are being developed. Several instrumental companies, already supply proteomics platforms for protein expression profiling studies utilizing liquid phase separation technology. Most of the proteome mapping activities, both in industry and academia, will make use of clinical material where healthy and sick individuals will be compared in selected patient groups looking for differential displays in protein expression. Clinical material such as tissue biopsies or sampled biofluids is commonly very limited. Consequently, microanalytical techniques will be of mandatory importance, enabling the optimal extraction of biological information from each sample. Chromatography has long since pioneered the miniaturization of analytical techniques and 2D-LC with microsized capillaries interfaced with ESI-MS/MS is currently a highly successful way of identifying proteomes, both in yeast [1–3] and mammalian cells [4]. Washburn et al. [3] made the first large-scale protein identification work using multidimensional capillary LC where they identified 1484 proteins of the yeast proteome. This group used tryptic peptides generated from whole yeast cell lysate and 'shotgun sequencing' by LC/MS/MS. In particular, tandem mass spectrometry methods (MS/MS), have shown a tremendous sequencing power.

* Corresponding author. *E-mail address:* thomas.laurell@elmat.lth.se.

References pp. 196–198

given biological system will ensure an encyclopedia of already verified protein identities. Varying levels of sequence coverage will be obtained in these studies, depending on both the abundance and the peptide sequences of the respective proteins.

As the availability of the biological sample is very limited, e.g. when analyzing protein expression in a biopsy from a pathological tissue, it is evident that the analytical techniques employed must be extremely sensitive and allow the handling of small sample volumes without any loss of analyte in the sample processing steps prior to analysis.

In response to these demands, microtechnology aimed at developing miniaturized chemical analysis systems has evolved as a strong candidate that very well provides the necessary tools. As silicon microtechnology has matured, it has spread its application area into the analytical chemistry field and currently this research field (micro Total Analysis Systems—μTAS or Lab-on-a-chip) is now demonstrating new approaches to high performance microscale analysis. The idea and concept of 'μTAS' was first presented almost two decades ago by Michael Widmer and showed the first proof of concept with his group in 1990 [21]. A recent paper by the Manz group made an overview of the 'micro Total Analysis Systems' [22,23].

This has further been developed and is nowadays commercialized for DNA and/or protein screening purposes, as integrated DNA and protein microchips, and capillary electrophoresis separation made on glass chips by Agilent Technologies and Caliper Inc., respectively.

To a large extent these developments have been driven by the need for new analytical tools for DNA analysis. As we currently are moving into the post-genome era and the proteome is on the agenda, new miniaturized approaches to perform protein analysis are gaining interest. In the following an analytical toolbox, based on microfluidic components, for the field of proteomics will be described. This toolbox provides novel and generic analytical solutions that are highly adaptable for analysis of various biomolecules, ranging from high to low abundant. The components are developed using silicon micromachining where the final microstructure material will be composed of either silicon, or where these units will form the basis for print templates, from which polymer-based microstructures are manufactured.

Our research group has focused on developing the current microtechnology platform to be interfaced with laser desorption ionization mass spectrometry (MALDI TOF-MS).

9.3. BASIC CONCEPTS OF PROTEIN DETERMINATION

Modern protein expression analysis studies utilizes both 1D-GE and 2D-GE, as well as multidimensional chromatography [24–28]. By the introduction of chromatography based protein platforms, it seems evident that a single protein platform is not sufficient to solve all issues that arise in biological studies undertaken in modern life science.

A proteomic analysis system has three parts, which in most cases form the basis for the proteomics analysis process; separation, mass spectrometry identification, and bioinformatics to identify proteins and address the functional relationship to the biological system that is being contemplated. In most cases, a separation is made of individual proteins of interest from complex mixtures such as cell supernatants, tissue samples, or from cellular organelles.

2D-GE separates proteins based on their molecular weight and isoelectric point. In the first dimension of 2D-GE, called isoelectric focusing (IEF), researchers load protein mixtures at one end of a polyacrylamide gel strip that contains a pH gradient. An electric field is applied to the gel and causes the proteins to migrate towards the other end of the strip. Individual proteins migrate until they reach their isoelectric point. In the second dimension, sodium dodecyl sulfate (SDS) polyacrylamide polymer networked gels are used. In this separation mechanism, the proteins are separated based upon size.

The combination of these two separation techniques is still the method providing the highest resolution, whereby intact proteins can be analyzed. 2D-GE-technology was originally developed in the early 1970s by O'Farell and Laemmli [29,30]. A schematic illustration of protein separation principles applied for global protein expression analysis is described in Fig. 9.1.

2D-GE is still the most widely used method for separating proteins. However, it can be foreseen that this will change in the near future. The main reason is that chromatographic and microfluidic techniques have the potential to be automated to a full extent providing unattended high capacity separations. This is not to say that 2D-GE will be out-performed, because it is in our belief that 2D-GE will remain the cornerstone of expression proteomics and continue to play an important role in proteomic studies.

To visualize the separated proteins, researchers treat the final SDS gels with various protein-binding stains such as fluorescent reagents e.g. CY-dyes, pre-labeled, or various post-gel fluorescent stains, silver staining techniques and Coomassie Blue staining.

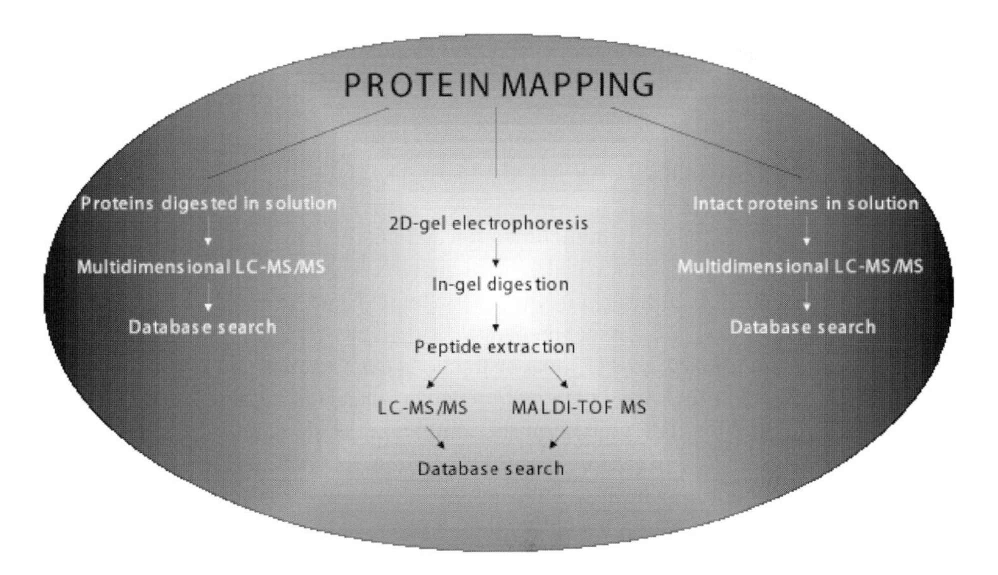

Fig. 9.1. Schematic illustration of protein separation principles applied for global protein expression analysis.

Alternatively, metabolic labeling can be applied, which utilizes a radioactive label in the initial mixtures followed by subsequent exposure of an X-ray film, (autoradiography). In both cases, the resulting image is a pattern of spots, each representing an individual protein or a group of overlapping proteins. Extremely high sensitivity can be obtained by metabolic labeling techniques. In in vitro studies, as little as a few thousand cells are sufficient to reach expression patterns where thousands of proteins can be determined. A limitation, though, is that an experimental step needs to be performed by the researcher in order to map the differential regulation; this would out-rule studies where, e.g. clinical material is to be investigated.

After staining, visualization is performed by image analysis where a selection and excision of individual protein spots from the gel is undertaken and the resulting gel plugs are both chemically and biologically degraded. Chemical degradation is performed by reduction and alkylation, followed by a tryptic digestion most often using an enzyme such as trypsin (in-gel digestion). The peptides are then extracted and prepared for identification by mass spectrometry.

In the past decade, mass spectrometry has emerged as the most important analytical tool for protein identification in proteomics studies, where various forms of mass spectrometry techniques are then used to determine the molecular masses of the resulting tryptic peptides.

Consequently, MALDI-TOF MS utilizes peptide mass fingerprinting, whereby the resulting peptide peak table is used in the database query. Peptide sequencing is applied utilizing MS/MS techniques where both electrospray ionization and laser desorption technologies can be applied to generate unique amino-acid sequences from protein digests [11]. Software developments within the informatics area have led to this becoming a highly powerful area, where various software algorithms are being used to compare the masses or sequences of these peptides with those in protein and gene sequence databases to determine the identity of the parent protein.

9.4. THE BENEFITS GAINED BY MINIATURIZATION

There are obvious reasons why miniaturization and microtechnology developments are gaining increased attention in analytical protein analysis development. As the biological material available for proteomic studies are very limited and precious, be that minute biopsies of pathogenic tissue, tissue preparations derived from laser microdissection, extracts/digests from excised 2D-gel spots or fractions from microliquid chromatography separations, it is evident that new analytical procedures needs to be developed that yield maximum information for each volume of the sample that is being analyzed.

There always seems to be a case of "we would like to see an order of magnitude lower level". These situations arise when one might try to address a biological situation where the interest is targeted maybe towards a single key protein, or a handful of proteins that form a biological entity of interest such as a pathway region, or a mechanism of action protein complexes.

In these situations, microtechnology and new miniaturized bioanalytical protocols may overcome the bottle neck of conventional macroscale proteomic methodology. As the dimensions of the assay formats are reduced, a number of benefits may be achieved. A few of these are listed below:

- reduced sample volumes⇒more and different assays may be performed on the same sample and ultimately samples that were too small earlier may now be analyzed
- reduced assay chemicals ⇒ the consumption of expensive assay compounds is dramatically reduced possibly allowing for high-throughput screening formats
- reduced dimensions that yield a faster mixing of compound by means of diffusion and thus a faster kinetic rate of reactions. The mean diffusion time for a molecule scales inversely to the $(\text{dimension})^2$. A compound that diffuses across the boundaries of a reaction vessel in 500 ms only needs 5 ms to perform the same act if the dimension of the vessel is reduced by a factor of 10
- surface area to volume ratio increases as the dimensions are reduced. This fact eventually favors that chemistry is performed at surfaces, e.g. immobilized chemistry becomes very efficient
- evaporation of a sample is speeded up considerably as the dimensions are reduced.

In the past, our group has put extensive effort into the development of micro/ nanostructures to improve proteomic analysis [31–53]. Topics such as high sample throughput, increased sensitivity, and low sample volume consumption have constantly been on the agenda. A special focus has been to develop new and more efficient ways for proteomic sample processing prior to MALDI-TOF MS and essentially bringing forth an attractive way of interfacing MALDI to any upstream analytical techniques, be that robotic sample handling, solid phase microextraction or on-line formats such as microliquid chromatography.

9.4.1. Silicon microfabrication

Our group has mainly developed microchip structures made from silicon. Commonly wet etching of monocrystalline ⟨100⟩- or ⟨110⟩-oriented silicon is utilised. The standard process steps are briefly described here. (a) The wafer is oxidized at high temperature, providing a protecting layer of silicon dioxide, followed by (b) a step where the formed silicon dioxide layer is coated with photoresist by spin coating. (c) A mask with the desired pattern is then placed in proximity to the wafer and is illuminated with UV-light. (d) The exposed wafer is subsequently developed, and (e) the remaining photoresist pattern is cured at an elevated temperature ca. 125 °C. (f) The silicon dioxide that now is unprotected is dissolved in buffered hydrofluoric (HF) acid. The HF etching stops as the underlying silicon is reached. (g) The remaining photoresist is then removed in the washing steps using acetone, ethanol and water. (h) Once this is done, the original photo mask pattern has been transferred to the silicon dioxide film on the wafer and (i) the silicon wafer is subsequently anisotropically wet etched usually using potassium hydroxide, KOH.

References pp. 196–198

Anisotropic wet etching of silicon provides extremely planar surfaces along the stop etch planes, which commonly define the achieved structure. If high surface area structures are desired, silicon may be treated in an electrochemical etch process, providing a highly porous layer at the surface of the silicon structure. Briefly, porous silicon is formed by anodizing a silicon wafer in a mixture of HF acid and ethanol. The obtained pore morphology may be tuned, e.g. by controlling the anodization voltage, current and illumination of the sample. In addition, fundamental properties of the semiconductor such as crystal orientation and dopant type influence the pore morphology. Fig. 9.2 shows an example of a microchannel array, with a channel width of 100 μm, which has been made porous to enable highly efficient enzymatic reactions, through immobilized enzymes, to be performed on-line in microfluidic systems.

9.4.2. Flow-through dispenser fabrication

The dispenser was fabricated in silicon using the above-described microfabrication. The flow-through microdispenser consists of two plates in which the inlet, outlet, flow-through channel and nozzle are defined. The two plates are joined by silicon direct bonding and silicone rubber tubes are attached to the inlet and outlet by silicone rubber adhesive. The inner diameter of the silicone rubber tubes is selected to fit standard 1/16″ tubes but other dimensions can also be used. The actuating component is a commercial piezo-electric multilayer element that generates the pressure pulse needed for ejecting droplets. The piezo-element is glued to a Plexiglass backing and in turn onto the microdispenser using epoxy adhesive. Fig. 9.3(a) shows a principal cross-section, length- and cross-wise of the microdispenser. Further details on the fundamental development of the dispenser can be found in Ref. [48]. Fig. 9.3(b) shows a series of dispenser generations as they have been

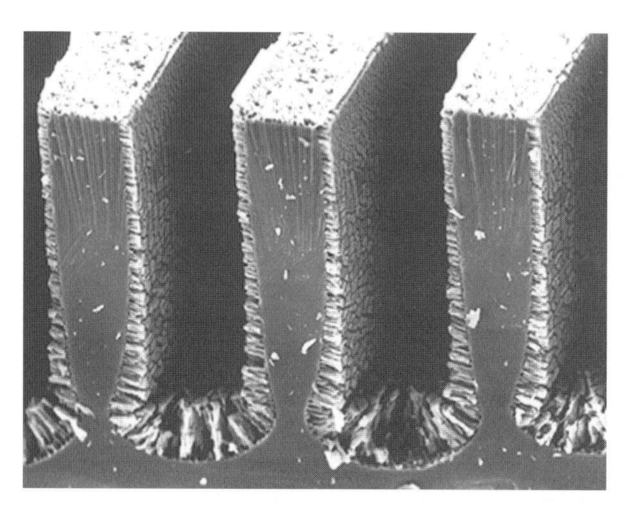

Fig. 9.2. SEM-image of a microchannel array covered with porous silicon. The vertical channels were initially crafted by anisotropic wet etching of ⟨110⟩-silicon whereafter the wafer was anodized in HF and ethanol, providing the highly porous surface layer on the channel walls.

(a)

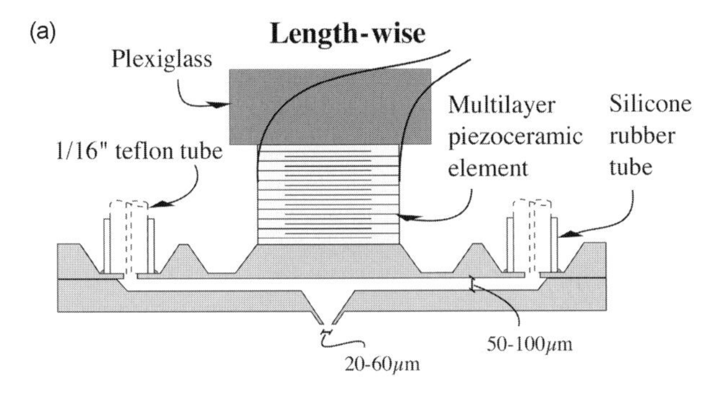

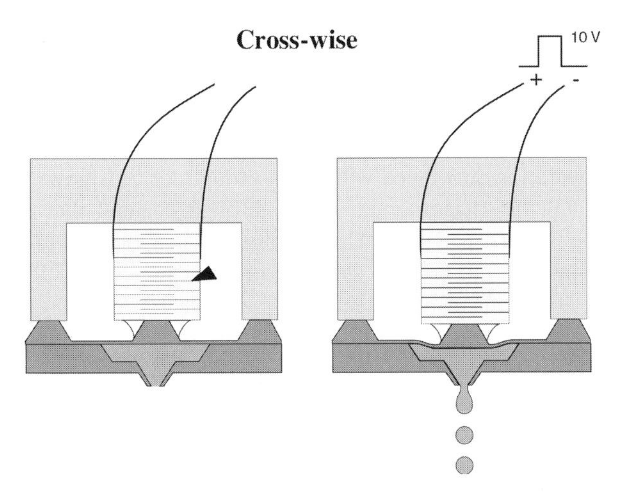

(b)

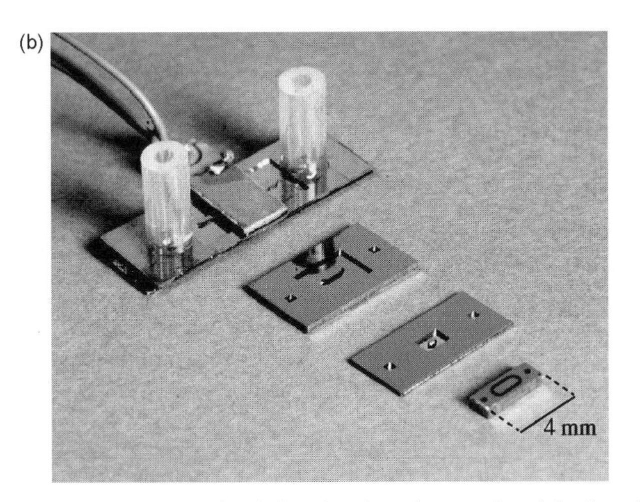

Fig. 9.3. (a) Cross-sectional view, length- and cross-wise of the flow through microdispenser and principal of operation. (b) A series of dispenser generations originating with an internal volume of 18 μl (upper left) and ending with an industry prototype having internal volume of 400 nl.

improved over the years. The smallest dispenser is the result of an industrial prototype version developed in collaboration with the biotech industry.

9.4.3. Dispenser operation

The piezo-dispensing technology provides high speed sample ejection of 100 picoliter droplets. The basics of piezo-electric dispensing originate from the inkjet printing area where such devices have been used since the mid-1970s. In order to use microdispensers in biochemical applications, careful attention must be paid to varying liquid parameters such as surface tension, viscosity, density and particle content, since these affect the droplet formation to a great extent. By matching the physical design of the microdispenser as well as the electrical pulse waveform for the piezo-ceramic actuator to the liquids used, it is possible to achieve the stable operation of a microdispenser for liquids with varying physical parameters. It is also important to keep the internal volumes of the units low to facilitate handling of small sample volumes and to reduce dispersion and avoid carry-over effects. The flow-through microdispenser operation developed within our group runs as follows.

The analytical flow is fed through the dispenser in a flow-through channel that is typically 8 mm long, 1 mm wide and 50–100 μm deep. This corresponds to a total volume of 400–800 nl. The volume from inlet to the nozzle, which is of great importance for analytical operation, is commonly 200 nl. The movement of a piezo-electric multilayer element is coupled to the wall opposing the nozzle and the channel volume is thereby slightly decreased when the piezo-element is activated by a voltage pulse, typically 10 V amplitude, 75 μs duration (Fig 9.3a). This causes the pressure in the channel to rise, which in turn accelerates the liquid in the nozzle and results in a droplet ejection (Fig. 9.4).

The size of the nozzle is typically 40 μm × 40 μm and the generated droplets have a diameter of ca. 50 μm, corresponding to a volume of 65 pl. The droplets can be ejected with a frequency of up to 5 kHz giving a maximum droplet flow rate of approximately 30 μl/min.

In 1998, we first presented the piezo-dispensing enrichment methodology for both peptides and proteins [31]. With a fast liquid evaporation, the sample will form multilayers on a very distinct small surface area. The scarce sample amounts are dispensed onto sample areas varying between 100 and 400 μm in diameter, thereby generating a high sample concentration on a small confined spot area. This on-spot sample enrichment forms the basis for a generic concept whereby, in principle, any sample can be dispensed and enriched onto a silicon high density target plate yielding an improved MALDI-TOF MS spectrum. This amplification principle is shown in the corresponding MS spectra in Fig. 9.5(a) and (b).

Fig. 9.5(a) shows the mass spectra of interleukin-8 (IL-8) when prepared using the dried droplet crystallization technique, commonly applied in most proteomics labs, and when deposited by microdispensing. The dispensed sample only used fractions (1/100) of the volume deposited by the dried droplet methodology yet giving orders of magnitude better than mass spectra. It can be noted that the dried droplet technique is also the crystal form that is deposited by all robotic sample handling instrumentation used for high-throughput protein identifications.

Fig. 9.4. Water droplet ejection from the microdispenser nozzle.

In Fig. 9.5(b), the piezo-dispensing enrichment is proven by sequentially amplifying IL-8. The increase in signal intensities found in the respective spectrum is shown with increasing number (100, 2000 or 8000) of microdroplets deposited onto the MALDI target plate. This clearly illustrates the amplification effect of the on-spot enrichment protocol. The dispenser has been improved in several areas since it was originally introduced in 1995. The introduction of a protruding nozzle greatly improved droplet directivity since the front surface wetting upon droplet ejection was minimized [34]. The total volume has also been reduced from the original 18 μl down to today's 400 nl. Therefore, it is now possible to handle sub-microliter sample volumes by the dispenser. The decreased volume has also led to a significant decrease in dispersion when the dispenser is inserted in a continuous flow system.

9.5. INTERFACING CAPILLARY LIQUID PHASE SEPARATIONS TO MALDI PEPTIDE MASS FINGERPRINTING

Our group has also developed the piezo-electric dispensing technology that can be interfaced with capillary liquid chromatography and MALDI-TOF MS [34,36,37,39,40,45]. The continuous mobile phase flow is fractionated on-line onto a high density MALDI-target plate. The obtained highly resolved chromatographic separation on the MALDI target will allow an efficient method of fast MS-screening and identification of the proteins present in

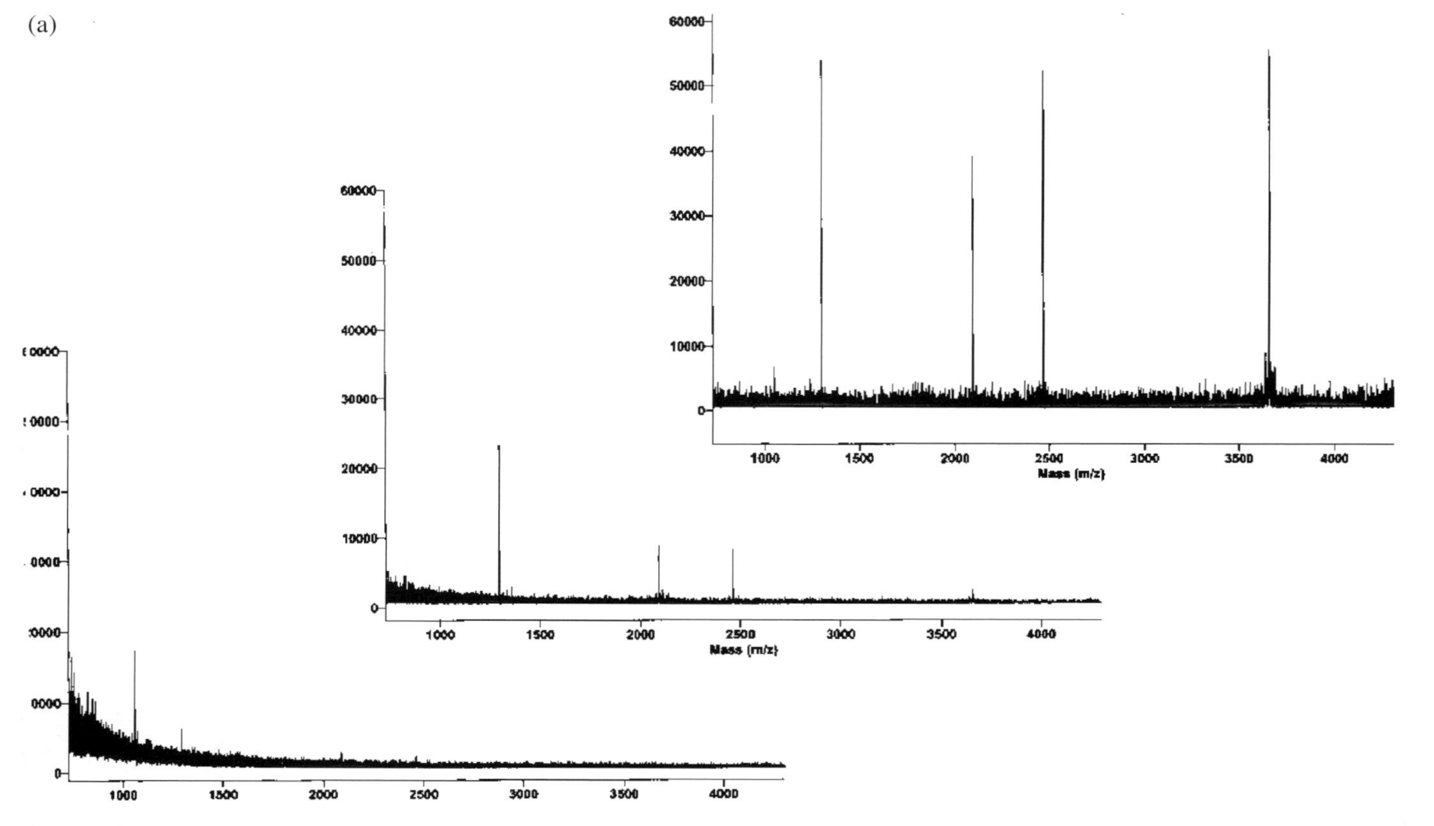

Fig. 9.5. (a) Mass spectra of a 5 nM standard petide mixture, Angitensin I, ACTH (1-17), ACTH(18-39) and ACTH (7-38). In the lower left mass spectra, 100 droplets were deposited, middle – 200 droplets and upper right – 8000 droplets. This illustrates the amplification effect that can be obtained by sequential dispensing and rapid sample evaporation. (b) A comparison of MALDI sample preparation using either (A) the dried droplet technique or (B) microdispensing and fast evaporation.

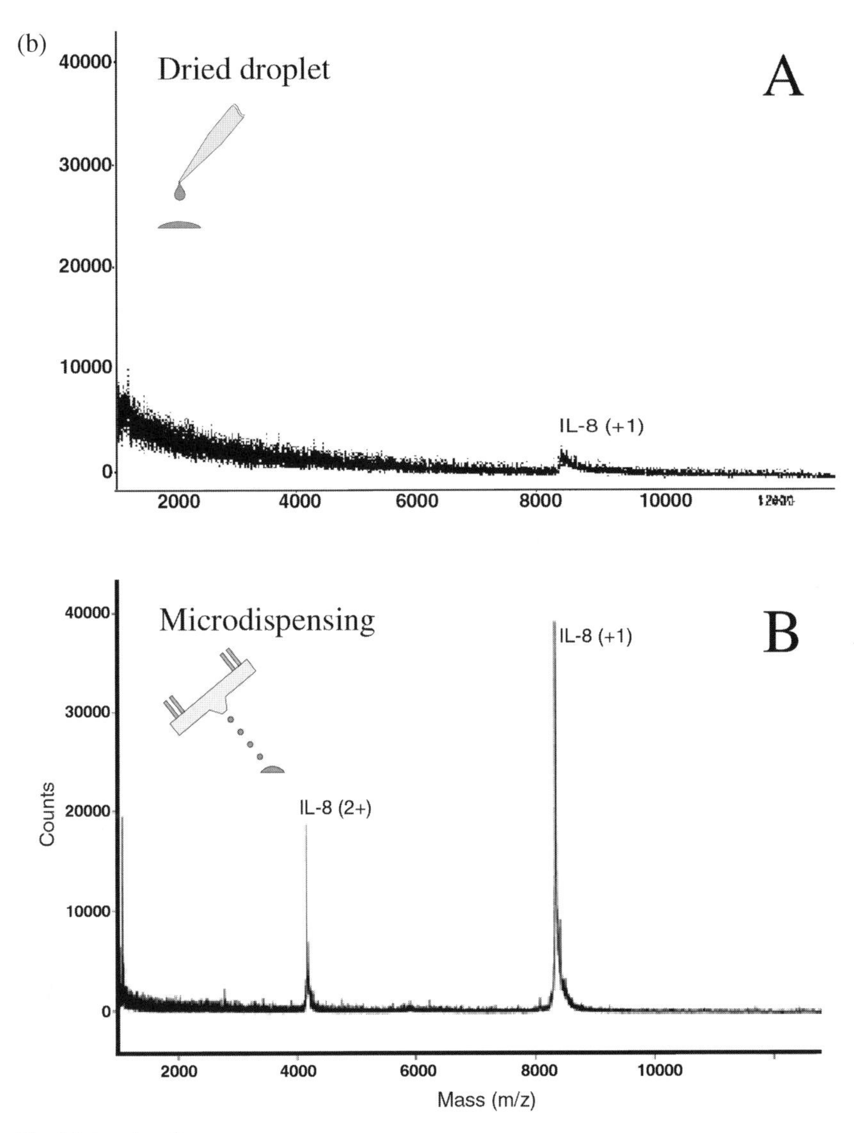

Fig. 9.5 (*continued*)

each biological sample. Here, the separated peptides are spotted by the dispenser onto a matrix/polymer thin-film MALDI-target plate [36,37]. The series of spots was then analyzed by MALDI followed by a merge of the obtained spectra producing a total peptide map of the protein.

In our set-up, the microdispenser forms the interface between the upstream liquid systems and the MALDI target plate. The liquid enters the dispenser and is ejected as droplets towards the target. In contrast to other microdispensers commercially available the one used here is of flow-through type where the droplets are ejected from a liquid passing the dispenser. This enables direct connection to continuous flow systems as well as simplifying the operation with respect to priming and changing of liquids, an important feature since all separations are made by gradient elutions.

The main benefit obtained by running capillary separations in many cases, even from sample spots that were previously separated by 2D-GE, is due to the improved sequence coverage. This can be compared when all the peptides are measured simultaneously. A drawback that occurs when all peptides are ionized and introduced in the TOF-flight tube is the ion suppression effect, where the mostly abundant peptides will be selectively ionized suppressing the ones at lower concentration levels. Fig. 9.6 shows an example where we have run the same sample using both direct MALDI analysis and capillary LC-separated peptides and thereafter enriched by piezo-dispensing onto the MALDI target plate. The additional masses displayed were found after running the samples by capillary LC interfaced with dispensing and spot enrichment. As shown in Fig. 9.6 the sequence coverage was increased significantly.

The operational stability of the interface is also very good. Stable dispensing is achieved over 5–10 h while cycling the eluent composition from aqueous/organic modifier 5:95 to organic/organic modifier 55:45.

9.6. INTEGRATED ON-LINE PROTEIN WORKSTATION

One essential part of protein identification with peptide mapping is to get a reproducible and sufficient digestion of the protein. A current bottleneck of this technique is the long incubation time required for digestion (6–24 h) when the protein is digested with a proteolytic enzyme in-solution. One way to address this problem is to use flow-through high capacity digestion by using immobilized enzymes that offer a high catalytic activity. Here we present work on enzyme reactors in microscale with methods such as enzyme

Dried droplet preparation

LC on-line to microdispensing

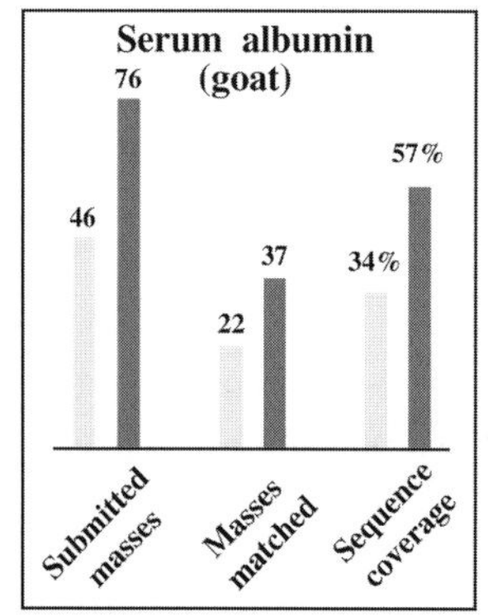

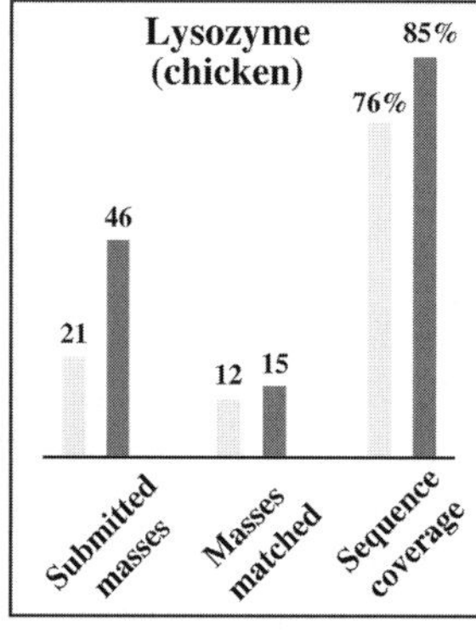

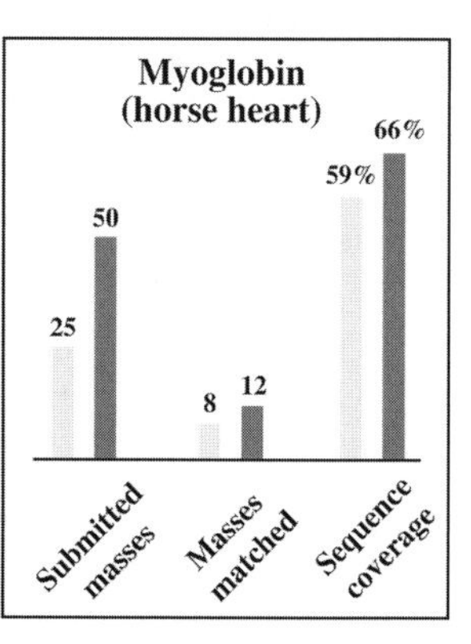

Fig. 9.6. Comparison of sequence coverage from resulting experiments where dried droplet sample preparation was compared with capillary liquid chromatography separation and MALDI fractionation.

activated beads packed in chip integrated microchannels, or monolithic high-surface are micro-chip reactors [47, 49].

Previously we have developed a continuous flow protein workstation whereby protein fraction studies can be conducted as outlined in Fig. 9.7 [32]. The platform holds three microstructure components whereby the integrated system generates advantages. (i) The large surface area enzyme digest reaction chamber, (ii) the piezo-dispenser from which the digested protein is spotted, and (iii) the high density MALDI target plate where the dispensed multilayer format generates the signal amplification in the subsequent MALDI read-out.

The figure illustrates a continuous flow system where the robotic sample introduction is followed by a microfluidic sample transfer into the catalytic microreactor where the protein digestion takes place. The immobilized enzyme microreactor (μ-IMER) commonly consists of a packed bed of porous beads or a porous silicon microreactor (c.f. Fig. 9.2) onto which the enzyme is covalently bound. The high surface area of the porous matrix compared with the small volume allows a fast digestion. Following the enzyme cleavage, the enzymatic products are eluted into the piezo-dispenser. Sample deposition is made onto the high density silicon target plates that are inserted in the mass spectrometer. See also section 9.8.2.

9.7. INTEGRATED MICROANALYTICAL TOOL-BOX COMPONENTS

Below will follow a presentation of each and every tool-box component that forms part of a newly developed integrated protein platform. The system set-up used for the proteomics applications is shown in Fig. 9.8.

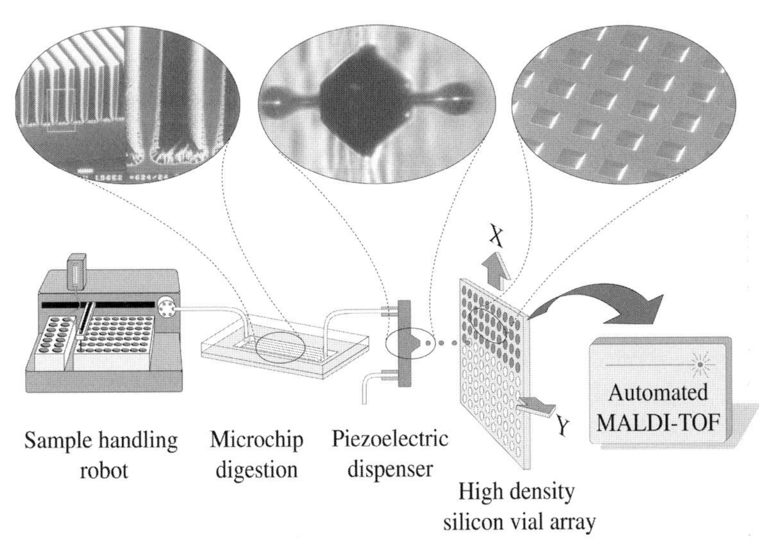

Fig. 9.7. Overview of microtechnology proteomics platform using silicon microstructures.

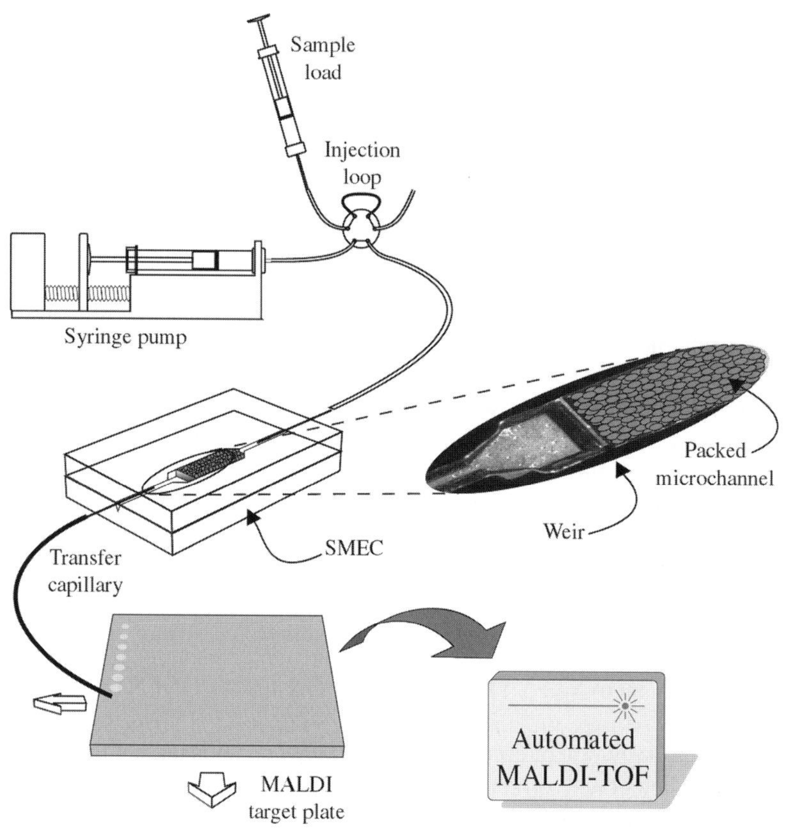

Fig. 9.8. Schematic overview of proteomics sample handling system using microextraction chip sample cleanup/enrichment.

9.7.1. Integrated on-chip microextraction system interfaced to peptide map fingerprinting

In order to increase sensitivity of the final mass spectrometry read-out, an enrichment step was built in our proteomics processing strategy where the final goal was to gain signal amplification factors ranging between 40 and 500 [35,42,50–52]. This has proven to have a profound influence on proteomics studies whereby quantitative analysis can be performed at lower abundance levels. The expression levels whereby quantitative proteomics studies are performed is in our opinion the most limiting factor to biological understanding of, e.g. pathophysiology progression.

In principle, we can make enrichments and purifications from a few microliter to hundreds of microliter sample volumes, depending on the sizes of biological materials available. The larger sample volumes typically arise in cases where we run multiple 2D-gel sets or fractionate preparative liquid chromatography fraction volumes. The processing steps made from these samples will need some extended sample loading and elution times, as compared with handling of a few microliter sample volumes. The extraction microstructure geometry was developed to hold a bead trapping geometry in the silicon microextraction chip (SMEC).

The extraction beads are loaded in front of the bead trap, retaining the beads that can be 20–100 μm in diameter [52,53]. We have found upon repeated packing of these microchannels that the packing density will be a maximum of 10 particles, aligned in parallel to each other.

The on-chip microextraction system described has been developed in order to analyze protein samples, regardless of whether they arise from 2D gels, chromatography fractions, affinity probes, or from any other isolated preparation. In the case of proteomic samples analyzed by 2D gels in biological studies, single or multi-spot analyses are performed by the on-chip microextraction chip. In cases where limited sample material is available multi-spot analysis is performed, i.e. the same spot from several gels are excised in order to generate larger amounts of protein that can be analyzed. In our experience, this will ensure a higher identification success. The drawback of this approach is that since typically three to six spots are being digested in the same 96-well vial position, which will result in a larger final sample volume. Consequently, processing of the larger volumes will take somewhat longer time when eluted through the on-chip microextraction.

In order to illustrate the benefits of using the silicon microextraction chip (SMEC) on-line in a flow-through format we used a model peptide mixture sample comprising Angiotensin I, ACTH Clip 1–17, ACTH Clip 18–39 and ACTH Clip 7–38 at a 10 nM concentration level as shown in Fig. 9.9.

It is obvious from these MS-spectra that the signal improvement and sensitivity gain is vital in order to enable low abundant protein identification. The lower detection limit for the peptide mixture was found to be below 1 fmol/μl level. Commonly, injection volumes ranging between 1 and 50 μl were investigated. We developed this system mainly to be used for sample clean-up/enrichment in the analysis of in-gel digested 2D-GE spots was used.

9.8. APPLICATIONS TO PROTEOMIC SAMPLES FROM HIGH SENSITIVITY DETERMINATIONS

As an example the identity of a sample originating from a 2D-gel spot as shown in Fig. 9.10 is correctly identified as ATP synthase. We made in-gel digestion of the protein spots with trypsin and the resulting peptide solutions were acidified through addition of 80 μl 0.1% TFA (100 μl total volume) and subjected to clean-up on the micro extractionchip.

The microchip structure was used to trap reversed phase chromatography media (Poros R2 beads) that facilitated the sample clean-up and subsequent trace enrichment of the corresponding ATP-synthase peptides generated from the protein digest step. Purification/enrichment of contaminated and dilute samples were thereby enabled to the MALDI-TOF MS analysis.

The identity of the ATP synthase was analyzed by a seed-layer sample preparation of the digested protein samples, and it was not possible to make a direct identification of the resulting peptide maps generated as shown in Fig. 9.10(A). However, after the microchip

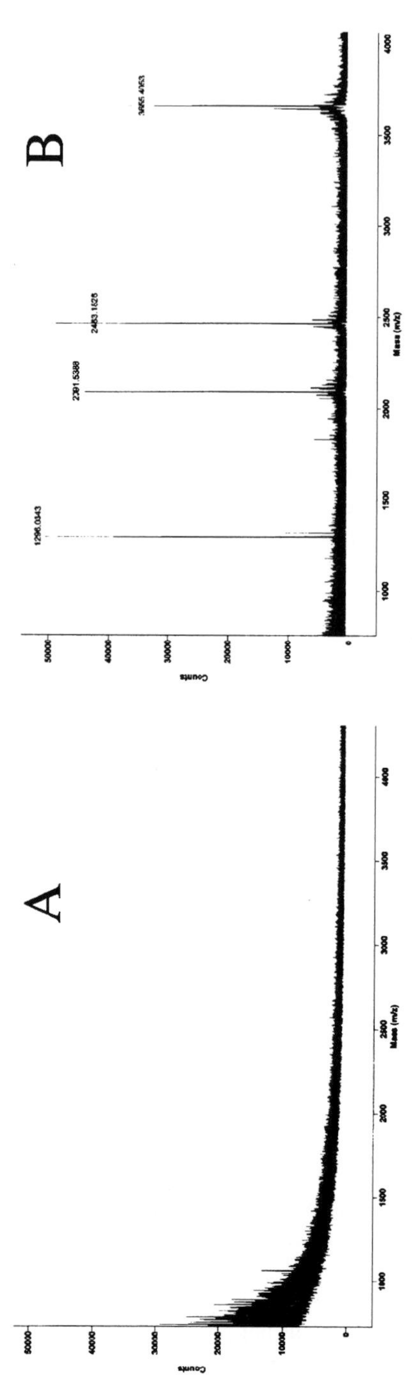

Fig. 9.9. MALDI spectra generated from experiments using the microextraction chip whereby signal amplification was achieved. (A) MALDI spectrum of Angiotensin I, ACTH Clip 1–17, ACTH Clip 18–39 and ACTH Clip 7–38. (B) Shows the corresponding mass spectra when the SMEC preparation was used.

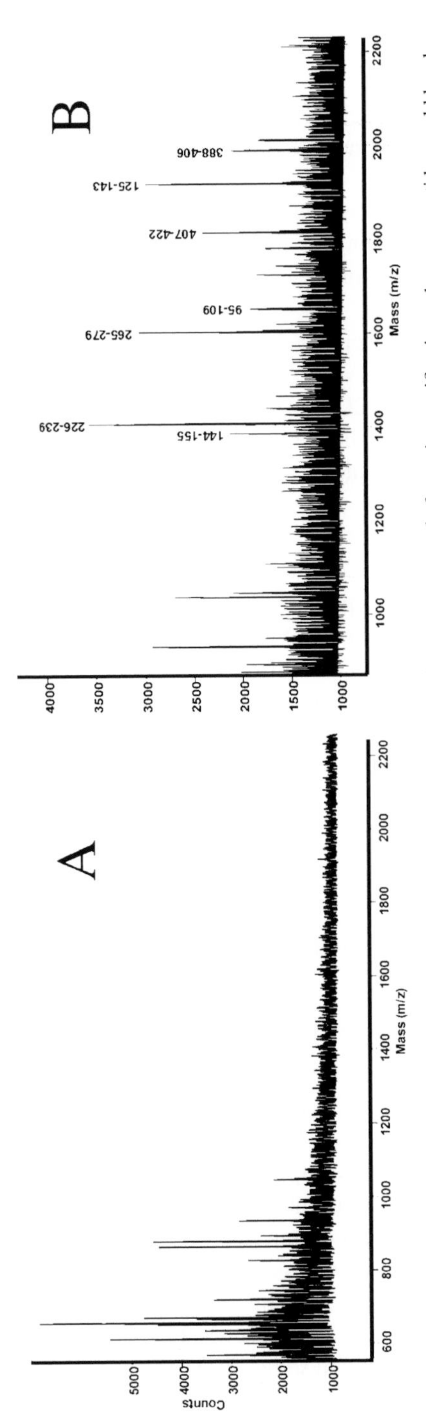

Fig. 9.10. MALDI mass spectra resulting from seed-layer sample preparation of the digest supernatant before micropurification, where no peptides could be observed (A). (B) The microchip sample cleanup/enrichment of the in-gel digested ATP synthase spot. The spot was correctly identified by database search as human ATP synthase (Swiss prot acc. no. P06576).

sample cleanup/enrichment seven peptides could correctly be ascribed and identified by database search from the corresponding MALDI spectrum in Fig. 9.10(B).

9.8.1. Analyzing targeted disease cells in proteomic studies

Discovery activities are currently focused on making protein expression studies that are directed towards the understanding of human diseases in order to find new ways for pharmaceutical intervention to result in successful treatment, and/or halt the progression of disease.

After we found that the conceptual principal was fulfilled utilizing high sensitivity microchip extraction with the SMEC, we used it for a large sample spot series on biological samples derived from 2D-PAGE gels [52].

The gel sample extracts ranged between 10 and 50 μl, and were loaded onto the SMEC, whereafter it was washed and eluted onto the MALDI target plate as discrete spots of 500 μm.

Typical proteomics samples were successfully analyzed by the use of microchip extraction and sample preparation is further shown in Fig. 9.11 where a Lipocortin V spectrum is shown with the corresponding peptide map fingerprint.

The annotation of both Lipocortin V- and ATP-synthase-expression was made in stimulated human primary cells. These diseased target cells were plated and cultivated under adherent conditions and harvested after 48 h stimulation. The regulative expression of a large number of proteins could be determined using the microchip extraction principle on 2D-gel based proteomic studies. Some of these findings were also proven by an earlier study, which could be used to overlay the identity maps made from the two studies [41,52].

9.8.2. On-line tryptic digestion sample preparation using the SMEC-microchip

An additional feature of the SMEC concept was illustrated by enzymatic digestion using trypsin-bound microparticles. The ability to easily pack microchip structures with the enzyme immobilized beads opens the possibility to use them as immobilized enzyme reactors (IMER) by the trapping of beads with immobilized enzyme.

The microchip SMEC was made by packing the microchip with POROS beads immobilized with trypsin. The microchip IMER was then used to digest a 1 μl sample of reduced and alkylated BSA at a concentration of 100 nM in 25 mM Tris, 1 mM $CaCl_2$, pH 8.0. A sample plug of 1 μl was injected and a carrier flow of digestion buffer transported the plug through the microchip IMER; the eluate was collected in an Eppendorf tube and then subjected to microscale sample cleanup/enrichment on a microchip packed with POROS 20 R2 as described earlier. The resulting mass spectra is presented in Fig. 9.12. Fully automated series' of proteomics analysis can thereby be made similar to that described in Fig. 9.7.

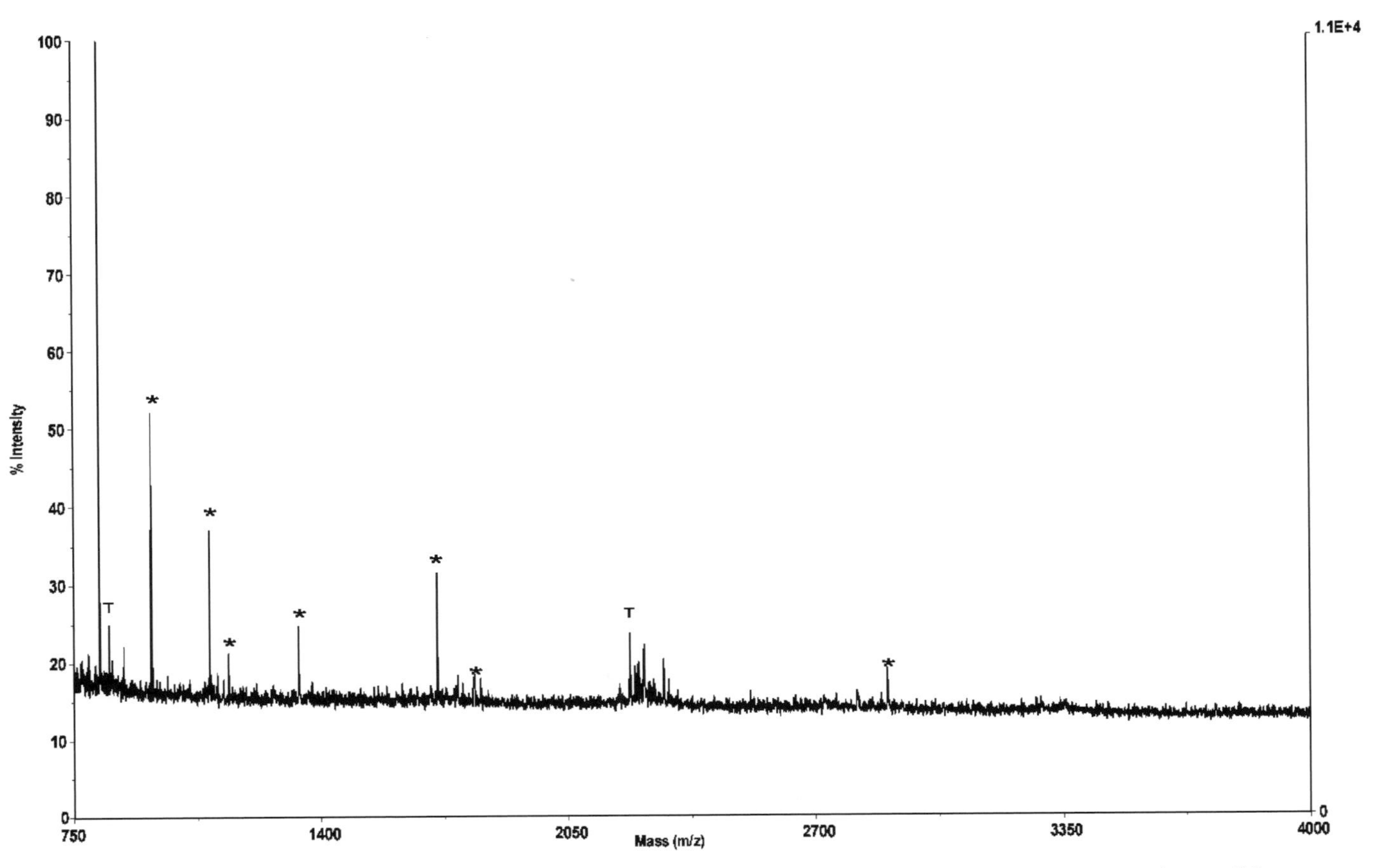

Fig. 9.11. MALDI spectrum from Lipocortin V generated from an excised 2D-gel spot that was separated originating from a human primary cell lysate.

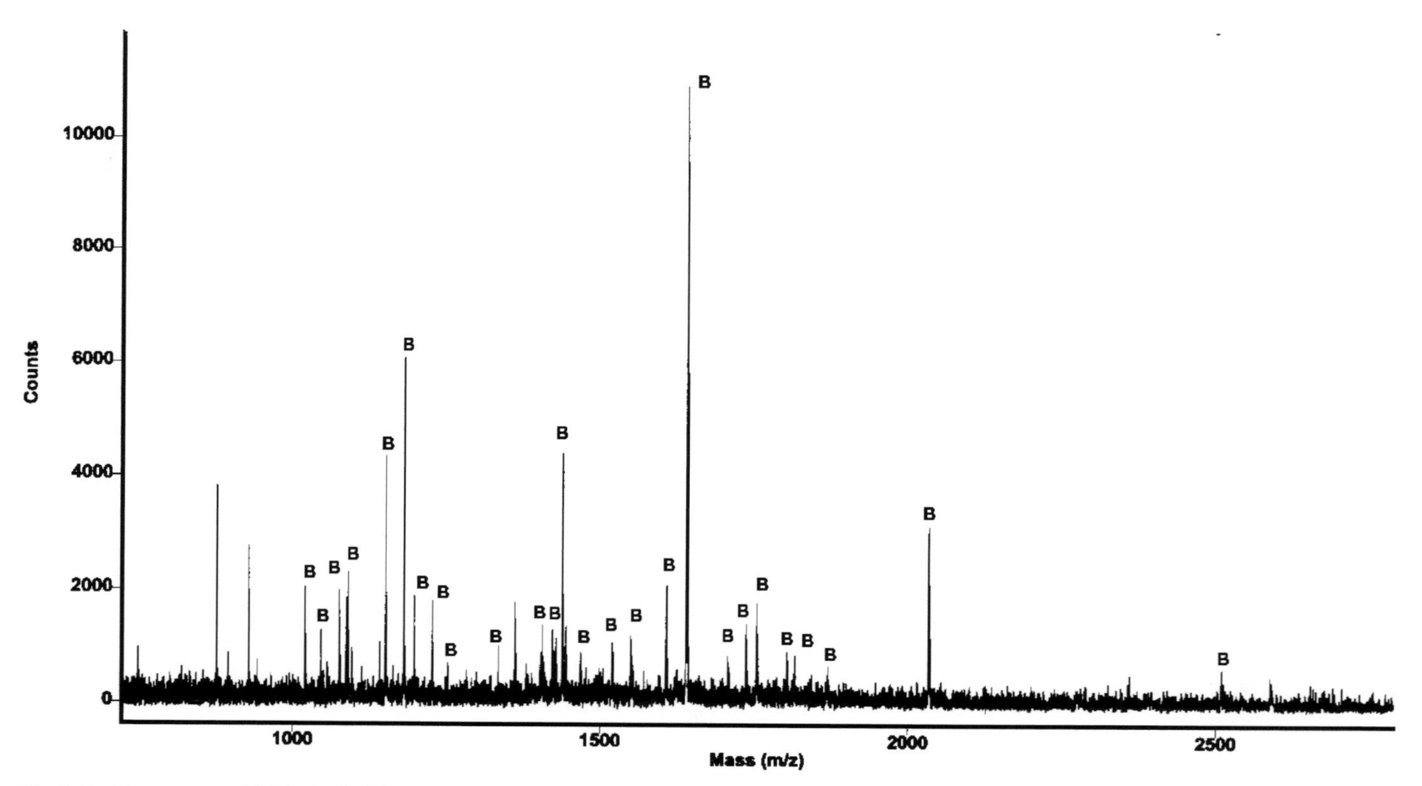

Fig. 9.12. Mass spectra of BSA obtained by performing enzymatic digestion in a microchip packed with trypsin immobilized beads.

9.9. IN-VIAL DIGESTION USING NANOVIAL MALDI-TARGET ARRAYS

A highly efficient technology for enzymatic digestion of proteins utilizing nanovial arrays interfaced to MALDI identification by peptide mass fingerprinting is outlined below.

It is based upon the dispensing of a protein solution with simultaneous evaporation. The protein (substrate) can be concentrated up to 300 times in a high density array of nanovials. The enzymatic kinetics will follow Michaelis–Menten kinetics and when using high substrate concentrations the digestion speed is dramatically increased. Therefore, the dispenser-aided nanodigestion is valuable for identification of low-level proteins (10–500 nM) as well as for automatic high efficiency digestions of medium to high level proteins performed for 0.2–10 min.

The developed silicon flow-through piezo-electric dispenser is adapted for low-volume and pre-concentrated samples in the nl–μl range and provides fast, accurate and contact-free sample positioning into the nanovials [51]. The steps in the in-vial digestion protein determination principle is illustrated in Fig. 9.13.

The nanovial piezo-electric microdispensing station is depicted where opposite to the dispenser (insert (b) in Fig. 9.14 showing the dispenser nozzle and a droplet),

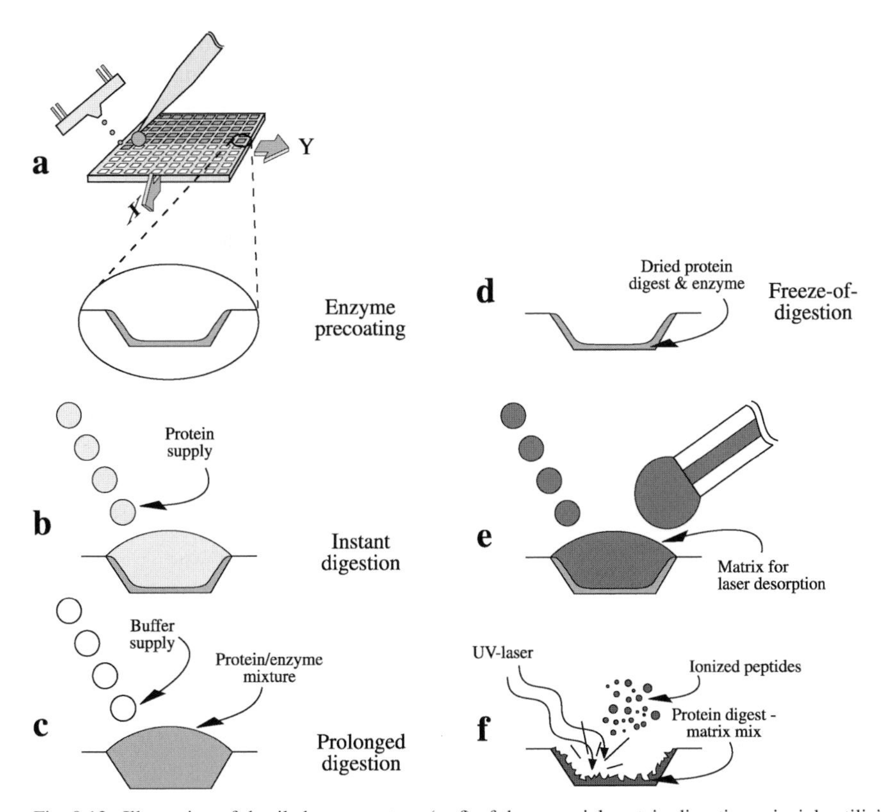

Fig. 9.13. Illustration of detailed process steps (a–f) of the nanovial protein digestion principle utilizing piezo-dispensing technology.

the MALDI-probe comprising the nanovial ((a) in Fig. 9.13) arrays were mounted on the motor driven *x–y* stage of the dispensing platform. Time and space increments of the *x–y* stage movements were controlled by LabView (National Instruments Inc., Austin, TX) routines developed in our group. The MALDI-probe was temperature controlled by high-power resistors enabling heating of the sample probe. This was to facilitate increased evaporation rate during the dispensing of liquids, to unfold the proteins, which facilitates enzymatic accessibility, as well as to enhance the kinetics during enzymatic catalysis. During all analyses the temperature was set to 40 °C, which is close to the optimal catalysis temperature of trypsin.

9.9.1. Dispenser-aided nanovial digestion

The protease (trypsin) was first added to the silicon chip either by dispensing into individual nanovials, or by covering all nanovials with 5–10 μl (10 mM) trypsin added by a pipette. When a defined amount of trypsin is desired, it can be dispensed in a burst of a 10–100 droplets at a rate of 10–15 Hz. The trypsin is usually dispensed in an automatic mode within minutes. Trypsin supply by pipette was even faster and more simple, but less accurate in terms of reproducible trypsin amount in each vial, as shown in Fig. 9.13.

As a second step in the method the protein sample was dispensed into the trypsin-coated nanovials. This was achieved by controlling the droplet frequency, and thus the digestion-volume. The proteolysis of the protein begins instantly and progresses as long as more sample is dispensed. This will be referred to as the instant digestion principle. In the performed experiments the digestion-times run between 10 s and 10 min.

In cases where the instant digestion time is inadequate for protein identification, an additional period of digestion is applied. This is done by the dispensing of 5–25 mM NH_4HCO_3 into the vials (see Fig. 9.13). When the volatile buffer is added to a protease and protein containing nanovial, the previously interrupted digestion-process is resumed. In this way, this additional step prolongs the digestion-time. In principle, the maintenance of catalytic activity in the chip vials can be prolonged, which is an important feature that allows flexibility for the digestion of most sample types. As the dispensing stops, the enzymatic catalysis terminates and immediately a freeze-of-digestion occurs. After complete digestion, the MALDI-matrix is applied into the nanovials with one of three different techniques investigated.

Fig. 9.14 shows the three alternatives whereby the protein sample as well as the digesting enzyme can be handled in the high density MALDI target plates.

(i) With the most preferable technique for adding matrix into the nanovials, the dispenser was used in the same manner as with enzyme and sample. Dispensing of the matrix was used in the experiments if not stated differently.
(ii) Alternatively, a GELoader-tip on a 10 ml pipette was used to manually apply the matrix-solutions by dipping a 5–20 nl droplets over each nanovial. This was only used for the bigger vials (400 × 400 mm).
(iii) As a third technique, different PTFE or silica capillaries were utilized together with a pump to add 2–5 nl matrix into the nanovials.

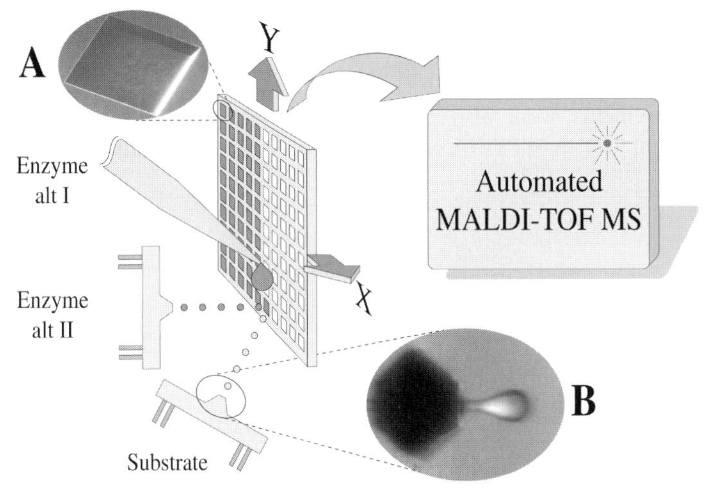

Fig. 9.14. Illustration of the three techniques whereby sample deposition and protein digestion was performed.

(iv) Subsequent to matrix-addition, MALDI-TOF analysis was performed either automatically or manually. By using internal calibration, the mass accuracy is typically 5–10 ppm for the database-matched peptides, although peptides with a mass deviating up to 30 ppm were accepted to be a match.

9.9.2. High speed in-vial protein catalysis

A spectrum resulting from an instant digestion of 50 fmol lys C (0.5 mM), which is digested in less than 75 s, is presented in Fig. 9.15(a). The nanodigestion resulted in 11

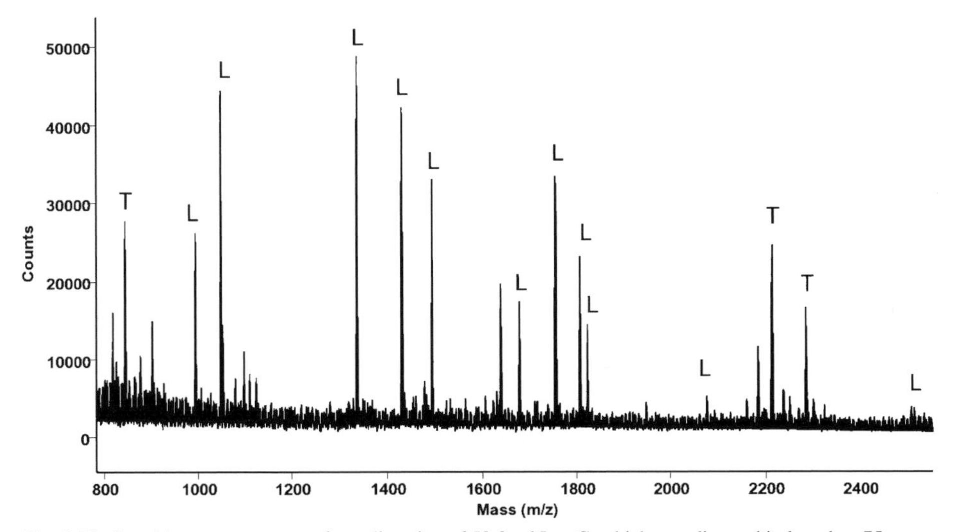

Fig. 9.15. Resulting mass spectrum from digestion of 50 fmol Lys C, which was digested in less than 75 s.

References pp. 196–198

clearly identified peptide fragments of Lys C corresponding to 62% of the protein sequence. Trypsin was used in the same amount as Lys C (50 fmol) and the matrix was added with the GELoader-tip technique.

High sensitivity measurements were also performed whereby the resulting mass spectrum formed digestion of 300 amol Lys C, which was digested in less than 5 minutes as shown in Fig. 9.16(A).

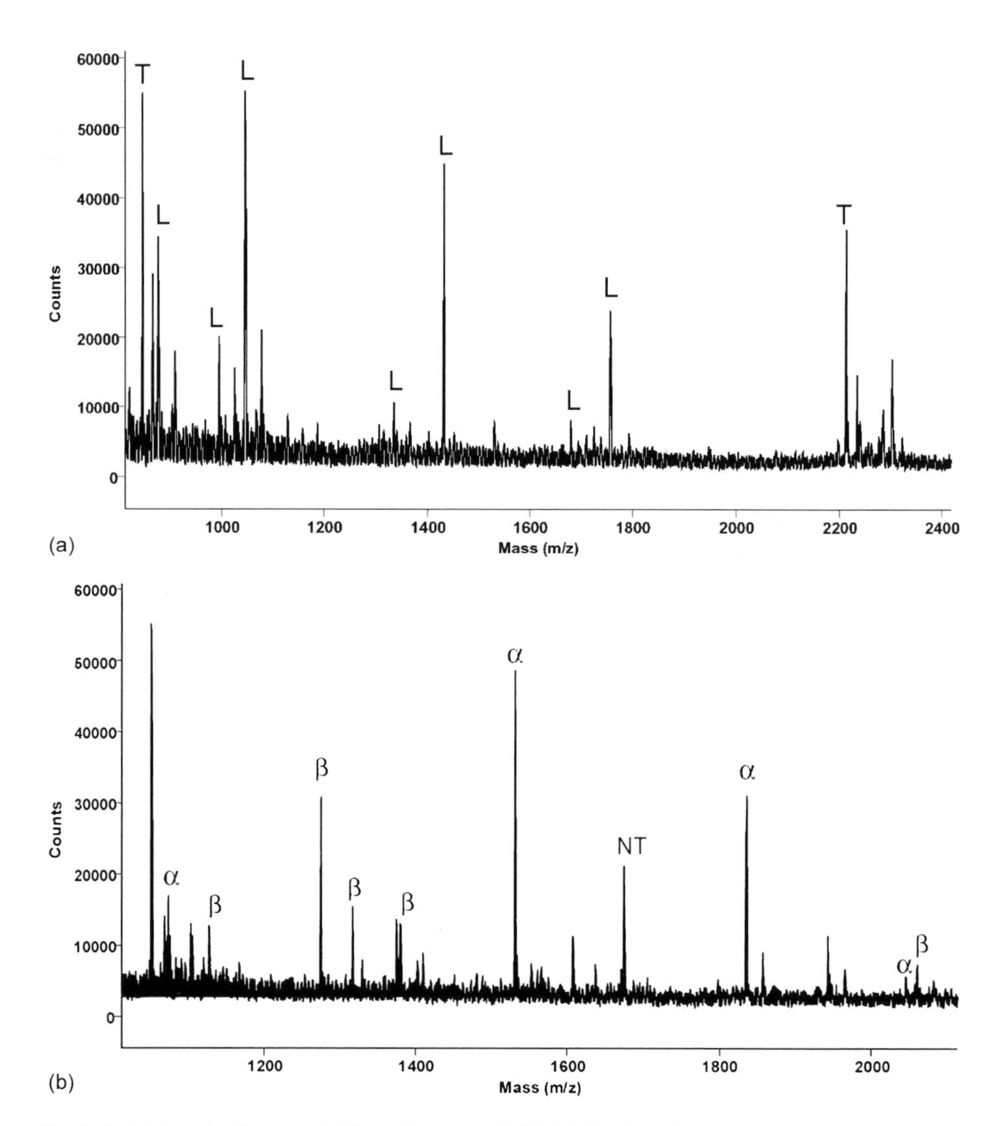

Fig. 9.16. (a) Nanovial digestion of 300 amol lysozyme C (10μM). The digestion time was 5 minutes. (b) Analysis of α-and β hemoglobin (10 μM), α-subunit with five peptides, 43% of amino-acid sequence coverage and β-subunit with four peptides and 31% of amino-acid sequence coverage, after only 30 s using the instant digestion technique.

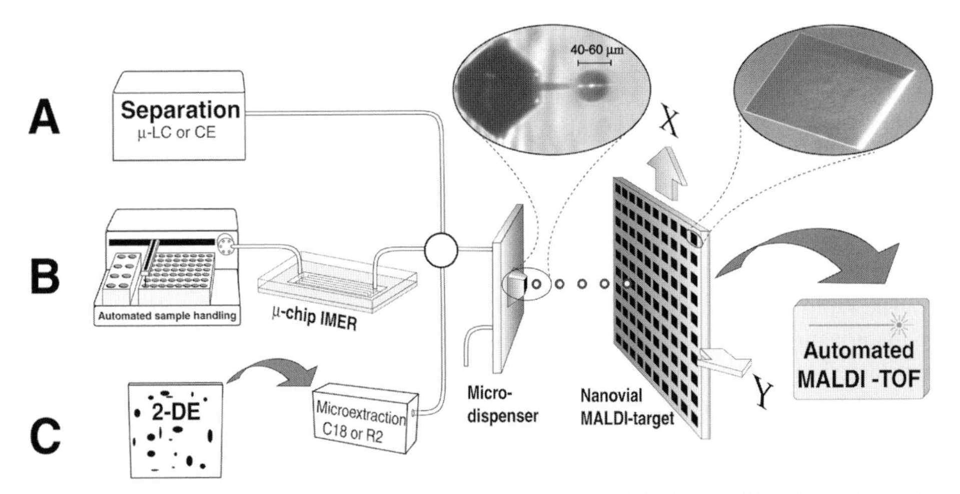

Fig. 9.17. Schematic compilation of platform developments performed in the Lund NanoTech Proteomics laboratories.

Fast catalysis of in-vial digestion was also demonstrated by the analysis of 10 μM hemoglobin for which no reduction or alkylation is needed. Protein identifications made by fast digestion methodology are shown in Fig. 9.16(b), where the spectrum shows peptides from the hemoglobin only after 30 s using the instant digestion technique.

9.10. COMPILED PLATFORM DEVELOPMENTS

The overall objective of our research group has been to develop automated protein analysis platforms that will allow high sensitivity and high speed protein analysis by MALDI MS. Fig. 9.17 shows an integrated solution to the concepts that we have outlined in this chapter, whereby microstructure developments utilizing silicon microtechnology to manufacture microstructures are being optimized for proteomic analysis.

We have in these projects emphasized that, regardless of the sample and/or separation technologies used, we will have to interface optimally for optimal protein identifications. Fig. 9.17 illustrates the strategy we currently use that encompasses (a) capillary liquid chromatography separations, (b) protein fractionation analysis, and (c) sample processing of 2D-gel spot analysis. The compiled platform presented covers most of the proteomics samples that are currently being processed in our group from various experimental expression studies.

9.11. CONCLUSIONS

Proteomics currently drives an increasing interest in protein analysis and the need for new technologies that enables better reading of protein expression. It is demonstrated that microfluidic tools can provide the means necessary for effective unattended automated protein identification and provide a welcome sensitivity increase in the analysis of peptides and proteins. The flow-through microdispenser is a key to flexibility that allows for the alteration of the presented platform configuration depending on the analytical situation. As biosciences progress, the development of miniaturized analytical technology will result in numerous new integrated chemical microsystems that can be expected to have a major impact on the life science research.

9.12. FUTURE PERSPECTIVES

The rapid development among the biotechnology companies in the world shows that miniaturization of analytical systems combined with intelligent use of biology is generally considered to be the strategy that will overcome the requirements of process speed for performing efficient proteome analysis.

The newly developed MALDI MS instrumentation capable of MS/MS-sequencing could benefit greatly from MALDI target plates that can facilitate the fractionation of an entire chromatographic peptide separation (i.e. thousands of sample spots). In view of this, separation of complex peptide mixtures and on-line nanofractionation of the samples, e.g. by microdispensing into nanovial arrays, may be an ideal solution [54]. The use of immobilized protease microreactors to digest intact proteins from 2D liquid chromatography may also provide an alternative approach to perform large-scale protein identification.

One of the most important post-translational modifications is protein phosphorylation, which many times has been directly linked to biological function, such as pathway signaling. By applying an automated in-vial dephosphorylation step to a nanovial fractionated peptide separation, potentially interesting phosphorylated peptides can be rapidly identified by the loss of 80 Da.

In view of this challenging future, our group is developing new microchip-based techniques to improve the current situation in protein analysis.

Mass spectrometry is one of the most rapidly growing analytical techniques in proteomics research due to its high sensitivity and ability to analyze small sample volumes ($\leq 1\ \mu l$). MALDI TOF-MS has become one of the working horses in every proteomics laboratory due to its sensitivity and the ease of automation. MALDI TOF-MS, however, still suffers from a poor interface, although new robotized sample handling protocols have improved this situation. Our group has therefore focused on developing a microtechnology based platform that interfaces with MALDI TOF-MS to a wide range of upstream sample handling and/or analytical techniques. Some of the major goals set forth in our research group for the close as well as the somewhat distant future will focus on the developments directed towards high sensitivity and high-throughput protein analysis with submicroliter sample consumptions.

9.13. REFERENCES

1. S.P. Gygi, B. Rist, S.A. Gerber, F. Turecek, M.H. Gelb and R. Aebersold, Nat. Biotechnol., 17 (1999) 994–999
2. P.A. Haynes and J.R. Yates III, Yeast, 17 (2000) 81–87
3. M.P. Washburn, D. Wolters and J.R. Yates III, Nat. Biotechnol., 19 (2001) 242–247

4. D.K. Han, J. Eng, H. Zhou and R. Aebersold, Nat. Biotechnol., 19 (2001) 946–951
5. A.L. McCormack, D.M. Schieltz, B. Goode, S. Yang, G. Barnes, D. Drubin and J.R. Yates III, Anal. Chem., 69 (1997) 767–776
6. Y. Oda, K. Huang, F.R. Cross, D. Cowburn and B.T. Chait, Proc. Natl Acad. Sci. USA, 96 (1999) 6591–6596
7. L. Pasa-Tolic et al. J. Am. Chem. Soc., 121 (1999) 7949–7950
8. J. Ji, A. Chakraborty, M. Geng, X. Zhang, A. Amini, M. Bina and F. Regnier, J. Chromatogr. B: Biomed. Sci. Appl., 745 (2000) 197–210
9. T.J. Griffin, D.K.M. Han, S.P. Gygi, B. Rist, H. Lee, R. Aebersold and K.C. Parker, Am. Soc. Mass Spectrom., 12 (2001) 1238–1246
10. J.R. Yates III, J. Proteome Res., (2002)
11. R. Aebersold and D.R. Goodlett, Chem. Rev., 101 (2001) 269–295
12. H. Zhou, J.A. Ranish, J.D. Watts and R. Aebersold, Nat. Biotechnol., 20 (2002) 512–515
13. M. Geng, J. Ji and F.E. Regnier, J. Chromatogr. A, 870 (2000) 295–313
14. R. Zang and F.E. Regnier, J. Proteome Res., 1 (2002) 139–149
15. M. Munchbach, M. Quadroni, M. Miotto and P. James, Anal. Chem., 72 (2000) 4047–4057
16. M.B. Goshe, T.D. Veenstra, E.A.T.P. Panisko, N.H. Conrads Angell and R.D. Smith, Anal. Chem., 74 (2002) 607–616
17. S.P. Gygi, G.L. Corthals, Y. Zhang, Y. Rochon and R. Aebersold, PNAS, 97 (2000) 9390–9395
18. S.J. Fey and P.M. Larsen, Curr. Opin. Chem. Biol., 5 (2001) 26–33
19. J.S. Andersen, C.E. Lyon, A.H. Fox, A.K.L. Leung, Y.W. Lam, H. Steen, M. Mann and A.I. Lamond, Curr. Biol., 12 (2002) 1–11
20. N.L. Anderson and N.G. Anderson, Electrophoresis, 19 (1998) 1853–1861
21. A. Manz, N. Grabner and H.M. Widmer, Sens. Actuators B-Chem., 1 (1990) 244–248
22. D.R. Reyes, D. Iossifidis, P.-A. Auroux and A. Manz, Anal. Chem., 74 (2002) 2623–2636
23. D.R. Reyes, D. Iossifidis, P.-A. Auroux and A. Manz, Anal. Chem., 74 (2002) 2637–2652
24. F. Lottspeich, Angew. Chem. Int. Ed., 38 (1999) 2476–2492
25. B.E. Chong, D.M. Lubman, F.R. Miller and A.J.M. Rosenspire, Rapid. Commun. Mass. Spectrom., 13 (1999) 1808–1812
26. D.B. Wall, M.T. Kachman, S. Gong, Hinderer, S. Parus, D.E. Miek, S.M. Hanash and D.M. Lubman, Anal. Chem., 72 (2000) 1099–2004
27. K.K. Wagner, K. Racaityte, K. Unger, T. Miliotis, L.-E. Edholm, R. Bischoff and G. Marko-Varga, J. Chromatogr. A, 893 (2000) 293–305
28. K.K. Wagner, K. Racaityte, R. Bischoff, G. Marko-Varga and K.K. Unger, Anal. Chem., (2003), in press
29. U.K. Laemmli, Nature, 227 (1970) 680–685
30. P.H. O'Farell, J. Biol. Chem., 250 (1975) 4007–4021
31. P. Önnerfjord, J. Nilsson, T. Laurell and G. Marko-Varga, Anal. Chem., 70 (1998) 4755–4762
32. S. Ekström, P. Önnerfjord, J. Nilsson, T. Laurell and G. Marko-Varga, Anal. Chem., 72 (2000) 286–293
33. S. Ekström, D. Ericsson, P. Önnerfjord, M. Bengtsson, J. Nilsson, G. Marko-Varga and T. Laurell, Anal. Chem., 73 (2001) 214–219
34. T. Miliotis, P. Önnerfjord, S. Kjellström, L.-E. Edholm, J. Nilsson, T. Laurell and G. Marko-Varga, J. Chromatogr. A, 886 (2000) 99–110
35. T. Laurell, J. Nilsson and G. Marko-Varga, Chromatogr. B, 752 (2001) 217–232
36. T. Miliotis, P.-O. Ericsson, G. Marko-Varga, R. Svensson, J. Nilsson, T. Laurell and R. Bischoff, J. Chromatogr. B, 752 (2001) 323–334
37. T. Miliotis, S. Kjellström, L.-E. Edholm, J. Nilsson, T. Laurell and G. Marko-Varga, J. Mass Spectrom., 35 (2000) 369–377
38. P. Önnerfjord, S. Ekström, J. Bergqvist, J. Nilsson, T. Laurell and G. Marko-Varga, Rapid Commun. Mass Spectrom., 13 (1999) 315–322
39. T. Miliotis, S. Kjellström, J. Nilsson, T. Laurell, L.E. Edholm and G. Marko-Varga, Rapid Commun. Mass Spectrom., 16 (2002) 117–126
40. T. Miliotis, J. Nilsson, G. Marko-Varga and T. Laurell, J. Neurosci. Meth., 109 (2001) 41–46
41. G. Marko-Varga, S. Ekström, G. Helldin, J. Nilsson and T. Laurell, Electrophoresis, 22 (2001) 3978–3983

42. S. Ekström, J. Nilsson, G. Helldin, T. Laurell and G. Marko-Varga, Electrophoresis, 22 (2001) 3984–3992
43. G. Marko-Varga, Sample preparation strategies for low volume determinations in in vitro studies of proteins using MADLI-TOF MS. Chromatographia, 49 (1999) 95–99
44. S. Ekström, G. Marko-Varga, J. Nilsson and T. Laurell, JALA, 5 (2000) 66–68
45. T. Miliotis, G. Marko-Varga, J. Nilsson and T. Laurell, JALA, 5 (2000) 69–71
46. T. Laurell, J. Nilsson and G. Marko-Varga, Trends Anal. Chem., 20(5) (2001) 225–231
47. M. Bengtsson, S. Ekström, J. Drott, A. Collins, E. Csöregi, G. Marko-Varga and T. Laurell, Phys. Status Solidi (a), 182 (2000) 495–504
48. T. Laurell, L. Wallman and J. Nilsson, J. Micromech. Microeng, 9 (1999) 369–376
49. M. Bengtsson, S. Ekström, G. Marko-Varga and T. Laurell, TALANTA, 56 (2002) 341–353
50. T. Laurell, G. Marko-Varga, S. Ekström, M. Bengtsson and J. Nilsson, Rev. Mol. Biotechnol., 82 (2001) 161–175
51. D. Ericsson, S. Ekström, J. Bergqvist, J. Nilsson, G. Marko-Varga and T. Laurell, Proteomics, 1 (2001) 1072–1081
52. S. Ekström, L. Wallman, J. Bergkvist, M. Löfgren, J. Nillson, G. Marko-Varga and T. Laurell, Proteomics, 2 (2002) 413–421
53. J. Bergqvist, S. Ekström, L. Wallman, M. Löfgren, G. Marko-Varga, J. Nilsson and T. Laurell, Proteomics, 2 (2002) 422–429
54. S. Martin, M. Vestal and K.R. Jonscher, Genomics Proteomics, 2 (2001) 56–59

G.A. Marko-Varga and P.L. Oroszlan (Editors)
Emerging Technologies in Protein and Genomic Material Analysis
Journal of Chromatography Library, Vol. 68

CHAPTER 10

Identification and Characterization of Peptides and Proteins Using Fourier Transform Ion Cyclotron Resonance Mass Spectrometry

M. PALMBLAD[a] and J. BERGQUIST[b,*]

[a]*Division of Ion Physics, Uppsala University, Box 534, SE-751 21 Uppsala, Sweden*

[b]*Department of Analytical Chemistry, Institute of Chemistry, Uppsala University, P.O. Box 599, SE-751 24 Uppsala, Sweden*

Abstract

Human body fluids have been rediscovered in the post-genomic era as great sources of biological markers and perhaps particularly as sources of potential protein biomarkers of disease. Analytical tools that allow rapid screening, low sample consumption, and accurate protein identification are of great importance in studies of complex biological samples and clinical diagnosis. Mass spectrometry is today one of the most important analytical tools with applications in a wide variety of fields. One of the fastest growing applications is proteomics, or the study of protein expression in an organism. Mass spectrometry has been used to find post-translational modifications and to identify key functions of proteins in the human body. In this chapter, we review the use of human body fluids as sources for clinical markers and present new data, showing the ability of Fourier transform ion cyclotron resonance (FTICR) mass spectrometry (MS) to identify and characterize proteins in four human body fluids: plasma, cerebrospinal fluid, saliva, and urine. The body fluids were tryptically digested without any prior separation, purification, or selection, and the digest was introduced into a 9.4 T FTICR mass spectrometer by direct infusion electrospray ionization (ESI). These samples represent complex biological mixtures, but the described method provides information that is comparable with traditional 2D-PAGE data. The sample consumption is extremely low, a few milliliters, and the analysis time is only a few minutes. It is, however, evident that separation of proteins and/or peptides needs to be included in the methodology, in order to detect low-abundance proteins and other proteins of biological relevance.

---

* Corresponding author. *E-mail address:* jonas.bergquist@kemi.uu.se.

10.1. ON DNA AND PROTEINS

Three key macromolecular players in all life on Earth are deoxyribonucleic acid (DNA), ribonucleic acid (RNA), and proteins. The DNA contains genes and the genes are the instructions, or blueprints, that are used to make proteins. Each protein performs its own specific tasks in the cell or body. RNA is the carrier of information between genes in the DNA and proteins. We say that genetic information is encoded in the DNA, transcribed into RNA (or mRNA, messenger RNA) and finally translated into the amino acid sequence of proteins (Fig. 10.1). This unidirectional flow of information is often referred to as 'the Central Dogma of molecular biology', stated by Crick in 1958 [1,2]. There are four 'letters' or bases in DNA and RNA, adenine (A), guanine (G), cytosine (C) and thymine (T). In RNA, thymine is replaced by uracil (U). These bases form triplets, or codons, and each codon corresponds to one out of 20 amino acids.

The Central Dogma states that the information (or more specifically the sequential information of nucleic or amino acid residues in nucleic acids and proteins, respectively) is transferred from nucleic acids to protein, from DNA to RNA to protein, and not in the reverse direction, from protein to nucleic acid. There are a few but notable exceptions to the unidirectional flow of information from DNA to RNA. In retroviruses, such as the human immunodeficiency virus (HIV), this information is transcribed from the viral RNA to the host organism DNA.

The translation of nucleic acid sequence to the amino acid sequence of proteins is defined by the genetic code (Fig. 10.2). If the DNA (or RNA) sequence is known, the genetic code will give the protein sequence, or at least a prediction of this. The point of mentioning these fundamental facts is that it has been, and still is, much less cumbersome to determine the oligonucleotide sequence of DNA than the amino acid sequence of proteins. The DNA is there in all the chromosomes in the cell nucleus (and a few other places, such as the mitochondria or chloroplast) and is easily extracted. Proteins are everywhere, inside and outside the cell, and they have very different physiochemical properties. Their concentrations range from less than one to millions of molecules per cell. This is

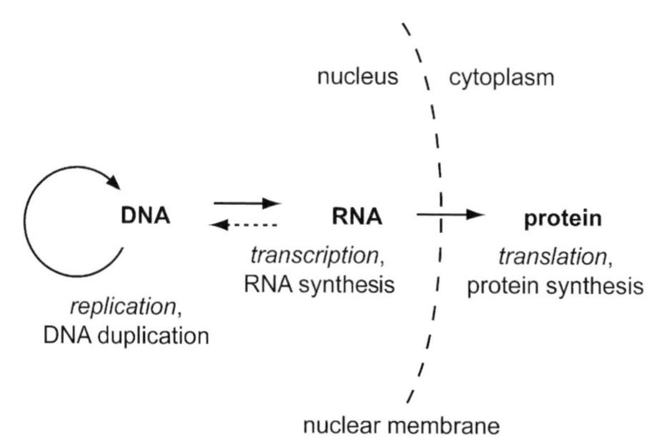

Fig. 10.1. The Central Dogma of molecular biology with arrows depicting the flow of residue sequential information. The dashed arrow represents the rare occasions where DNA is synthesized from an RNA template.

TTT Phe	TCT Ser	TAT Tyr	TGT Cys
TTC Phe	TCC Ser	TAC Tyr	TGC Cys
TTA Leu	TCA Ser	TAA **Ter**	TGA **Ter**
TTG Leu	TCG Ser	TAG **Ter**	TGG Trp
CTT Leu	CCT Pro	CAT His	CGT Arg
CTC Leu	CCC Pro	CAC His	CGC Arg
CTA Leu	CCA Pro	CAA Gln	CGA Arg
CTG Leu	CCG Pro	CAG Gln	CGG Arg
ATT Ile	ACT Thr	AAT Asn	AGT Ser
ATC Ile	ACC Thr	AAC Asn	AGC Ser
ATA Ile	ACA Thr	AAA Lys	AGA Arg
ATG **Met**	ACG Thr	AAG Lys	AGG Arg
GTT Val	GCT Ala	GAT Asp	GGT Gly
GTC Val	GCC Ala	GAC Asp	GGC Gly
GTA Val	GCA Ala	GAA Glu	GGA Gly
GTG Val	GCG Ala	GAG Glu	GGG Gly

Fig. 10.2. The genetic code defines the mapping of the DNA sequence into protein sequence, where a code word of three nucleic bases, or a codon, is translated into one amino acid residue in a protein. One codon (ATG) defines the starting point of translation (always a methionine, Met), and one of three alternative codons terminates (Ter) translation (TAA, TAG, and TGA).

a fundamental challenge in proteomics. In fact, most proteins that we know anything about, at least with respect to sequence or function, are predictions from DNA and RNA sequences, essentially using the genetic code, some knowledge on what genes look like and comparison with similar proteins that have been characterized biochemically.

The picture is complicated by the fact that in eukaryotes, such as humans, the transcribed RNA sequence is often edited, or spliced, generating a protein sequence that is not a straightforward translation of the DNA sequence. Proteins are also modified post-translationally in a way that is not trivial to predict from the DNA sequence. Examples of such modifications are phosphorylations ($-O-PO_3$) of serine, threonine and tyrosine, often involved in regulation of biological activity on the protein level, and glycosylation, carbohydrates bound to serine and threonine (*O*-glycosidic linkage) or asparagine (*N*-glycosidic).

Regardless of the modifications, and this is an important point, there are always stretches of the protein whose sequences are correctly predicted and these stretches are sufficiently long to uniquely identify the protein among all other proteins in the organism. By identification we mean that the protein is associated with a certain gene, or part of a chromosome. The roles of most genes/proteins are unknown today.

This identification can be made by looking at a small part of the protein sequence, or alternatively, the patterns, or fingerprints, resulting from enzymatic digestion of the protein. Both the sequence 'tags' (typically 6 amino acid residues or more) and fingerprints of enzymatic digestion can be routinely generated using mass spectrometry and are specific enough to uniquely identify any protein. This chapter will describe how mass spectrometry can be used to generate both of these types of information for the purpose of identifying and characterizing proteins in complex biological samples.

The first complete DNA sequence of an 'organism', the bacteriophage ϕX174, was completed in 1977, soon followed by the human mitochondrial genome and phage λ. The first bacterial genome, *Haemophilus influenzae*, was completed in 1995 [3] and the first eukaryotic genome, *Saccharomyces cerevisiae*, or baker's yeast, was completed in 1996 [4]. Rapid advances in automated, multiplexed, DNA sequencing machines in the 1990s made it possible to complete a draft sequence of the human genome by the year 2000 [5,6].

Sanger successfully sequenced the first protein by determining the amino acid sequence of insulin, a work completed in 1953 [7]. Today, many methods are available for sequencing proteins, some of which are based on mass spectrometry. A general problem in protein sequencing is sensitivity—the amount of protein necessary to obtain a complete protein sequence. Mass spectrometric methods can provide excellent sensitivity, but have difficulties generating the complete sequence of even a small protein. Often, several techniques have to be combined in a procedure that is far from routine.

In the post-genomic era, focus has shifted to find methods to approach the other molecular 'datasets', such as the proteome (the set of all proteins in an organism) and the metabolome (all other small, organic molecules in an organism). The systematic studies of all these molecular datasets bear the 'ics' suffix, such as 'proteomics', the study of the proteome. Table 10.1 summarizes the four principal molecular datasets, how they are approached today and to what degree of success. Other important parameters are the time and space dependencies of these sets. The genome is (with just a few exceptions) constant throughout the organism and its lifetime. The other 'omes' vary from cell to cell and from one moment to the next, posing formidable analytical challenges.

Gene expression is routinely analyzed on the RNA level by microarray techniques [8,9], but protein abundance cannot simply be predicted from the corresponding mRNA abundance [10]. Much effort is now spent on developing analogous techniques for proteome analysis (for a review, see Stoll et al. [11]). Still, two-dimensional (2D)

TABLE 10.1

THE MOLECULAR DATASETS IN BIOLOGY, WHAT THEY MEAN AND HOW THEY ARE STUDIED (P. KRAULIS, PERSONAL COMMUNICATION)

Dataset	Definition	Measurement	Time-dependence
Genome	The complete DNA sequence of the genetic material in the organism	Routine, completeness (CGE, DNA arrays)	Generation
Transcriptome	The complete set and levels of mRNA transcripts in a cell or tissue of the organism	Routine, near-completeness (arrays)	Seconds–minutes
Proteome	The complete set and levels of proteins in a cell or tissue, including modifications and complexes	Not quite routine (2D-PAGE, MS, protein arrays, fusions with tag- or reporter-proteins,...)	Seconds–years
Metabolome	All metabolites and their concentrations. Metabolites are here defined as small organic molecules, not protein, RNA or DNA	Hard (MS, NMR, specific chemical reagents,...)	<Second–years

isoelectric focusing and size separation in sodium dodecyl sulfate (SDS) in a polyacrylamide gel electrophoresis (2D-PAGE) [12,13] is the workhorse for protein expression analysis. The proteins on a 2D-PAGE gel can be identified through in-gel tryptic digestion and mass spectrometry, also called peptide mass fingerprinting (PMF) [14–19], possibly using partial sequence tags from tandem mass spectrometry [20–23]. Proteins in simple mixtures, such as proteins co-migrating in 2D-PAGE can also be identified by PMF if the mass accuracy is sufficiently high [24]. 2D-PAGE has inherent limitations in speed and dynamic range [36] and is biased against what proteins can be detected [25]. Hydrophobic analytes, such as membrane proteins, pose severe challenges in 2D electrophoresis [26–28]. These can be partially circumvented using strategies based on liquid chromatography or capillary electrophoresis [36,25].

What kind of information are we looking for and what tools do we need? Primarily, we need a general method to detect and identify the proteins. We may wish to quantify the expression of the proteins to correlate levels of a specific protein or proteins to a particular physiological phenomenon that we are studying. We also wish to characterize the proteins themselves in more detail, for example, determining their amino acid sequences and localizing, characterizing and quantifying post-translational modifications important for structure and function. We may also be interested in protein–protein interaction, and 'charting the protein complexome', i.e. the protein–protein interaction networks [29–31]. Mass spectrometry has played a central role in these efforts so far.

Kelleher, McLafferty and co-workers have applied a 'top-down' approach in the analysis of proteins and mixture of proteins [32,33]. In this approach, intact proteins are analyzed by ESI mass spectrometry, identified and characterized by MS/MS. The alternative, or 'bottom-up' or 'shotgun' approach, digests all proteins before analysis by MS, LC/MS or CE/MS. This approach has been applied by Smith et al. in the analysis of microbial and mammalian proteomes [34–37].

10.2. ON BODY FLUIDS AS SOURCES FOR CLINICAL MARKERS

In recent years, several investigators have emphasized the usefulness of the detection and identification of specific proteins in various body fluids as markers of metabolism, as well as their structural and/or regulatory roles in different disorders (for references see below under each separate body fluid). It is of general interest to find a technique that allows for simultaneous observation of many different proteins in body fluids in the search for new diagnostic markers of diseases and physiological correlates of behavior. The overall accepted approach to study the total protein expression of a cell, tissue, organ, or organism under particular physiological circumstances relies today only upon a coordinated use of 2D-PAGE [12,13], image analysis, mass spectrometric protein identification, and database mining/bioinformatics [14–19]. There is a need for proteomic approaches that can provide more information than traditional 2D-PAGE analysis. The ability of mass spectrometry to characterize complex biological samples without any complicated pre-purification is an attractive alternative to established time- and sample-consuming techniques. Current mass spectrometric techniques have developed into important tools in screening for clinical biomarkers, and should be investigated as a possible technique for total protein analysis.

10.2.1. Plasma

An individual weighing 75 kg has about 6 l of blood (i.e. about 8% of the body weight). About 60% of the blood is a straw-colored liquid called plasma, which is 90% water. The other 40% of the blood that is not plasma is made up of the red blood cells (erythrocytes), white blood cells (leukocytes; granulocytes, lymphocytes and monocytes), and platelets. With the exception of the oxygen and carbon dioxide carried by hemoglobin, most of the molecules needed by the individual cells, as well as waste products from the cells, are carried along dissolved in the plasma. In addition, the plasma contains plasma proteins; these differ from the other molecules carried in the plasma in that they are not nutrient for or waste products of tissue cells, but function in the bloodstream itself. Plasma proteins are of three major types: albumin, whose main function is to maintain the plasma hyperosmotic to the interstitial fluid; globulins, a key group of proteins that act to defend the body against foreign invaders; and fibrinogen, which is responsible for blood clotting. Plasma from which fibrinogen (together with some other components) has been removed as a result of clotting is called serum.

A number of proteins are today screened by electrophoretic plasma protein analysis (agarose gel electrophoresis), and are used as clinical markers, e.g. cerebrospinal fluid leakage (transthyretin), dehydration, inflammation, malignant tumors (albumin), inflammation, liver status, hemoglobin catabolism, spleen status, hemolytic processes (haptoglobin), and anemia (transferrin).

It is well known that protein analysis of plasma and serum may provide valuable information on a large number of functions in the body. The analysis of plasma for diagnostic purposes has been described in numerous publications and books [38,39]. Most of these proteins play their main role in the extracellular space, and often have a molecular weight above 45 kDa (e.g. human serum albumin 67 kDa). In the plasma, a large number of intracellular proteins are present, due to the leakage from cells into the interstitial fluid.

The leakage of these proteins, many of which are enzymes, is a result of normal cell catabolism. Plasma levels of specific proteins, however, can be used as diagnostic markers because, at various pathological conditions, the leakage together with cellular death dramatically increases the extracellular levels. The plasma proteins have many important functions in the circulation, e.g. stabilizing the salt–water balance and transportation of low-molecular weight compounds (for instance, hydrophobic substances, e.g. fatty acids or substances that are toxic in free form, e.g. bilirubin). There are also a number of smaller proteins bound to plasma proteins to avoid excretion by glomerular filtration in the kidney (e.g. retinol-binding protein bound to transthyretin [44,45], and globulin dimers bound to haptoglobin [46]). Furthermore, many of these proteins act as protectors of various systems such as extracellular protease inhibitors, immunoglobulins, complement factors, and coagulation factors. Most of the high-abundant plasma proteins are synthesized by the hepatocytes in the liver (except for the immunoglobulins, which are synthesized by the B-lymphocytes).

Other cells synthesize less-abundant proteins (e.g. α_1-antitrypsin by the macrophages, factor VIII and von Willebrand's factor by the endothelium [39]). The protein pattern is thus expected to vary greatly in time and space within the plasma samples.

Although the levels of different plasma proteins have a natural variation, the human serum albumin (HSA) level is strictly regulated by the colloidosmotic pressure (salt–water balance) feedback mechanisms. This regulation results in a very low inter- as well as intra-individual relative standard deviation below 7% (note that the albumin level is reduced with $< 10\%$ when lying down and during pregnancy [39]). Other protein levels are regulated by, e.g. steroid hormones (such as ceruloplasmin and α_1-antitrypsin [47]) or by acute-phase reactions (e.g. C-reactive protein, α_1-antichymotrypsin, α_1-antitrypsin, orosomucoid, haptoglobin, complement factors C3 and C4, C4b-binding protein, fibrinogen, and factor VIII [39]), resulting in large variations during acute (e.g. pneumonia) as well as chronic (e.g. rheumatoid arthritis) processes.

10.2.2. Cerebrospinal fluid

The cerebrospinal fluid (CSF) is in continuum with the extracellular fluid of the central nervous system, and therefore reflects the biochemical environment of the brain. CSF is mainly formed in the *choroid plexus* at a rate of 0.3–0.4 ml/min (see Ref. [174]). The total volume of the fluid is 100–150 ml in adults, which means that the CSF is totally replaced three or four times every day.

Diagnostic markers for patients with spinal muscular atrophy were found as early as 1976 by Kjellin and Stibler [48], and in 1977, Delmotte and Gonsette [49] found CSF protein abnormalities in patients with multiple sclerosis. The number of disorders that have been screened for CSF proteins is very large. Especially different forms of cerebral inflammatory processes with intrathecally (in the brain) produced immunoglobulins [50], different forms of amyloidosis [51,52], psychiatric disorders like schizophrenia [53,54], dementia such as Alzheimer's disease (AD), and vascular dementia [55,56] have rendered a lot of interest. However, the determination of brain-specific proteins present in low concentration in CSF is complicated because more than 80% of CSF proteins originate from the plasma [57,58]. A number of strategies have been explored to circumvent this problem, using different forms of chromatography, electrophoresis, and combinations thereof [53,59,60].

10.2.3. Saliva

Saliva is a watery secretion produced by three pairs of large salivary glands plus numerous minute glands, the bucal glands, which are located in the mucosal lining of the mouth. The secretion of saliva is controlled by the autonomic nervous system. On average we produce 1–1.5 l of saliva every 24 h. The saliva, which contains mucus, lubricates the food so that it can be swallowed easily. Saliva is slightly alkaline, owing to the presence of sodium bicarbonate. In humans, saliva also contains a digestive enzyme (amylase), which

References pp. 235–240

begins the breakdown of starches. Like all digestive enzymes, amylase works by hydrolysis. Protein analysis of saliva in clinical diagnosis is attractive because saliva offers a simple, rapid, and non-invasive method for the short- and long-term monitoring of pathological disorders and drug therapy. However, the fluid collection must be clearly defined due to variations in saliva composition and flow rate. A number of specific proteins in saliva are used as clinical markers. Among these proteins are phenotypic variants of α-amylase as indicators of inheritance of autosomal co-dominant alleles, disorders such as cystic fibrosis [61], and diabetic ketoacidosis [62]. Beeley reviewed clinical applications of saliva proteins in 1991 [63].

10.2.4. Urine

Urine is the liquid waste filtered from the blood by the kidney and stored in the bladder pending elimination through the urethra. The analysis of human urine is one of the oldest diagnostic procedures known to medicine. The collection of urine samples is in most cases achieved by the natural, non-invasive method, and the amounts are often sufficient for any kind of analysis. By studying proteins in urine, it is possible to follow conditions that cause glomerular damage. The porous walls of the glomerular capillaries in the kidney normally allow only proteins with a molecular weight (M_w) < 65 kDa into the tubular system, whereas larger proteins stay in the blood stream. The tubular cells subsequently reabsorb the smaller proteins, and only low concentrations of proteins are normally present in the urine. Most of the renal diseases, as well as some extra-renal, may, however, be seen as proteinuria—defined by a loss of proteins into the urine. Marshall and Williams recently reviewed the clinical significance of analyzing urine, and listed a large number of clinical applications, including various renal diseases, diabetes, and Bence-Jones proteinuria [45].

10.3. ON FTICR MASS SPECTROMETRY

Mass spectrometry (MS) is the analysis or separation of charged particles, such as atoms and molecules, based on their mass-to-charge ratios. Most mass spectrometers are comprised of three principal parts. Ions are produced in an *ion source* and subsequently separated or analyzed in a *mass analyzer* before detection by a *detector* (Fig. 10.3). All of these, or at least the analyzer and detector, are inside a vacuum system. The cyclotron principle, the core of FTICR mass spectrometry, was introduced by E.O. Lawrence in

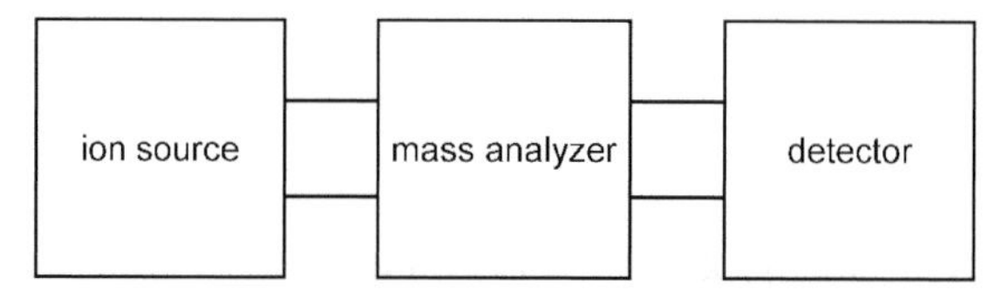

Fig. 10.3. A generic mass spectrometer can be divided into three principal parts—an ion source, a mass analyzer and a detector. A modern instrument also includes a computer to retrieve and analyze data.

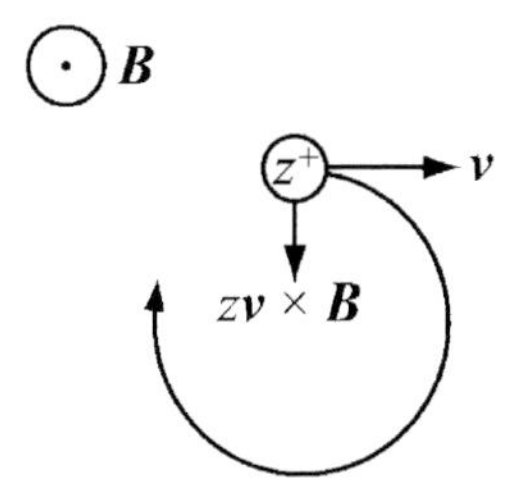

Fig. 10.4. The force and resulting motion of a positive ion of charge z, in a magnetic field **B** (direction out of the paper).

1931 [64] and in 1936. F.M. Penning added an electrostatic confinement perpendicular to the magnetic field [65], a construction that later became known as the 'Penning trap'.

The use of ion cyclotron resonance for mass measurements was pioneered by Hipple, Sommer and Thomas in the late 1940s [66,67] as a precise method to determine the Faraday constant. Over the following 15 years, ICR-MS instrumentation was greatly improved, rapidly expanding its applications. High resolution and sensitivity became hallmarks of ICR and the technique was used with particular success in studies of gas-phase ion chemistry. Varian Associates introduced the first commercial ICR-MS in 1965 [68]. The next major step was taken in 1970 by R.T. McIver with the introduction of the 'trapped ion analyzer cell', combining trapping and measurement in one cell (trap). In 1974, Comisarow and Marshall significantly improved ICR mass spectrometry by building the first FTICR mass spectrometer [69], analogously to previous FT spectroscopies, particularly FTNMR [70,71]. Marshall, Comisarow and others have since continuously improved FTICR mass spectrometry, increasing the mass range, mass resolving power and sensitivity of the technique [72,73].

10.3.1. Ion cyclotron resonance

An ion moving in the presence of a spatially uniform magnetic field, **B**, is subject to a force **F**,

$$\mathbf{F} = m\frac{\mathrm{d}v}{\mathrm{d}t} = zv \times \mathbf{B} \tag{10.1}$$

where m, z, and v are the mass, charge and velocity of the ion, respectively. (Fig. 10.4)

If $v_{\perp}$ is the velocity component perpendicular to **B** and B_0 the magnetic flux density, then the angular acceleration is this velocity squared divided by the radius of the planar motion, r, is:

$$\frac{mv_{\perp}^2}{r} = zv_{\perp}B_0 \tag{10.2}$$

The angular velocity ω around **B** is $v_{\perp}/r$ and substituting $v_{\perp}$ in Eq. (10.2):

$$m\omega^2 r = z\omega r B_0 \tag{10.3}$$

or

$$\omega = \frac{zB_0}{m} \tag{10.4}$$

Eq. (10.4) is also known as the cyclotron equation, and ω the unperturbed ion cyclotron frequency. Note that the cyclotron frequency is not dependent on the radius (or velocity) of the ion but on the magnetic field and the mass-to-charge ratio. If B_0 is stable and ω can be accurately determined, then m/z can also be determined with high accuracy. It turns out that these conditions can be met in practice.

To measure the angular frequency ω for an ion, the cyclotron motion is excited by applying a radiofrequency signal with the same frequency across two opposite electrodes. This produces a *coherent* cyclotron motion, i.e. all ions are in phase with another. The ions induce an image current also in two other electrodes, offset by 90° from the first pair. After the excitation frequency is turned off, this image current is recorded. A mass spectrum is obtained by scanning the excitation frequency and plotting the response, the ion cyclotron resonance, as a function of mass-to-charge ratio according to Eq. (10.4). In reality, the picture is more complicated due to static electric fields from the trapping electrodes (in front of and behind the plane in Fig. 10.5) that cause a lower frequency *magnetron* motion to be superimposed on the cyclotron motion. However, this can also be compensated for calibrating spectra.

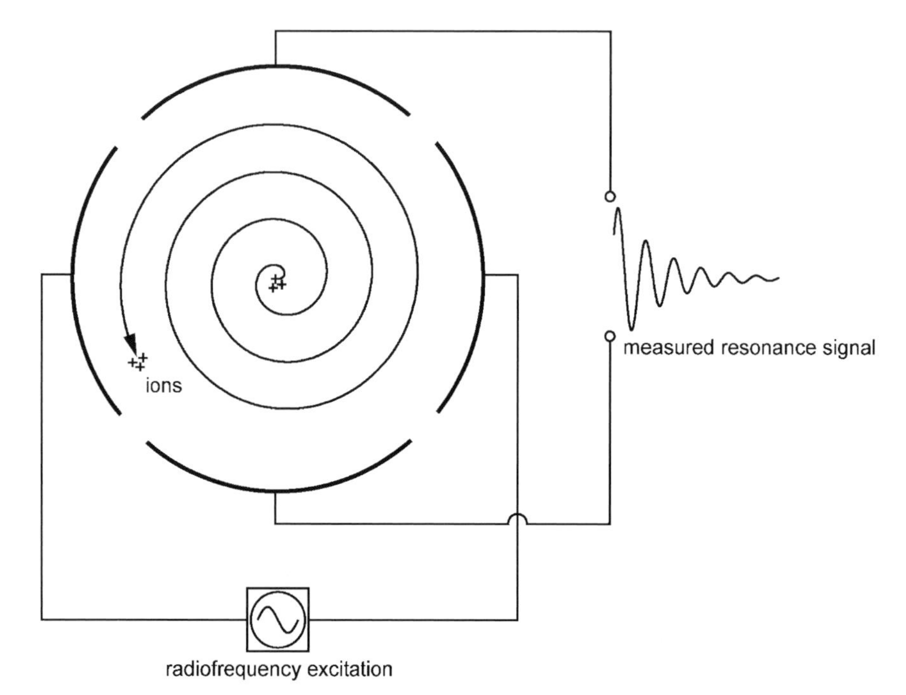

Fig. 10.5. An external radiofrequency pulse applied across two electrodes (right and left) excites the cyclotron motion of resonating ions. These ions then induce an image current in two detector plates (top and bottom) that can be detected. The response as a function of applied frequency is converted to a mass spectrum by Eq. (10.4).

10.3.2. Fourier transform spectroscopy

Before computers and rapid analog-to-digital converters were generally available, most types of spectroscopies (whether optical, infrared (IR), Raman, nuclear magnetic resonance (NMR), electron spin resonance (ESR) or ion cyclotron resonance (ICR)) were operated in *scanning* mode, this means either (or both) the excitation frequency or the magnetic field was scanned and only a single resonant frequency recorded at any given time during the experiment. The detectors were undiscriminating towards signals at different frequencies, and the frequencies had to be separated *before* the detector. This problem was first solved two centuries ago by the French mathematician Fourier [74]. The *Fourier transform* (FT) relates a time domain signal, $f(t)$, to the frequency spectrum, $f(\nu)$ as:

$$f(\nu) = \mathrm{F}[f(t)] = \int_{-\infty}^{\infty} f(t)\mathrm{e}^{-2\pi \mathrm{i}\nu t}\,\mathrm{d}t \tag{10.5}$$

In FT spectroscopies, energy (signal) is recorded as a function of time, converted to a digital form and stored in a computer. The (discrete) signal sampled from time $t = 0$ to time $t = T$ is then converted to a frequency domain spectrum in the computer, typically by the fast Fourier transform (FFT) algorithm [75]. The discrete version of the Fourier transform can be written (omitting a scale factor):

$$f(\nu_n) = \sum_{k=0}^{N-1} f_k\,\mathrm{e}^{-2\pi \mathrm{i}nk/N} \tag{10.6}$$

where

$$f_k = f(t_k), \qquad t_k = \frac{kT}{N}, \qquad k = 0,\ldots,N-1 \tag{10.7}$$

and

$$\nu_n = \frac{n}{T} \tag{10.8}$$

The transform [Eq. (10.6)] has a real (absorption) and an imaginary (dispersion) component. If all $f(\nu_n)$ were in phase we could extract the absorption spectrum from the measured signal, but due to a frequency-dependent phase shift it is often more convenient to look at the phase-independent magnitude spectrum, $M(\nu_n)$:

$$M(\nu_n) = \sqrt{\mathrm{Re}(f(\nu_n))^2 + \mathrm{Im}(f(\nu_n))^2} \tag{10.9}$$

In some cases, the phase shift can be compensated and the absorption spectrum retrieved [76]. The magnitude-mode spectrum is a factor $\sqrt{3}$ broader than the absorption-mode spectrum for a purely Lorentzian line [77].

For an excellent introduction to FT spectroscopies see, 'Fourier Transforms in NMR, Optical and Mass Spectrometry' by Marshall and Verdun [77].

The Fourier transform had already been applied to NMR and IR spectroscopy before it was successfully coupled to ion cyclotron resonance by Comisarow and Marshall in 1974 [69,70,78]. FTICR and FTNMR are homologous techniques and they have 'a broad range of conceptual, physical, experimental, and data reduction characteristics' in common [71]. The practical advantages of FTICR over scanning-mode ICR are significant. The signal-to-noise ratio and acquisition speed increase dramatically because all frequencies are detected all the time. This is also known as the multichannel or Fellgett's advantage in Fourier transform methods. The speed increases a factor of 10 000 and the sensitivity a factor of 100 [73].

For structural characterization and identification, tandem mass spectrometry (MS/MS) techniques have proved very efficient. In a typical MS/MS experiment, a precursor ion is selected and subjected to conditions sufficient to induce fragmentation of the molecule. The fragments are then analyzed in the mass spectrometer. There are several means to provide the conditions for fragmentation, the most commonly employed being collision-induced dissociation (CID) [79]. Recently, electron-capture dissociation (ECD) [80] in FTICR mass spectrometry has attracted much attention and has been applied to sequence polypeptides and characterize post-translational modifications [29,81–91].

10.3.3. Ionization techniques

Two breakthroughs that proved to have a major impact on biological mass spectrometry came in the 1980s with the development of matrix-assisted laser desorption/ionization (MALDI) and the application of electrospray ESI. MALDI was discovered by Karas and Hillenkamp at the University of Frankfurt, Germany [92]. A related concept was developed independently by Tanaka and co-workers at the Shimadzu Corporation in Kyoto, Japan [93]. These techniques have significant improvements in sensitivity over previous laser desorption methods. MALDI is also a very 'soft' ionization method.

In ESI, pioneered by Fenn and co-workers at Yale University [94] and independently by Aleksandrov et al. in Leningrad [95], charged droplets are dispersed from a capillary, or needle, in a high electric field. As solvent evaporates, the Coulombic repulsion overcomes the surface tension and droplets fission [96,97]. Eventually, gas-phase ions are produced, although some of the details in this process remain to be unveiled. ESI is a very gentle ionization method, capable of transferring even large non-covalent complexes such as ribosomes [98,99] and viral particles [100] intact into the gas-phase.

The fundamental principles of electrospray is still a matter of investigation but it has been shown that charged droplets formed by electrospray undergo a recursive asymmetric fission process, where smaller and smaller droplets are formed [96,97]. Charged droplets rupture when the charge approaches the Rayleigh limit [101], where Coulomb repulsion equals the surface tension,

$$q^2 = 8\pi^2 \varepsilon_0 \gamma d^3 \tag{10.10}$$

where q is the charge, ε_0 the electrical permittivity of the surrounding, γ the liquid surface tension and d the diameter of the droplet.

A pictorial view of an electrospray ion source is shown in Fig. 10.6. High voltage, typically a few kilovolts over a distance of a few millimeters, is applied between

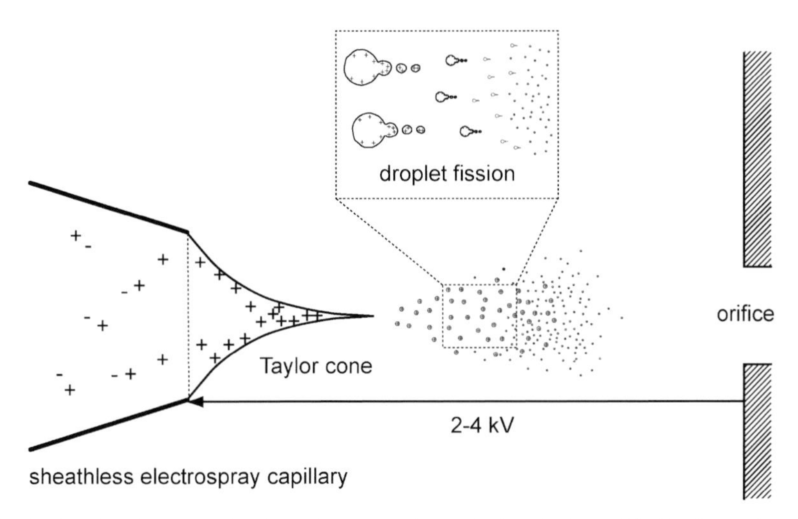

Fig. 10.6. Application of a strong electric field separates charges in the solution and liquid of opposite net charge is pulled out into a Taylor cone. The liquid is disrupted, where the Coulombic repulsion exceeds the surface tension, a process that is repeated recursively in smaller and smaller droplets as solvent evaporates.

the electrospray liquid and the orifice of the mass spectrometer. An electrospray ion source is a constant-current electrochemical cell, causing redox reactions and ionization of neutral electroactive molecules in the electrosprayed solution. This makes the study of redox chemistry intrinsically difficult. The formed ions may hinder or assist the detection of the analyte species, depending on properties of the analyte and solution.

Since the sample is at atmospheric pressure and the liquid flow is not negligible, instruments coupled to ESI sources need larger vacuum pumps than instruments with vacuum ion sources. The first successful combination of ESI and FTICR mass spectrometry was reported by Henry et al. in 1989 [102]. When combining an ion source such as ESI with a mass analyzer requiring ultra-high vacuum it is beneficial to use flow rate as low as possible to reduce the gas load. A lower flow rate also reduces contamination of the interface. The ion yield is only flow-sensitive at very low (increasing with flow) [103] or very high (decreasing) flow rates. In the common operating range, the yield depends mostly on the concentration of analyte (and presence of interfering contaminants). For a review of the central concepts related to ESI, see a recent paper by Cole [104].

Although many other ion sources, including MALDI, have been coupled to FTICR mass spectrometers, the combination of ESI and FTICR mass spectrometry is particularly powerful due to the generation of multiple charged ions inherent in the electrospray process and that trapping efficiency and resolving power increases with charge (of a given mass) in an FTICR mass spectrometer. Vestal has recently reviewed ionization techniques in mass spectrometry [105]. For further reading on mass spectrometry in biology, the recent reviews by Burlingame [106] and Matsuo and Seyama [107] are recommended.

10.4. EXPERIMENTAL APPROACHES

This chapter will describe some of the more interesting findings of clinically relevant proteins in human body fluids and give a demonstration of the applicability of liquid separations and electrospray ionization (ESI) Fourier transform ion cyclotron resonance (FTICR) mass spectrometry (MS) in the screening of human body fluids for potential protein biomarkers of disease. To illustrate the concept behind this approach to protein identification it is useful to compare with the traditional route including 2D-PAGE, automated spot picking, tryptic digestion and protein identification by PMF and/or sequence tags generated by MS/MS. The alternative approach presented here scrambles the information by digesting all proteins simultaneously; separating these by direct infusion or a suitable liquid separation method coupled to the FTICR mass spectrometry (Fig. 10.7) and identifies the proteins by highly accurate mass measurement.

These two methods should be regarded as complementary, each having benefits and disadvantages over the other. This chapter describes how such a thorough study has been made in human plasma, cerebrospinal fluid, saliva, and urine. The unsurpassed mass-resolving power and mass accuracy are prominent features of FTICR-MS that are

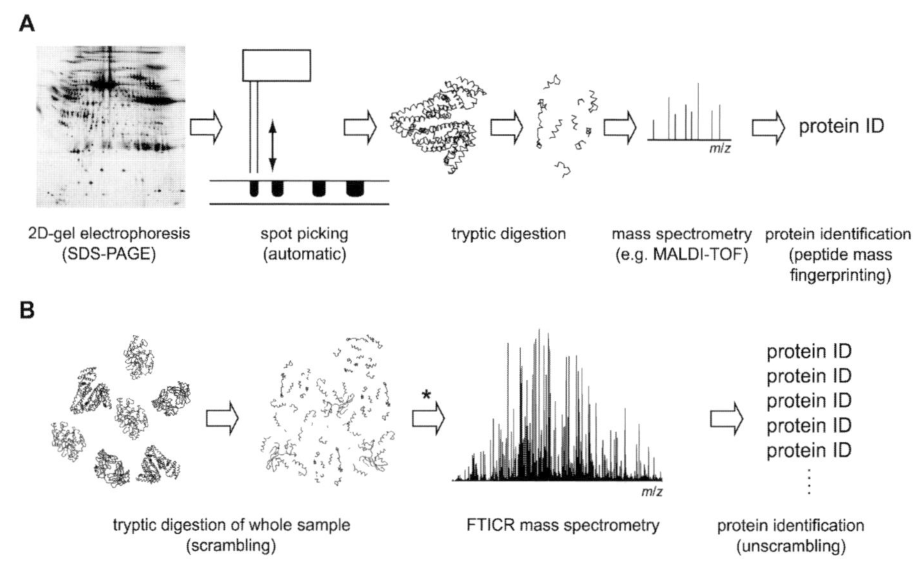

Fig. 10.7. The picture shows two different approaches to proteomics involving mass spectrometry. The established technique (A) is based on 2D-gel electrophoresis, spot picking and tryptic digestion before mass spectrometry (typically MALDI-TOF) and protein identification by PMF using one of several available algorithms, such as MASCOT [40], ProFound [41,42] or MS-Fit [43]. The second approach (B), or 'shotgun' proteomics, begins by digesting all components in the sample at once, then separates the scrambled peptides by (FTICR) mass spectrometry and retrieves a list of identified proteins by unscrambling the available information. For complex mixtures, a liquid separation method such as LC or CE can be inserted at the $*$. A similar approach has also been taken by Smith et al. [36,37] in the analysis of bacterial and yeast proteomes.

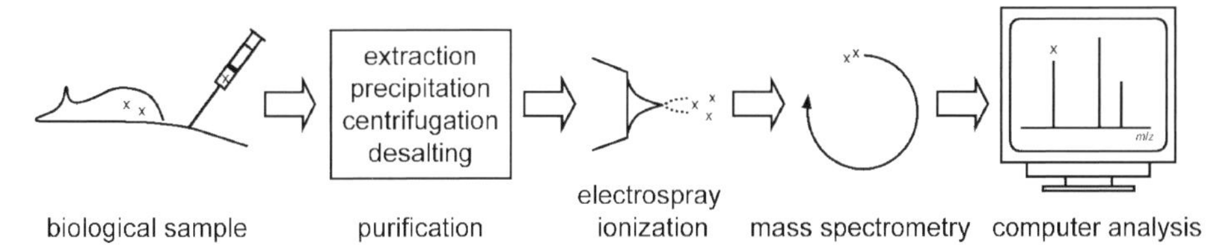

Fig. 10.8. Flow scheme view of the analysis of a biological sample by ESI mass spectrometry. The purification is more or less elaborative depending on the tissue or fluid sampled. Liquid separation methods such as LC or CE can be conveniently used on-line with the electrospray mass spectrometer.

necessary when analyzing such complex biological samples. The throughput and low sample consumption are also important features.

The mass spectrometric analysis of a biological sample, or any sample, does not start or end with the acquisition of a mass spectrum. In fact, this is often the easiest and fastest step in the analysis using modern commercial instruments. Samples in the real world are often complex and as rich in contaminants as they are in substances of interest. These contaminations could for one reason or another interfere with ionization or mass spectrometric measurements. Different ionization techniques have different spectra of sensitivities to such contaminants. ESI, for instance, is notoriously sensitive to surfactants and salts. One or more stages of sample preparation or purification are often necessary to remove unwanted contaminants or to reduce the complexity of the sample to facilitate the mass spectrometric analysis.

The successful acquisition of a mass spectrum does not mark the end of any analysis. Even a single mass spectrum contains an enormous amount of information, usually counted in megabytes. Still, this number diminishes in comparison with 2D, or time-series, of spectra whose information contents are in the gigabyte range. It is evident that efficient computer algorithms are as important a tool as any of the preceding physiochemical techniques, in order to extract useful information out of mass spectrometric measurements. This chapter will also describe automatic methods for the analysis of raw data as well as statistical methods for peptide identification.

Fig. 10.8 describes a minimal experimental flow scheme from sample to final analytical result.

10.4.1. Sample purification

It is often necessary to remove salts and other low-molecular weight compounds from solutions of peptides or proteins. This can be achieved with high efficiency (99% or better) and good recovery (80% or more) using a micro-column packed into the tip of a disposable plastic pipette tip, such as the ZipTip$^{®}_{C18}$ (Millipore Corp., Marlborough, MA, USA) with a bed volume of 0.2 μl. The tips were first wetted using 50% acetonitrile (ACN) (v/v) in H_2O and a micropipette. The column was equilibrated using 1% acetic acid (HAc) in H_2O. Each of these steps was repeated three times. The sample was loaded by filling and emptying the tip 10 times in the sample, then desalted by washing with 1% HAc in H_2O 3–5 times.

(Phoenix, AZ, USA). The MAPTAC coating generates a constant positive inner capillary surface, thus reversing the electroosmotic flow [119]. The electrospray end of the capillary was tapered [113] and coated with polyimide and particular gold [114] or graphite [115] as described in Section 10.4.2.

The CE instrumentation was assembled in-house and consisted of a high voltage power supply (Bertan, Hicksville, NY, USA) operated in constant negative voltage mode. The CE separation voltage was set at -30 kV relative to the grounded electrospray end of the capillary. The current was ~40 μA during the separation. The CE buffer used for the experiments consisted of 12.8 mM HAc, 0.2 mM ammonium acetate (NH_4Ac) in 25:75 (v/v) ACN/deionized H_2O at pH 3.0 [120].

The buffer was filtered and degassed by ultrasonication. The capillary was equilibrated with 4–5 capillary volumes of buffer prior to sample injection. The sample was injected using hydrodynamic injection at a pressure of 1.7 bar for 56 s, injecting ~40 nl onto the capillary, followed by injection of buffer (1.7 bar, 56 s). The migration times for the last eluting components in the samples were 10–15 min under these conditions.

10.4.4. Electrospray ionization

In the experimental setup used in the experiments presented here, the electrostatic voltage was applied at the liquid end via a conductive outer coating on a fused silica capillary, tapered to a sharp tip [114,115,121]. The sample was pushed through a 30 cm, 25 μm i.d. fused silica capillary by applying a pressure of ~1.3 bar at the sample end in a pressurized container (Fig. 10.11). The flow rates in this system are 30–100 nl/min, or in/near the nanospray regime [104].

A micrograph of the two types of coated capillaries used and a commercial nanospray needle is shown in Fig. 10.12. The inset shows an operating graphite-coated electrospray capillary.

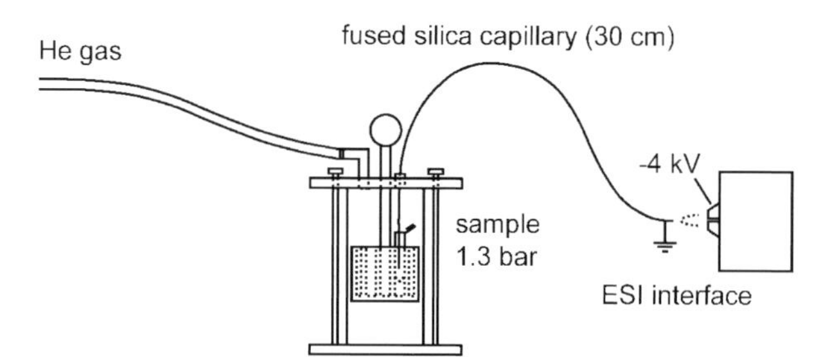

Fig. 10.11. The home-built pressurized direct infusion electrospray apparatus. The sample is kept inside a pressurized container and pressure of ~1.3 bar is sufficient to push the sample through a 30 cm, 25 μm i.d. fused silica capillary. The electrospray end is coated with a conductive layer and kept at ground potential, whereas a negative potential of 2–4 kV is applied at the mass spectrometer orifice.

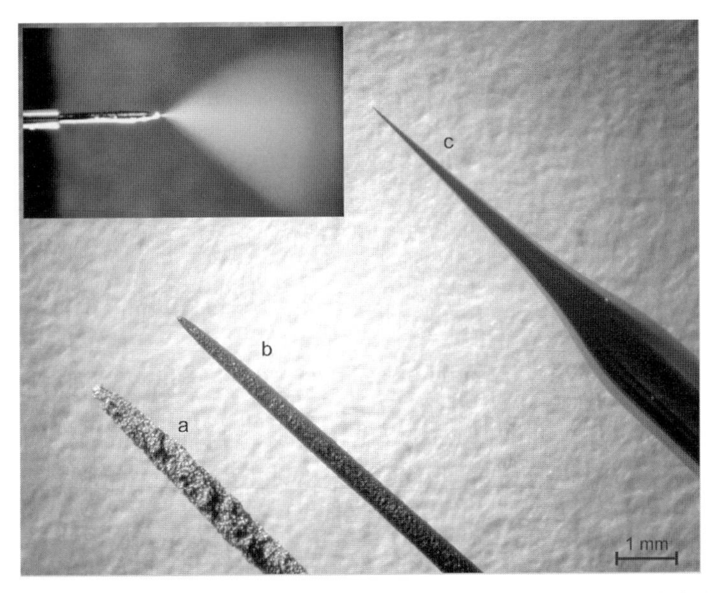

Fig. 10.12. Micrograph of electrospray needles: gold/polyimide 'fairy dust' (a) and graphite/polyimide 'black dust' (b) coated fused silica capillaries and a New Objective PicoTip™ metal-coated nanospray needle (c). The inset shows a spraying graphite-coated capillary (inset picture courtesy of M. Wetterhall).

10.4.5. Mass spectrometry

A general layout of the Bruker 9.4 T FTICR Mass Spectrometer is shown in Fig. 10.12. A 160 mm bore 9.4 T central-field, passively shielded, superconducting Nb_3Sn magnet (Magnex Scientific Ltd, Oxford, UK) is cooled in a cryosystem by liquid He to 4.2 K. The horizontal bore of the magnet is aligned with a metal frame supporting a vacuum system, ion sources and a gas inlet system on tracks. From the main vacuum chamber a titanium tube leads into the center of the magnet, where it ends with a flange holding and INFINITY™ cell. The vacuum system is a differentially pumped ultra-high vacuum system capable of sustaining a base pressure of below 4×10^{-10} mbar using turbomolecular pumps backed by mechanical rough pumps.

The instrument is controlled and data is retrieved and analyzed on a UNIX or Windows workstation. The instrumentation includes a pulse-shaping system for generation of frequency shots, correlated frequency sweeps, on-resonance and sustained off-resonance irradiation (SORI) for CID experiments [122,123], pre-amplifier and electronics for direct and heterodyne mode detection. Outside the cell is an electron gun for internal electron impact (EI) ionization and chemical ionization (CI) of volatile substances and this gun can also be used for ECD [80,124,125].

The ESI source (Analytica, Branford, CT, USA) is equipped with an inlet glass capillary (15 cm in length, an inner diameter of 0.5 mm and platinum capping at both ends). A further improvement to the electrospray source was the addition of a computer-controlled shutter [126]. This shutter essentially closes the spectrometer, and no more ions are accumulated in the hexapole ion trap. This allowed a controlled investigation of multipole-storage assisted dissociation (MSAD) of peptides [127,128].

For SORI and other experiments that require introduction of gas, two computer controlled fast valves for gas inlet are provided in this instrument. By opening an inlet valve connecting the cell and a volume of collision gas, like Ar or Xe, at a pressure of 5 mbar, for ~10 ms, the pressure in the cell is raised to a level suitable for SORI experiments ($\sim 10^{-7}$ mbar).

This instrument and its performance are described in detail in a recent paper [129]. Another 9.4 T FTICR instrument has also been described by Senko et al. [130].

10.4.6. Mass spectrometry, peptide mass fingerprinting and information

Mass spectrometry measures the ratio of mass to electric charge, m/z. Larger molecules, such as peptides, frequently become multiply charged in ESI. This means that one molecule can appear at different m/z in a mass spectrum, each corresponding to a different z. The first step after the transformation of the time-domain signal is therefore to deconvolute or reduce [131] the mass spectrum to retrieve the individual masses.

The mass of a protein is not well defined in the sense that each protein molecule has the same mass. The natural abundances of heavier isotopes, such as ^{13}C and ^{34}S, contribute to an isotopic distribution characteristic for each molecule. The mass difference between isotopic peaks is around 1.00 Da and the difference in m/z around $1.00/z$. This simple information can be used to find the charge state of a particular cluster of peaks.

There are several algorithms available to deconvolute mass spectra. The simplest methods use only information on the m/z of peaks, to find charge states, or isotopic distributions, and remove them from the list of peaks being processed. If the isotopic distributions are known or can be estimated a more thorough approach can be taken, where a model for the isotopic distribution of a species of such mass (give or take a few mass units) is fitted to the experimental data [131–134].

The isotopic distributions themselves contain potentially valuable information. One such case is isotope-labeling experiments, where differentially labeled samples and controls are analyzed and relatively quantitated in the same spectrum using isotope-coded affinity tags [135].

In addition to the mass degeneracy due to isotopes, the mass of a protein is further degenerated as a function of variability in sequence, e.g. different splice forms or post-translational processing or modifications.

Instead of using the mass of the whole protein for identification, the protein can be cleaved or fragmented into smaller pieces with well-defined masses. The standard method is enzymatic digestion using trypsin. This enzyme cleaves specifically C-terminals of the basic residues arginine and lysine, except for N-terminals of proline. The measured peptide masses from a tryptic digest are compared with calculated sets of masses for different protein sequences or libraries of mass spectra in a database, an approach called peptide mass fingerprinting (PMF) [14]. To aid protein identification, these peptides can be fragmented in vacuo to generate short sequence tags [20]. Sequences of 5–6 amino acids in length are often specific for a certain protein in the database. The sequence tag approach is

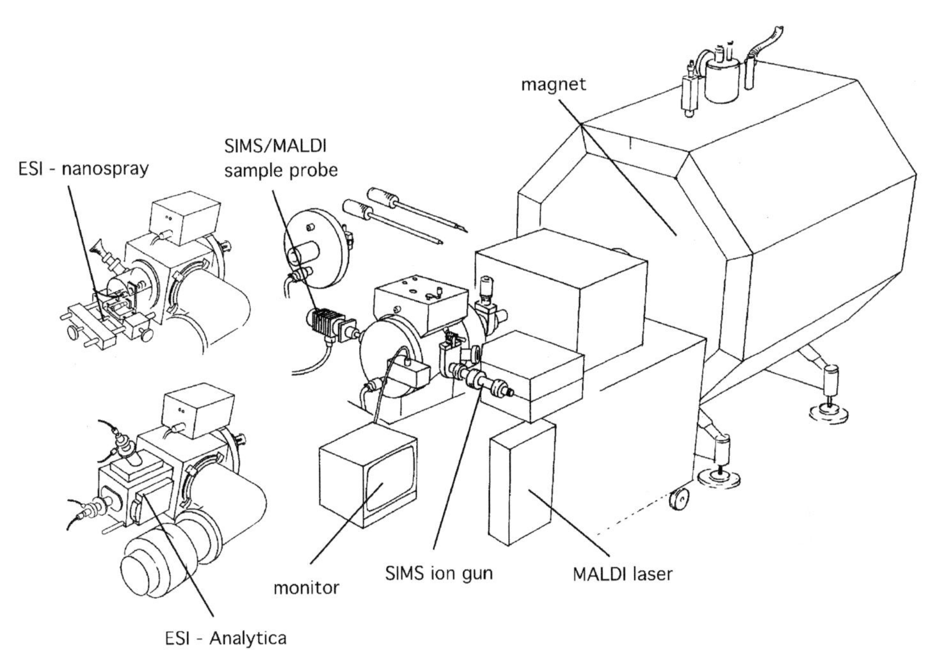

Fig. 10.13. General layout of the Bruker 9.4 T FTICR mass spectrometer with interchangeable ion sources (figure courtesy of G. Baykut, Bruker Daltonics). The experiments presented in this chapter were performed using either of the ESI sources.

more error-tolerant and can identify peptides with unpredicted post-translational modifications or errors in the database sequence.

To get an idea of the statistical significance for certain identification it is appropriate to compare with random peptide mass matches from sequences derived from an unrelated organism. The random matches from the *S. cerevisiae* proteome can be used to calculate the probability (or risk) for a random match (false positive) when identifying a human protein and thus calculate the significance [136]. The biologist may object that *Homo sapiens* and *S. cerevisiae* are related organisms, but the evolutionary relationship must be much closer and protein sequence homologies near 100% to generate significant 'cross-talk' in PMF, i.e. common tryptic peptides in distantly related human and yeast proteins. It has been shown that PMF can distinguish between myoglobin from horse and zebra, differing in only one residue, i.e. having sequence homology >99% [137].

There are other means to discriminate between true and random matches. For instance, the *pattern* of matching peptides in the sequence may also differ between true and random matches. In general, true matches have non-random distributions of peptides in the protein sequence (Fig. 10.14). The non-randomness of the distribution can be quantified, for instance by an adaptation of the runs test [138]. In the test the distributions of the lengths of contiguous peptide *spatial* runs in the protein sequence are compared with those calculated for random matches and the likelihood of

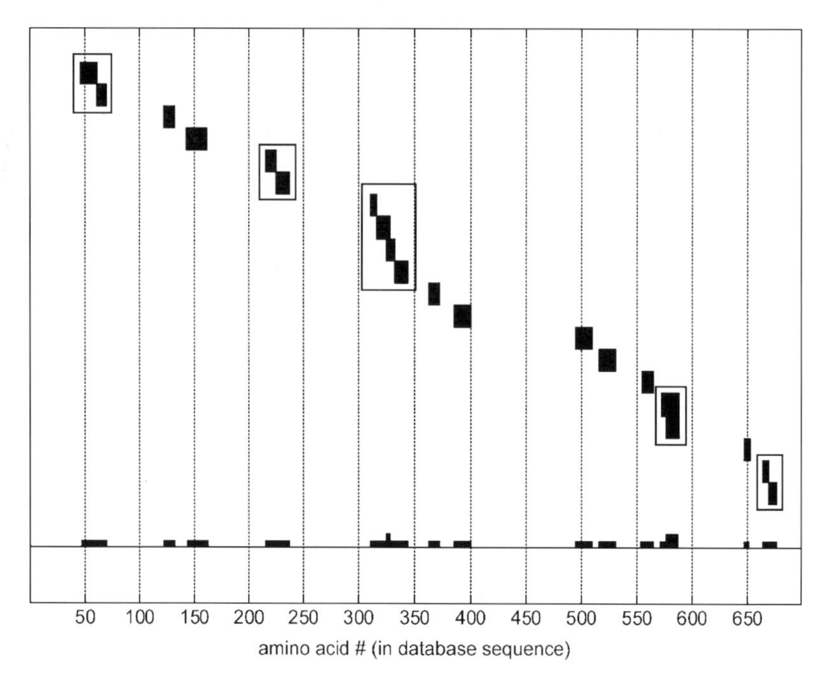

Fig. 10.14. Non-random distribution of tryptic peptides in a peptide mass fingerprint. Regions of contiguous matching peptides are indicated. The pattern, or non-randomness, can be used to further discriminate between random and true matches in PMF.

a measured peptide distribution occurring at random. The random distributions can be simulated or estimated from data of unrelated proteins.

Another way of quantifying this 1D pattern is to look at the distributions of peptides with zero, one or two close neighbors in the sequence, here simply called the *connectivity* in a peptide map.

This choice of spatial pattern descriptors of peptide maps is supported by the empirical notion that true peptide matches are often contiguous. This pattern also appears as a factor in the Bayesian model used in the ProFound algorithm [41].

Two-dimensional spectra, whether from liquid chromatography or capillary electrophoresis or any other type of separation, contain more information than the masses and intensities alone. The chromatographic retention or electrophoretic migration time contains information on the physio-chemical properties of the peptides, such as hydrophobicity in RPC and size and charge in standard capillary zone electrophoresis (CZE). A model of chromatographic retention [139–143] or electrophoretic migration [144–153] can be fitted to experimental data from known proteins. These models are then used to predict the retention time for candidate peptides in a database. The candidates can then be ranked based on combining deviation of calculated mass from measured mass and predicted retention from measured retention by calculating a total χ^2 value for all candidates within a certain mass interval.

Hydrophobicity is one of the most important physico-chemical characteristics of amino acids and probably the most poorly defined term. Cornette et al. has reviewed and compared 37 different hydrophobicity scales [154].

To illustrate the use of a predictor of retention, a simple model of retention in reversed-phase, also discussed by Hodges and co-workers [140–143], can be adapted to the chromatographic separations used here. Peptides from abundant proteins in complex mixtures can be used as internal standards to calculate a 'retention coefficient' for each amino acid according to:

$$T_{\text{retention}} = \sum_{i=1}^{20} N_i C_i + T_0 \tag{10.11}$$

where the C_i are the retention coefficients for the 20 amino acids and there are N_i amino acids i. The C_i values could be expected to correlate with hydrophobicity of the amino acid side-chains in RPC. The C_i and T_0 can be fitted to experimental data, retention times $T_{\text{retention}}$, for a large number of peptides ($\gg 20$). Eq. (10.11) also allows for a linear dependence of retention time on peptide size, as a constant term added to each C_i. In addition to a sufficiently large number of peptides in the predictor training set to avoid overfitting, each amino acid should also be represented by several peptides in this training set.

Most models of electrophoretic mobility of peptides are non-linear functions of peptide charge and molecular weight [151]. The molecular weight is easily calculated for each peptide candidate, and the charge, or distribution of charges, is estimated from measured pK_a values of amino acids in N and C-termini and side chain pK_a values. The sum of the electrophoretic mobility and electroosmotic flow at a certain electric field determines the electrophoretic migration.

The methodology presented in this chapter scrambles the information by digesting all proteins simultaneously, separating these by direct infusion or a suitable liquid separation method (LC or CE) coupled to FTICR mass spectrometry (Fig. 10.7) and identifies the proteins by highly accurate mass measurement and possibly the complementary types of information from ECD, patterns in tryptic digestion, retention time or electrophoretic mobility discussed above.

10.5. PROTEINS IN MIXTURES AND HUMAN BODY FLUIDS

Before resorting to liquid separation techniques it is of interest to estimate how many proteins can be identified by tryptic digestion and direct infusion into the mass spectrometer. This is not only a convenient figure-of-merit to compare mass spectrometers or experimental protocols, but have real importance, as gel spots are often heterogeneous [24] and SDS-PAGE is denaturing and does not preserve protein complexes. To identify proteins in a purified multi-protein complex with PMF, it is also of value to know how large complexes can be characterized by direct infusion of a total tryptic digest. First, three proteins, equine cytochrome *c* (12 kDa), equine myoglobin (17 kDa) and bovine serum albumin (66 kDa), were digested, pooled and analyzed together.

The mass spectrum (Fig. 10.15) contains about 1300 peaks, which could be reduced to 330 unique masses [137]. A general database search (containing sequences from all

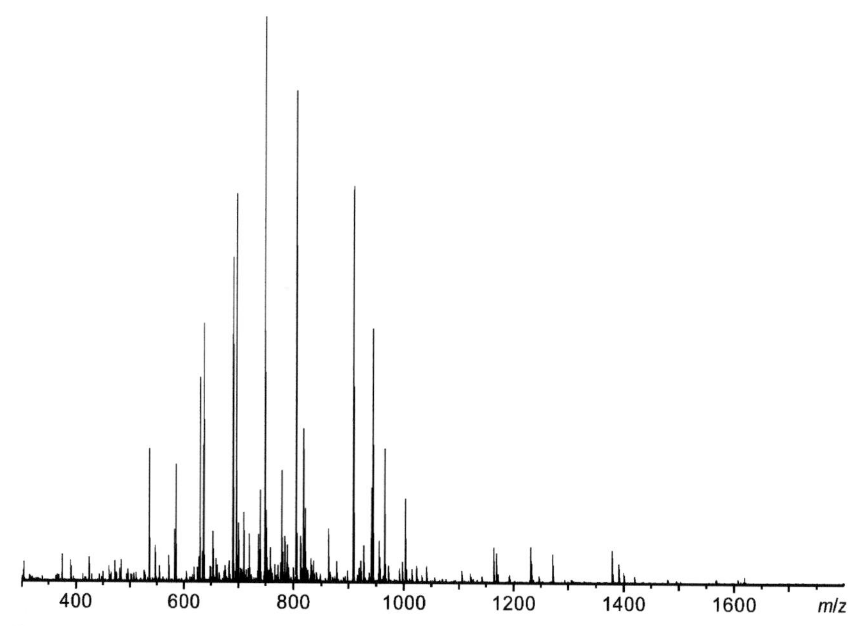

Fig. 10.15. FTICR mass spectrum of a mixture of tryptic digests of cytochrome *c*, myoglobin and albumin. A total of 1294 peaks could be reduced to 573 isotopic clusters and 330 unique masses. Searching a database containing protein sequences from all available organisms identified the correct proteins from the correct species. For this spectrum, 2048 scans were accumulated and 512 K data points acquired.

species) finds the three proteins from the correct species. Ingendoh and co-workers later performed similar experiments using MALDI and a 7 T FTICR instrument with similar success [155]. Three or four proteins are by no means a limit for what can be analyzed through direct infusion FTICR mass spectrometry, which will be shown in this chapter. In general, the organism and tissue or cell type are all known, and the databases used for searching for matches with peptide mass fingerprints are more specific and thus smaller.

Another use for direct infusion of multi-protein tryptic digests is a first rapid screen of the most abundant proteins in a complex biological sample. We have demonstrated this possibility by analyzing the total tryptic digests of proteins in four human body fluids of clinical interest: plasma, cerebrospinal fluid (CSF), urine, and saliva [156]. The protein profiles are known to be similar in plasma, CSF and urine, and the most abundant proteins are also found with the highest significances (lowest probabilities for random match) [156]. At most about 10 significant matches are found by direct infusion of the digested body fluids (in saliva, typically only salivary amylase could be found by direct infusion).

The tryptic digestion had been carried out to near completion, meaning only a small number of missed cleavage sites in the proteins. Most peptides were found between m/z 600 and 1000 (Fig. 10.15). The 800–900 $m/3$ region of this spectrum is shown in Fig. 10.16 to illustrate the spectral peak density. The unit for mass-to-charge ratio is called thomson, denoted Th, where 1 Th = 1 Da/1 elementary charge.

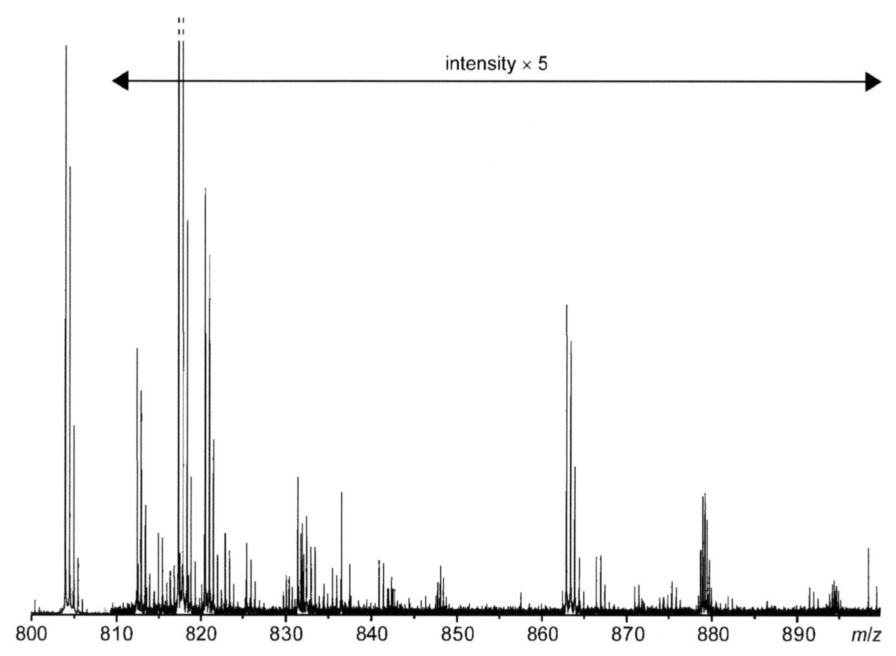

Fig. 10.16. Magnification of the region 800–900 Th of the FTICR mass spectrum in Fig. 10.15 illustrating the peak density (~ 1 peak/Th).

The resolving power of the 9.4 T FTICR instrument is sufficient to resolve peptides with mTh differences, for example, an albumin and a myoglobin tryptic peptide with a mass-to-charge difference 0.0045 Th could be resolved in this routinely acquired broadband spectrum without tuning for maximum resolution (Fig. 10.17).

The results from analysis of the three-protein mixture shows that at least a small number of proteins can be identified simultaneously with direct infusion and FTICR mass

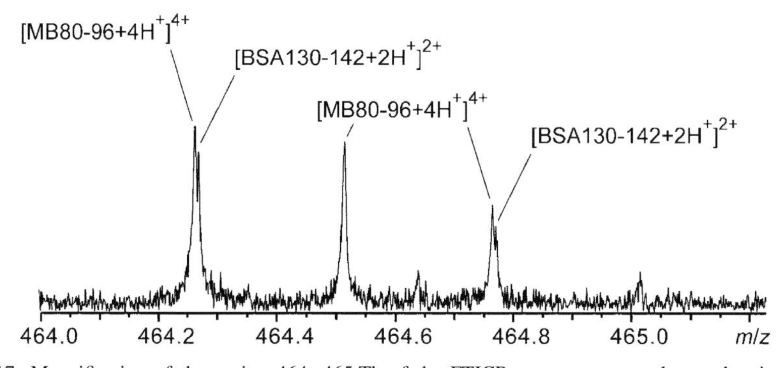

Fig. 10.17. Magnification of the region 464–465 Th of the FTICR mass spectrum above, showing partially resolved isotopic peaks from quadruply charged horse myoglobin (MB) 80–96 (CHHEAELKPLAQSHATK), 464.2464 Th (calc.) (monoisotopic), 464.2445 Th (exp.) and doubly charged bovine serum albumin (BSA) 136–142 (YLYEIAR), 464.2509 Th (calc.), 464.2504 Th (exp.).

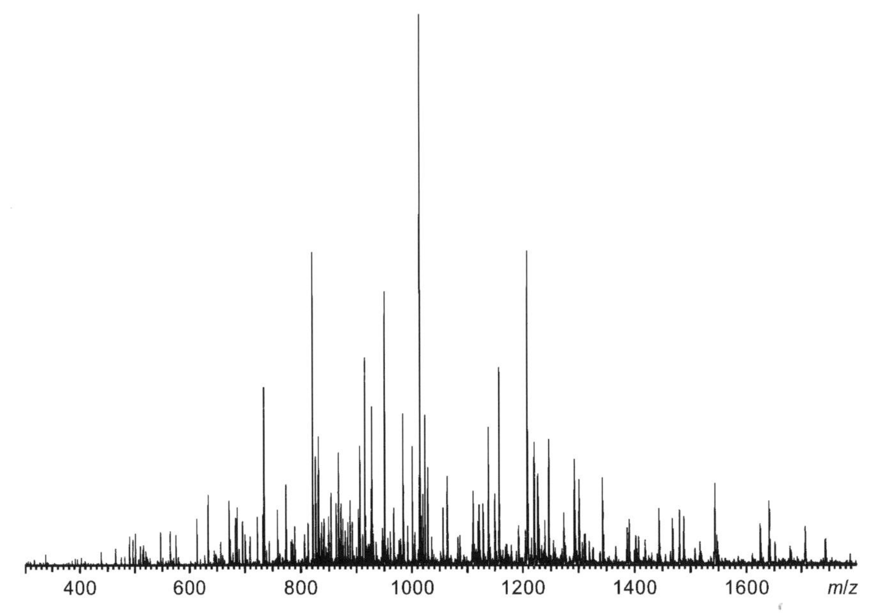

Fig. 10.18. 'World-record' FTICR mass spectrum of a human plasma tryptic digest containing about 5500 peaks that could be reduced to 1300 unique peptide masses. Direct database searches in MASCOT or ProFound returned only the most abundant protein, serum albumin, but subsequent analysis identified a number of the more abundant proteins with high statistical significances.

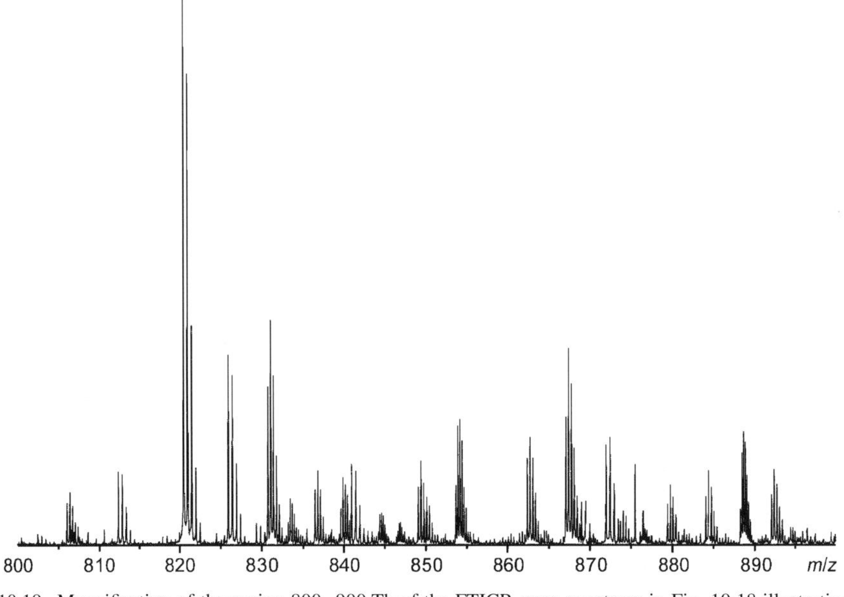

Fig. 10.19. Magnification of the region 800–900 Th of the FTICR mass spectrum in Fig. 10.18 illustrating the peak density (~5 peaks/Th).

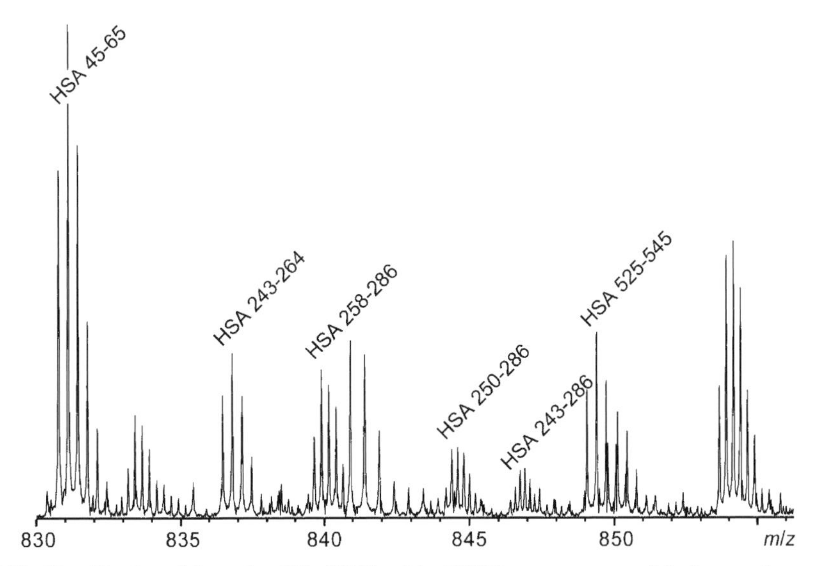

Fig. 10.20. Magnification of the region 830–856 Th of the FTICR mass spectrum of the human plasma tryptic digest showing identified HSA fragments.

spectrometry. A biological protein extract is far more complex than this three-protein mixture, illustrated by the resulting spectrum of a plasma tryptic digest in Fig. 10.18.

To illustrate the spectral peak density and compare this with the three-protein mixture, the same region 800–900 Th is shown in Fig. 10.19. In the range 600–1000 Th the peak density is ~5 peaks/Th. Fig. 10.20 shows a further magnification into the most dense region and the isotopic clusters putatively identified as HSA fragments by the analysis software. Despite the broadband resolving power of the FTICR mass spectrometer, this sample is so complex that there is no baseline for most of the spectrum.

Accurate mass determination is paramount in PMF and mass spectrometric protein identification [34,157,158].

The mass accuracy can be estimated by looking at the mass measurement error distribution for the most abundant protein in plasma, human serum albumin (HSA) (Fig. 10.21).

The centered distribution in Fig. 10.21 is mainly the result of true matches, whereas the matching peptides outside ± 10 ppm are due to false matches. The true matches are drawn from a much more narrow distribution than the false matches. This distribution is the signal to be detected in PMF. The absolute number of matching peptides within any chosen cut-off (e.g. ± 10 ppm) does not suffice for protein identification. Instead, this number should be compared with the expected number of random matches given the experimental data and error cut-off. This comparison can be done by taking the ratio of the likelihood of the measured distribution being drawn from a random matching distribution and the likelihood of being drawn from a convolution of a random and true matching distribution. The ratio of convoluted signal (true matches) and background (false matches) should be large enough to discriminate between true and false matches. The distribution of matching peptides in

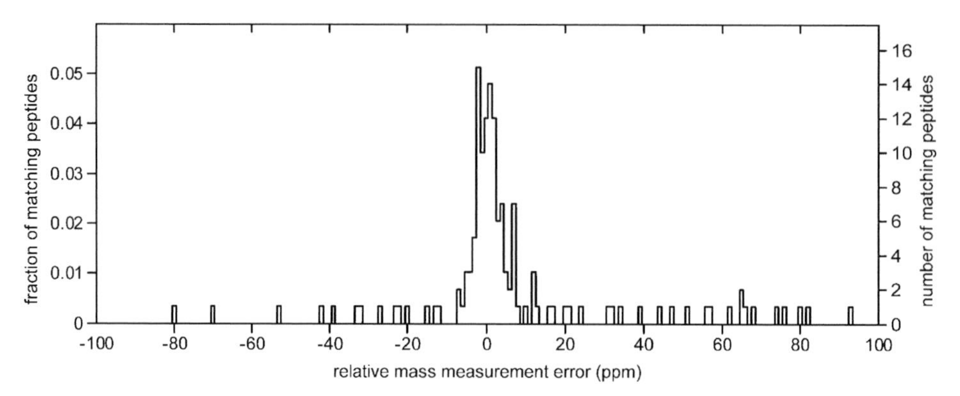

Fig. 10.21. Relative mass measurement error distribution for HSA in plasma in the mass spectra above.

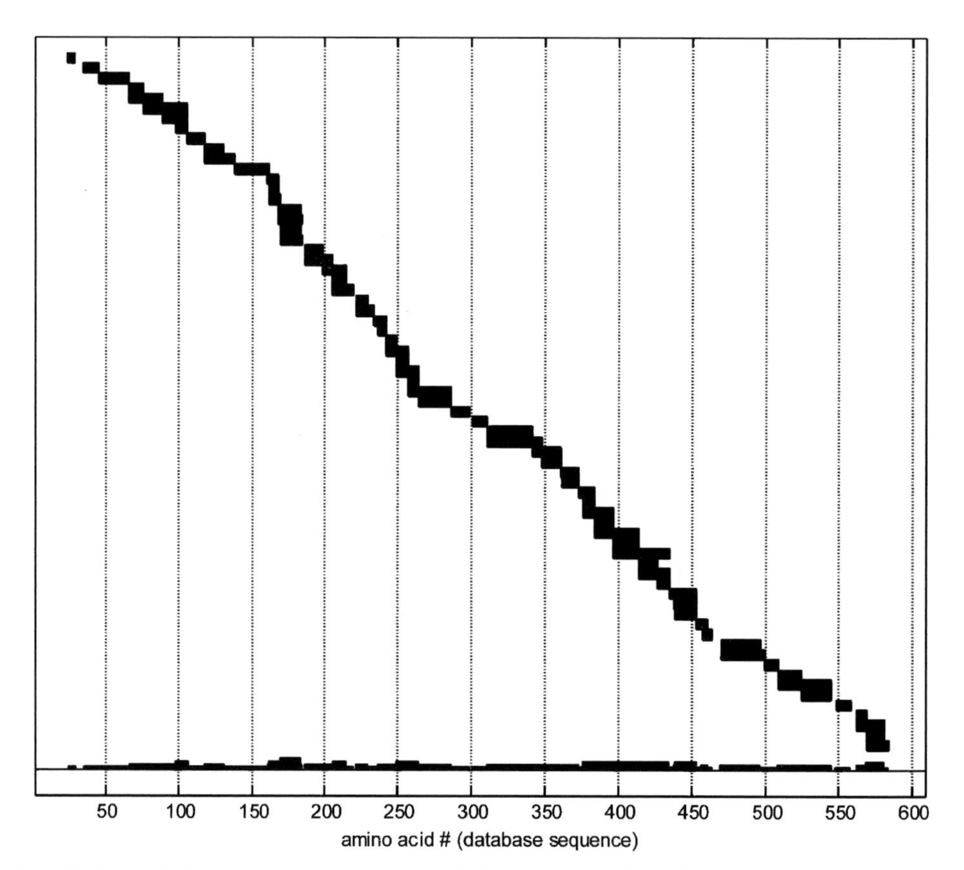

Fig. 10.22. Distribution of 70 matching peptides within 10 ppm, covering 92% of the mature protein sequence, of HSA in a human plasma tryptic digest. Residues 1–18 and 19–24 are signal peptide and propeptide and not expected to show up in this graph. The bottom line is the degree of sequence coverage by overlapping peptides.

TABLE 10.2

MATCHES BETWEEN THE 150 SEQUENCES IN THE A PRIORI SELECTED DATABASE, BFPDB (FOR BODY FLUID PROTEIN DATABASE) AND DIRECT INFUSION ESI-FTICR MASS SPECTRA OF FOUR BODY FLUID TRYPTIC DIGESTS RUN UNDER IDENTICAL CONDITIONS (A DIFFERENT PLASMA DIGEST FROM SHOWN SPECTRA). SEQUENCES IN BOLDFACE ARE ≥99% SIGNIFICANT; OTHERS ARE 95–99% SIGNIFICANT WHEN TESTED ALONE. AT MOST 1–2 SEQUENCES OF THOSE ≥99% COULD BE FALSE MATCHES

Plasma	Cerebrospinal fluid	Saliva	Urine
Serum albumin	**Serum albumin**	**Amylase alpha 1A**	**Serum albumin**
Immunoglobulin G	**Transferrin**	**Apolipoprotein A-IV**	**Uromodulin**
Immunoglobulin A	**Cystatin C**		**Haptoglobin**
Fibrinogen alpha	**Immunoglobulin G**	Fibrinogen alpha	**Cytokeratin**
Transferrin	**Perlecan**	Complement component B	**AMBP protein**
	Alpha-1-antitrypsin	Complement component 9	**Alpha-1-microglobulin**
Plasma kallikrein		Immunoglobulin G	**Epidermal growth factor**
Complement component C4	Complement component C4	Complement complement C3	**Apolipoprotein D**
Haptoglobin	Prostaglandin D2 synthase	Vitronectin	
Cystatin C	Alpha-2-glycoprotein 1		Complement component C4
	Haptoglobin		Alpha-1-acid glycoprotein 1
	Fibulin 1 isoform D		Protein Z
	Actin gamma		Prostaglandin D2 synthase
	Alpha-1-antichymotrypsin		Alpha-2-HS glycoprotein

the database sequence of HSA (± 10 ppm) can be illustrated by a 'sequence coverage' plot (Fig. 10.22). Note that the first 24 amino acids constitute signal and propeptides and are not expected in the released form of HSA.

The likelihood ratios for mass measurement errors and peptide connectivity of 150 body fluid proteins in an a priori selected database were compared with 4861 *S. cerevisiae* proteins as negative controls [156]. Table 10.2 shows the matching proteins and their statistical significances calculated by comparison with the yeast sequences in four analyzed body fluids. A small number of the most abundant proteins could be identified already in the direct infusion spectra. Although not many proteins could be routinely identified in whole body fluid tryptic digests, the method could nonetheless be used as an initial screening of biological samples in general to identify the most abundant proteins. Further optimization is possible; one option would be to remove the highly abundant HSA, essentially doubling the relative abundance of all other proteins (except a few that are bound to HSA).

Coupling of liquid separations such as LC [159–161] or CE [36,117,162–166] to FTICR mass spectrometry significantly increases the capacity for analyzing complex samples. Primarily, this can be assigned to the reduced ion suppression and increased dynamic range compared with direct infusion. Belov, Gorshkov, Smith and co-workers have further increased the dynamic range by loading the FTICR analyzer cell after selection in a quadrupole ion trap, called 'active dynamic range expansion' [167–169].

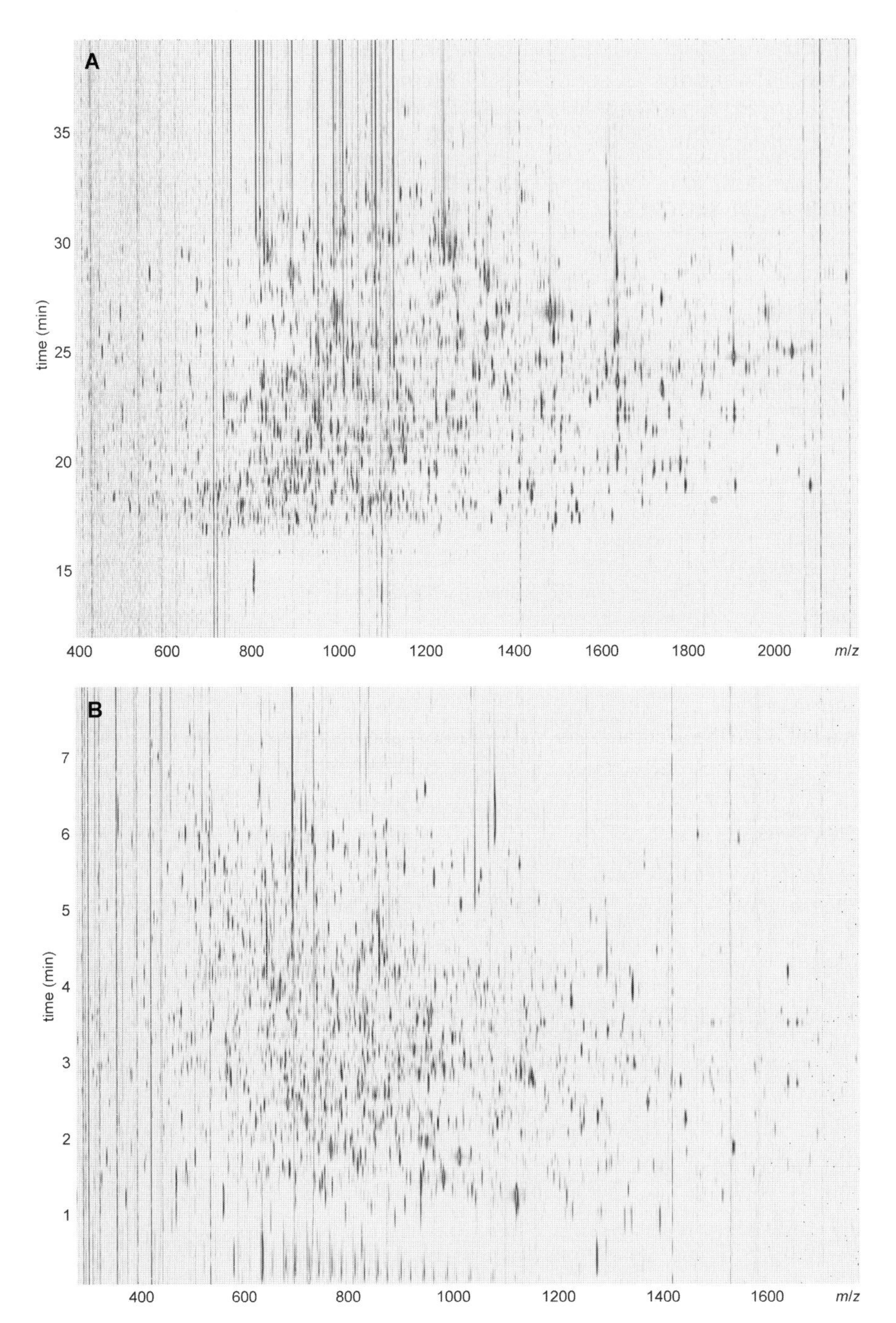
A
35
30
25
20
15
time (min)
400
600
800
1000
1200
1400
1600
1800
2000
m/z
B
7
6
5
4
3
2
1
time (min)
400
600
800
1000
1200
1400
1600
m/z

Smith et al. have also used this quadrupole to induce fragmentation to generate MS/MS type data [170]. A suitable figure-of-merit to compare direct infusion and different on- and off-line separations and mass spectrometry is how much information can be obtained in a given time and/or with a given amount of sample. The simplicity and reliability of direct infusion and the time needed to re-equilibrate LC columns and CE capillaries between each run should also be taken into account (or alternatively, several LC or CE systems could be coupled to the same mass spectrometer). Fig. 10.23 shows an LC-FTICR mass chromatogram and a CE-FTICR electropherogram of CSF tryptic digests.

The total number of data points in 2D datasets such as those illustrated in Fig. 10.23 is the number of data points in the m/z domain times the number of spectra acquired, i.e. $\sim 128-512$ kbyte times 256–512 spectra = 32–256 Mbyte raw data generated in each run.

When searching the a priori defined body fluid protein database, a significantly larger number of proteins could be identified compared with the direct infusion spectra of CSF with tryptic digests (Table 10.3). This comparison with direct infusion experiments demonstrates that separation methods enhance the ability of FTICR mass spectrometry to analyze complex biological samples, even given the relatively modest performance of the LC and CE equipment used here. Fig. 10.24 shows the distributions of likelihood-ratio based scores for the 150 sequences in the a priori defined database and the 4861 sequences in the yeast database.

Here, all identifications have been made on mass measurement errors (standard PMF) and a probabilistic notion of the non-randomness of the distribution of tryptic peptides in the sequence coverage, i.e. the amount of overlap and contiguity of these peptides.

In addition to the accurate mass measurement and distribution of peptides in the protein sequence, containing information on the non-random behavior of trypsin, protein structure, post-translational modifications and database sequence errors, there is also information on the physio-chemical properties of the individual peptides. This is primarily hydrophobicity in RPC [139–143] and size and charge in capillary zone electrophoresis [144–153]. A predictor of retention time according to Eq. (10.11) was 'trained' by least squares fit to 103 measured retention times for tryptic peptides of serum albumin and transferrin in CSF. HSA contains only one tryptophan residue and should not be used alone to train a predictor based on amino acid composition, especially since tryptophan is also one of the most hydrophobic residues. Table 10.4 shows the retention coefficients found by fitting to Eq. (10.11). In this separation, the pH of the mobile phases was ~ 3. The pH has been shown to be very important in reversed-phase chromatographic retention and electrophoretic migration of peptides [153,172].

Fig. 10.23. (color panel, opposite side). LC-FTICR (A) and CE-FTICR (B) analysis of cerebrospinal fluid tryptic digests. LC and CE data was acquired with skillful assistance of M. Ramström and M. Wetterhall, respectively. Peaks were picked automatically by an AURA macro run in XMASS. The macro performed time-domain convolution with squared sine functions of period equal to the sampling time followed by peak picking using the function supplied with the XMASS software and centroid fitting to each peak. The generated peak report files were merged and analyzed by computer programs ESIMSA and DATACOMP. The logarithmically scaled intensities (color) and the plots were generated by sifting out 4/5 (A) or 9/10 (B) of the data points (keeping the maximum in each interval) using an AWK [171] script before plotting in MATLAB.

TABLE 10.3

PROTEINS IDENTIFIED BY MASS MEASUREMENT ERROR AND PEPTIDE CONNECTIVITY IN A CEREBROSPINAL FLUID DIGEST BY DIRECT INFUSION, LIQUID CHROMATOGRAPHY AND CAPILLARY ELECTROPHORESIS ESI-FTICR MASS SPECTROMETRY FROM A 150-SEQUENCE A PRIORI SELECTED DATABASE. PROTEINS IN BOLDFACE ARE ≥99% SIGNIFICANT, OTHERS ARE 95–99% SIGNIFICANT. AT MOST 1–2 OF THE PROTEINS ON THE 99% SIGNIFICANCE LEVEL CAN BE EXPECTED TO BE RANDOM MATCHES AND 4–6 IN THE 95–99% INTERVAL (REDUNDANT SEQUENCES NOT SHOWN IN TABLE)

Direct infusion	LC-FTICR	CE-FTICR
Serum albumin	**Serum albumin**	**Serum albumin**
Transferrin	**Transferrin**	**Immunoglobulin G**
Cystatin C	**Complement component 9**	**Transferrin**
Immunoglobulin G	**Complement component C4**	**Transthyretin (pre-albumin)**
Perlecan	**Coagulation factor IX**	**AMBP protein**
Alpha-1-antitrypsin	**Complement component C1S**	**Plasma kallikrein**
	Apolipoprotein A-II	**Uromodulin**
Complement component C4	**Alpha-2-HS glycoprotein**	**Vitamin D-binding protein**
Prostaglandin D2 synthase	**Complement component D**	**Hemopexin**
Alpha-2-glycoprotein 1	**Prostaglandin D2 synthase**	**Immunoglobulin M**
Haptoglobin	**Vitamin D-binding protein**	**Fibulin 1 isoform A**
Fibulin 1 isoform D	**Lactotransferrin**	**Alpha-1B-glycoprotein**
Actin gamma	**Plasma retinol-binding protein**	**Apolipoprotein A-IV**
Alpha-1-antichymotrypsin	**Vitronectin**	**Alpha-2-macroglobulin**
	Statherin	
	Clusterin	Protein z
	Alpha-1-antichymotrypsin	Alpha-2-HS glycoprotein
	Apolipoprotein A-IV	Coagulation factor XI
	Alpha-2-macroglobulin	Fibrinogen alpha chain
	Immunoglobulin M	Alpha-1-antitrypsin
	Serum paraoxonase/arylesterase	Complement component C4
	Immunoglobulin G	Factor IX
	Fibrinogen alpha	Complement component 9
		Fibrinogen gamma chain
	Antithrombin III	Apolipoprotein C-III
	Hemopexin	Clusterin
	Fibulin 1 isoform A	Cystatin C
	Complement subcomponent C1R	Cytokeratin
	Cystatin C	Protein sap1
	Actin beta	Alpha-2-HS glycoprotein
	Alpha-1B-glycoprotein	Alpha-1-acid glycoprotein 2
	Glial fibrillary acidic protein	
	Alpha-2-HS glycoprotein	
	Leucine-rich alpha-2-glycoprotein	
	Transthyretin (pre-albumin)	
	Alpha-1-acid glycoprotein 2	
	Immunoglobulin A	
	Actin gamma	
	AMBP protein	
	Plasma kallikrein	
	S-100 calcium-binding protein A7	

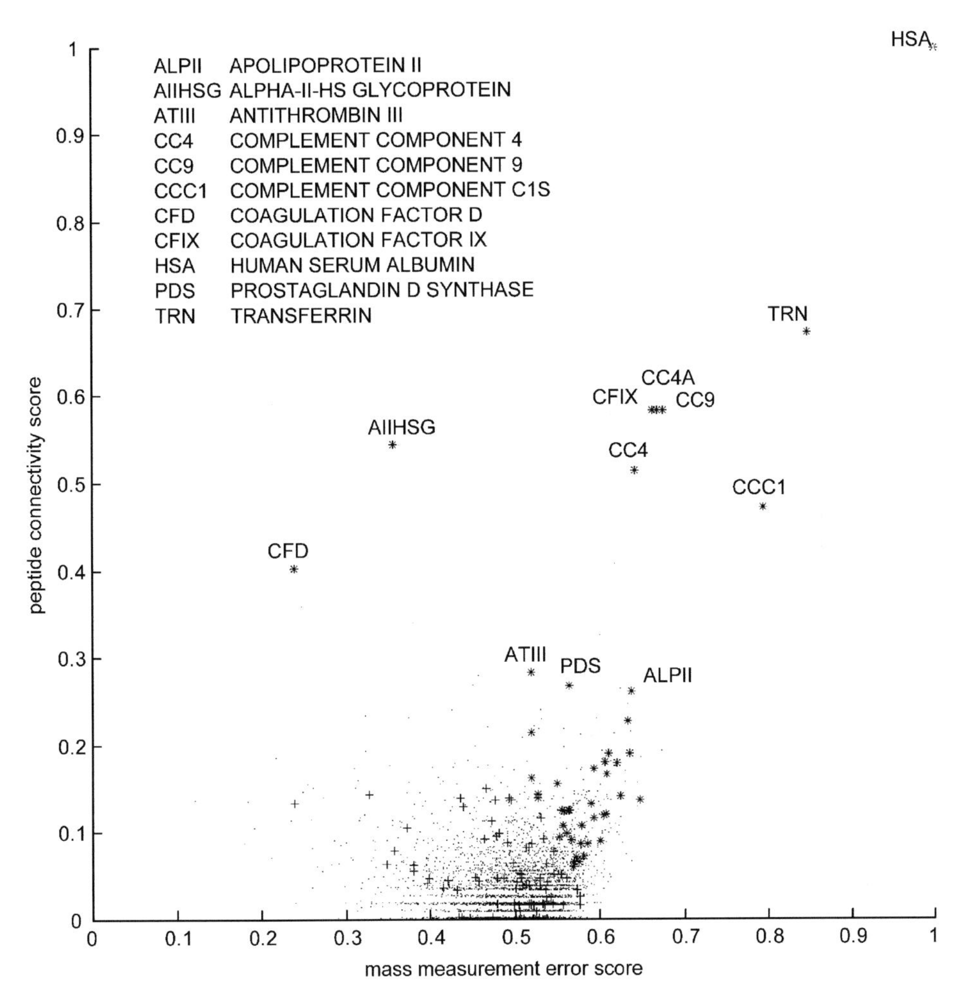

Fig. 10.24. Distribution of mass measurement errors and peptide connectivity likelihood scores for 150 body fluid proteins (+, ∗) in an LC-FTICR mass spectra of CSF compared to 4861 *S. cerevisiae* sequences (·). The LC-FTICR dataset contained 70 204 peaks, which could be reduced to 16 296 isotopic clusters and 6551 unique masses. Asterisks (∗) denote 95% significant matches.

Fig. 10.25 shows the correlation between predicted and measured retention for the 103 peptides of serum albumin and transferrin in the LC-FTICR mass spectra identified by accurate mass measurement alone. This is contrasted with the correlation between predicted retention for 622 peptides from 100 yeast proteins showing only a weak correlation to the measured retention for the randomly matching peptides in CSF (Fig. 10.25B). This weak correlation is due to a correlation between peptide size and retention time. In itself, the information on retention time is not sufficient to identify proteins in complex mixtures, but it is complementary to the mass measurement and peptide connectivity and can hence be used to resolve ambiguities and increase the

TABLE 10.4

RETENTION COEFFICIENTS IN MASS SPECTRUM NUMBERS (#) AND MINUTES FOR THE 20 AMINO ACIDS DERIVED FROM TRYPTIC PEPTIDES OF HSA AND TRANSFERRIN IN A CSF TRYPTIC DIGEST ANALYZED BY REVERSED-PHASE LC-FTICR MASS SPECTROMETRY AND ACIDIC (pH ~ 3) MOBILE PHASES. AS EXPECTED, THESE HAVE A POSITIVE CORRELATION WITH HYDROPHOBICITY

Amino acid	Retention coefficient (#)	Retention coefficient (min)
Arginine (Arg or R)	−4.02	−0.76
Serine (Ser or S)	−3.75	−0.71
Lysine (Lys or K)	−3.47	−0.66
Asparagine (Asn or N)	−2.87	−0.54
Glutamic acid (Glu or E)	−1.39	−0.26
Aspartic acid (Asp or D)	0.23	0.04
Glycine (Gly or G)	1.53	0.29
Threonine (Thr or T)	1.94	0.37
Alanine (Ala or A)	2.18	0.41
Histidine (His or H)	3.00	0.57
Proline (Pro or P)	5.10	0.97
Methionine (Met or M)	5.19	0.98
Glutamine (Gln or Q)	5.37	1.02
Cystein (Cys or C)	6.98	1.32
Leucine (Leu or L)	12.04	2.28
Valine (Val or V)	12.88	2.44
Phenylalanine (Phe or F)	14.13	2.68
Isoleucine (Ile or I)	14.24	2.70
Tyrosine (Tyr or Y)	14.67	2.78
Tryptophan (Trp or W)	24.69	4.68
(T_0)	(25.17)	(4.77)

significance of protein matches in LC-MS of whole sample tryptic digests. The same approach could also be applied in CE-MS experiments.

Combining information from retention/migration time predictions, accurate mass measurements and non-randomness of matching peptides in peptide mass fingerprints in a total χ^2-score improves the significances of proteins found and the number of proteins identified with a sufficient statistical significance. This predictor is based on a very simple model of retention and the accuracy could be expected to improve if more advanced features are incorporated. One such improvement would be to use a secondary structure predictor (a review was recently published by Rost [173]) to estimate the number of intramolecular hydrogen bonds and solvent accessibility to amino acid side-chains. These are likely to play an important role for peptide behavior in RPC.

A computer software was written that combined analysis of raw data with PMF and database searches. The raw data analysis consisted of apodization by time-domain convolution of a Gaussian or a half-period sine function, automated peak picking and deconvolution to unique masses and their retention times, i.e. maxima for all selected ion chromatograms. In the LC-FTICR mass chromatogram discussed here, 70 204 peaks were found and reduced to 16 296 isotopic clusters and 6551 unique peptide masses. The whole

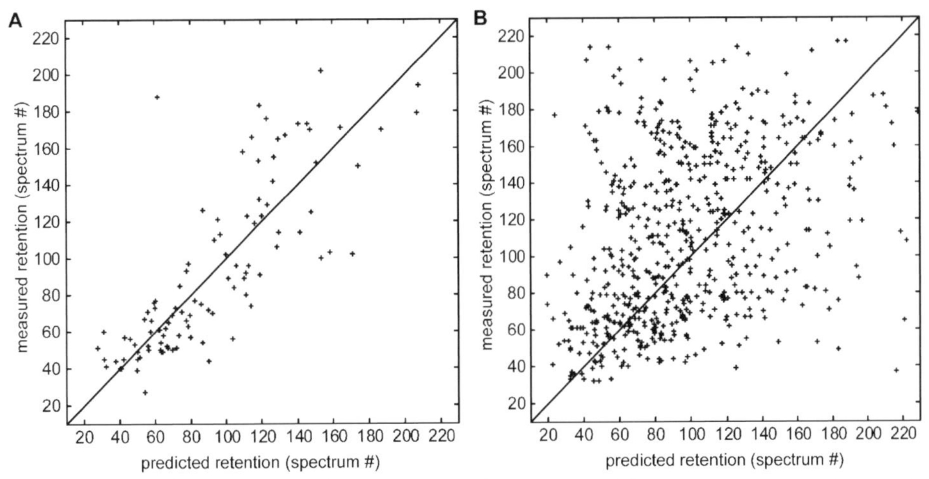

Fig. 10.25. Correlation between predicted and observed retention for HSA and transferrin tryptic peptides (A) and random peptide matches from 100 yeast proteins (B).

process took a few seconds on a standard PC and matching to protein databases took about 1 s/100 sequences.

10.6. CONCLUSIONS

The resolving power and accuracy of the FTICR-MS technique facilitated the screening of complex biological samples without any extensive pre-purification. The tremendous amount of information (more than 5500 peaks in a single spectrum) that this procedure reveals may well be used for fingerprint analysis and for screening of clinically relevant markers. The megabyte amounts of raw data generated in the described analyses require efficient computer algorithms for processing spectra, and statistically robust methods for searching protein sequence databases to identify and characterize the peptides and proteins in complex biological samples. In mass spectra of the intact body fluids, none or very few proteins could be identified, whereas in the digested samples several of the abundant proteins were characterized with high sequence coverage (between 50 and 95% with mass measurement errors below 5 ppm). Thus, this approach may become an important complement to ordinary tools not only in clinical diagnosis but also in proteomics and functional genomics. Fig. 10.26 summarizes the flow of information from sample to identified proteins. Present work includes the optimization of capillary electrophoresis, FTICR MS; capillary electrochromatography, FTICR MS; liquid chromatography, FTICR MS; and liquid chromatography, ECD-FTICR MS. These hyphenated techniques typically generate 10 000–100 000 peaks in each analysis. The separation methods will be applied in the

73. A.G. Marshall, C.L. Hendrickson and G.S. Jackson, Mass Spectrom. Rev., 17 (1998) 1
74. J.B.J. Fourier, Théorie analytique de la chaleur, Chez Firmin Didot, Père et Fils, Paris, 1822
75. J.W. Cooley and J.W. Tukey, Math. Comput., 19 (1965) 297
76. B.A. Vining, R.E. Bossio and A.G. Marshall, Anal. Chem., 71 (1999) 460
77. A.G. Marshall and F.R. Verdun, in: Fourier Transforms in NMR, Optical, and Mass Spectrometry, Elsevier, New York, 1990
78. M.B. Comisarow and A.G. Marshall, Fourier Transform Ion Cyclotron Resonance Spectroscopy Method and Apparatus, Nicolet Technology Corporation, USA, 1976, Patent US3937955
79. R.N. Hayes and M.L. Gross, Methods Enzymol., 193 (1990) 237
80. R.A. Zubarev, N.L. Kelleher and F.W. McLafferty, J. Am. Chem. Soc., 120 (1998) 3265
81. E. Mirgorodskaya, P. Roepstorff and R.A. Zubarev, Anal. Chem., 71 (1999) 4431
82. R.A. Zubarev, D.M. Horn, E.K. Fridriksson, N.L. Kelleher, N.A. Kruger, M.A. Lewis, B.K. Carpenter and F.W. McLafferty, Anal. Chem., 72 (2000) 563
83. A. Stensballe, O.N. Jensen, J.V. Olsen, K.F. Haselmann and R.A. Zubarev, Rapid Commun. Mass Spectrom., 14 (2000) 1793
84. D.M. Horn, Y. Ge and F.W. McLafferty, Anal. Chem., 72 (2000) 4778
85. K.F. Haselmann, B.A. Budnik and R.A. Zubarev, Rapid Commun. Mass Spectrom., 14 (2000) 2242
86. F.W. McLafferty, D.M. Horn, K. Breuker, Y. Ge, M.A. Lewis, B. Cerda, R.A. Zubarev and B.K. Carpenter, J. Am. Soc. Mass Spectrom., 12 (2001) 245
87. S.D. Shi, M.E. Hemling, S.A. Carr, D.M. Horn, I. Lindh and F.W. McLafferty, Anal. Chem., 73 (2001) 19
88. D.M. Horn, K. Breuker, A.J. Frank and F.W. McLafferty, J. Am. Chem. Soc., 123 (2001) 9792
89. K. Håkansson, H.J. Cooper, M.R. Emmett, C.E. Costello, A.G. Marshall and C.L. Nilsson, Anal. Chem., 73 (2001) 4530
90. Y. Ge, B.G. Lawhorn, M. ElNaggar, E. Strauss, J.H. Park, T.P. Begley and F.W. McLafferty, J. Am. Chem. Soc., 124 (2002) 672
91. S.K. Sze, Y. Ge, H. Oh and F.W. McLafferty, Proc. Natl Acad. Sci. USA, 99 (2002) 1774
92. M. Karas and F. Hillenkamp, Anal. Chem., 60 (1988) 2299
93. K. Tanaka, H. Waki, Y. Ido, S. Akita, Y. Yoshida and T. Yoshida, Rapid Commun. Mass Spectrom., 2 (1988) 151
94. M. Yamashita and J.B. Fenn, J. Phys. Chem., 88 (1984) 4451
95. M.L. Aleksandrov, L.N. Gall, V.N. Krasnov, V.I. Nikolaev, V.A. Pavlenko and V.A. Shkurov, Dokl. Akad. Nauk SSSR, 277 (1984) 379
96. P. Kebarle and L. Tang, Anal. Chem., 65 (1993) A972
97. A. Gomez and K. Tang, Phys. Fluids, 6 (1994) 404
98. D.R. Benjamin, C.V. Robinson, J.P. Hendrick, F.U. Hartl and C.M. Dobson, Proc. Natl Acad. Sci. USA, 95 (1998) 7391
99. A.A. Rostom, P. Fucini, D.R. Benjamin, R. Juenemann, K.H. Nierhaus, F.U. Hartl, C.M. Dobson and C.V. Robinson, Proc. Natl Acad. Sci. USA, 97 (2000) 5185
100. G. Siuzdak, B. Bothner, M. Yeager, C. Brugidou, C.M. Fauquet, K. Hoey and C.M. Chang, Chem. Biol., 3 (1996) 45
101. L. Rayleigh, Phil. Mag. Ser. 5, 14 (1882) 184
102. K.D. Henry, E.R. Williams, B.H. Wang, F.W. McLafferty, J. Shabanowitz and D.F. Hunt, Proc. Natl Acad. Sci. USA, 86 (1989) 9075
103. T. Covey, in: Biochemical and Biotechnological Applications of Electrospray Ionization Mass Spectrometry, A.P. Snyder (Ed.), American Chemical Society, Washington, DC, 1995, p. 21
104. R.B. Cole, J. Mass Spectrom., 35 (2000) 763
105. M.L. Vestal, Chem. Rev., 101 (2001) 361
106. A.L. Burlingame, R.K. Boyd and S.J. Gaskell, Anal. Chem., 70 (1998) 647R
107. T. Matsuo and Y. Seyama, J. Mass Spectrom., 35 (2000) 114
108. X. Geng and F.E. Regnier, J. Chromatogr., 296 (1984) 15
109. C.C. Stacey, G.H. Kruppa, C.H. Watson, J. Wronka, F.H. Laukien, J.F. Banks and C.M. Whitehouse, Rapid Commun. Mass Spectrom., 8 (1994) 513

110. J. Frenz, W.S. Hancock, W.J. Henzel and C. Horváth, Reversed phase chromatography in analytical biotechnology of proteins. HPLC of Biological Macromolecules: Methods and Applications, Marcel Dekker, New York, 1990, p. 145
111. R.D. Voyksner, Combining liquid chromatography with electrospray mass spectrometry. Electrospray Ionization Mass Spectrometry, R.B. Cole (Ed.), Wiley, New York, 1997, p. 323
112. K.B. Tomer, Chem. Rev., 101 (2001) 297
113. D.R. Barnidge, S. Nilsson, K.E. Markides, H. Rapp and K. Hjort, Rapid Commun. Mass Spectrom., (1999) 994
114. D.R. Barnidge, S. Nilsson and K.E. Markides, Anal. Chem., (1999) 4115
115. S. Nilsson, M. Wetterhall, J. Bergquist, L. Nyholm and K.E. Markides, Rapid Commun. Mass Spectrom., 15 (2001) 1997
116. J.A. Olivares, N.T. Nguyen, C.R. Yonker and R.D. Smith, Anal. Chem., 59 (1987) 1230
117. J.C. Severs, S.A. Hofstadler, Z. Zhao, R.T. Senh and R.D. Smith, Electrophoresis, 17 (1996) 1808
118. J.P.E. Landers, in: Handbook of Capillary Electrophoresis, 2nd ed., CRC Press, Boca Raton, FL, 1997
119. J.F. Kelly, L. Ramaley and P. Thibault, Anal. Chem., 69 (1997) 51
120. J. Samskog, M. Wetterhall, S. Jacobsson and K. Markides, J. Mass Spectrom., 35 (2000) 919
121. M. Wetterhall, S. Nilsson, K.E. Markides and J. Bergquist, Anal. Chem., 74 (2002) 239
122. R.B. Cody and B.S. Freiser, Int. J. Mass Spectrom. Ion Phys., 41 (1982) 199
123. J. Gao, Q. Wu, J. Carbeck, Q.P. Lei, R.D. Smith and G.M. Whitesides, Biophys. J., 76 (1999) 3253
124. J. Axelsson, M. Palmblad, K. Håkansson and P. Håkansson, Rapid Commun. Mass Spectrom., 13 (1999) 474
125. Y.O. Tsybin, P. Håkansson, B.A. Budnik, K.F. Haselmann, F. Kjeldsen, M. Gorshkov and R.A. Zubarev, Rapid Commun. Mass Spectrom., 15 (2001) 1849
126. J. Axelsson, K. Håkansson, M. Palmblad and P. Håkansson, Rapid Commun. Mass Spectrom., 13 (1999) 1550
127. K. Sannes-Lowery, R.H. Griffey, G.H. Kruppa, J.P. Speir and S.A. Hofstadler, Rapid Commun. Mass Spectrom., 12 (1998) 1957
128. K. Håkansson, J. Axelsson, M. Palmblad and P. Håkansson, J. Am. Soc. Mass Spectrom., 11 (2000) 210
129. M. Palmblad, K. Håkansson, P. Håkansson, X. Feng, H.J. Cooper, A.E. Giannakopulos, P.S. Green and P.J. Derrick, Eur. Mass Spectrom., 6 (2000) 267
130. M.W. Senko, C.L. Hendrickson, L. Pasa-Tolic, J.A. Marto, F.M. White, S. Guan and A.G. Marshall, Rapid Commun. Mass Spectrom., 10 (1996) 1824
131. D.M. Horn, R.A. Zubarev and F.W. McLafferty, J. Am. Soc. Mass Spectrom., 11 (2000) 320
132. M.W. Senko, S.C. Beu and F.W. McLafferty, J. Am. Soc. Mass Spectrom., 6 (1995) 229
133. M.W. Senko, S.C. Beu and F.W. McLafferty, J. Am. Soc. Mass Spectrom., 6 (1995) 52
134. Z. Zhang and A.G. Marshall, J. Am. Soc. Mass Spectrom., 9 (1998) 225
135. S.P. Gygi, B. Rist, S.A. Gerber, F. Turecek, M.H. Gelb and R. Aebersold, Nat. Biotechnol., 17 (1999) 994
136. J. Eriksson, B.T. Chait and D. Fenyö, Anal. Chem., 72 (2000) 999
137. M. Palmblad, M. Wetterhall, K. Markides, P. Håkansson and J. Bergquist, Rapid Commun. Mass Spectrom., 14 (2000) 1029
138. J. Bradley, Distribution-Free Statistical Tests, Prentice-Hall, Englewood Cliffs, NJ, 1968
139. J.L. Meek, Proc. Natl Acad. Sci. USA, 77 (1980) 1632
140. M.T. Hearn, M.I. Aguilar, C.T. Mant and R.S. Hodges, J. Chromatogr., 438 (1988) 197
141. R.S. Hodges, J.M. Parker, C.T. Mant and R.R. Sharma, J. Chromatogr., 458 (1988) 147
142. C.T. Mant, N.E. Zhou and R.S. Hodges, J. Chromatogr., 476 (1989) 363
143. C.T. Mant, T.W. Burke, N.E. Zhou, J.M. Parker and R.S. Hodges, J. Chromatogr., 485 (1989) 365
144. P.D. Grossman, J.C. Colburn and H.H. Lauer, Anal. Biochem., 179 (1989) 28
145. T.A. van de Goor, P.S. Janssen, J.W. van Nispen, M.J. van Zeeland and F.M. Everaerts, J. Chromatogr., 545 (1991) 379
146. M. Castagnola, L. Cassiano, R. Rabino, D.V. Rossetti and F.A. Bassi, J. Chromatogr., 572 (1991) 51
147. M. Castagnola, L. Cassiano, I. Messana, G. Nocca, R. Rabino, D.V. Rossetti and B. Giardina, J. Chromatogr. B: Biomed. Appl., 656 (1994) 87

148. A. Cifuentes and H. Poppe, J. Chromatogr. A, 680 (1994) 321
149. A. Cifuentes and H. Poppe, Electrophoresis, 16 (1995) 516
150. I. Messana, D.V. Rossetti, L. Cassiano, F. Misiti, B. Giardina and M. Castagnola, J. Chromatogr. B: Biomed. Sci. Appl., 699 (1997) 149
151. A. Cifuentes and H. Poppe, Electrophoresis, 18 (1997) 2362
152. M. Castagnola, D.V. Rossetti, M. Corda, M. Pellegrini, F. Misiti, A. Olianas, B. Giardina and I. Messana, Electrophoresis, 19 (1998) 1728
153. M. Castagnola, D.V. Rossetti, M. Corda, M. Pellegrini, F. Misiti, A. Olianas, B. Giardina and I. Messana, Electrophoresis, 19 (1998) 2273
154. J.L. Cornette, K.B. Cease, H. Margalit, J.L. Spouge, J.A. Berzofsky and C. DeLisi, J. Mol. Biol., 195 (1987) 659
155. A. Ingendoh, M. Witt, J. Fuchser and G. Baykut, High mass accuracy MALDI FT-ICR MS for protein identification out of mixtures of enzymatic digests, in: 49th ASMS Conference on Mass Spectrometry and Allied Topics, Chicago, Il, May 27–31, 2001
156. J. Bergquist, M. Palmblad, M. Wetterhall, P. Håkansson and K.E. Markides, Mass Spectrom. Rev., 21 (2002), 2
157. R.A. Zubarev, P. Håkansson and B.U.R. Sundqvist, Anal. Chem., 68 (1996) 4060
158. C. Masselon, G.A. Anderson, R. Harkewicz, J.E. Bruce, L. Pasa-Tolic and R.D. Smith, Anal. Chem., 72 (2000) 1918
159. Y. Shen, N. Tolic, R. Zhao, L. Pasa-Tolic, L. Li, S.J. Berger, R. Harkewicz, G.A. Anderson, M.E. Belov and R.D. Smith, Anal. Chem., 73 (2001) 3011
160. L. Li, C.D. Masselon, G.A. Anderson, L. Pasa-Tolic, S.W. Lee, Y. Shen, R. Zhao, M.S. Lipton, T.P. Conrads, N. Tolic and R.D. Smith, Anal. Chem., 73 (2001) 3312
161. T.L. Quenzer, M.R. Emmett, C.L. Hendrickson, P.H. Kelly and A.G. Marshall, Anal. Chem., 73 (2001) 1721
162. S.A. Hofstadler, F.D. Swanek, D.C. Gale, A.G. Ewing and R.D. Smith, Anal. Chem., 67 (1995) 1477
163. S.A. Hofstadler, J.C. Severs, R.D. Smith, F.D. Swanek and A.G. Ewing, Rapid Commun. Mass Spectrom., 10 (1996) 919
164. L. Yang, C.S. Lee, S.A. Hofstadler, L. Pasa-Tolic and R.D. Smith, Anal. Chem., 70 (1998) 3235
165. P.K. Jensen, L. Pasa-Tolic, G.A. Anderson, J.A. Horner, M.S. Lipton, J.E. Bruce and R.D. Smith, Anal. Chem., 71 (1999) 2076
166. P.K. Jensen, L. Pasa-Tolic, K.K. Peden, S. Martinovic, M.S. Lipton, G.A. Anderson, N. Tolic, K.K. Wong and R.D. Smith, Electrophoresis, 21 (2000) 1372
167. M.E. Belov, E.N. Nikolaev, G.A. Anderson, H.R. Udseth, T.P. Conrads, T.D. Veenstra, C.D. Masselon, M.V. Gorshkov and R.D. Smith, Anal. Chem., 73 (2001) 253
168. M.E. Belov, M.V. Gorshkov, K. Alving and R.D. Smith, Rapid Commun. Mass Spectrom., 15 (2001) 1988
169. M.E. Belov, G.A. Anderson, N.H. Angell, Y. Shen, N. Tolic, H.R. Udseth and R.D. Smith, Anal. Chem., 73 (2001) 5052
170. M.E. Belov, M.V. Gorshkov, H.R. Udseth and R.D. Smith, J. Am. Soc. Mass Spectrom., 12 (2001) 1312
171. A.V. Aho, B.W. Kernighan and P.J. Weinberger, in: The AWK Programming Language, Addison-Wesley, Reading, MA, USA, 1988
172. V. Sanz-Nebot, I. Toro, F. Benavente and J. Barbosa, J. Chromatogr. A, 942 (2002) 145
173. B. Rost, J. Struct. Biol., 134 (2001) 204
174. A.L. Betz, G.W. Goldstein and R. Katzman, Basic Neurochemistry: Molecular, Cellular and Medical Aspects, 5th ed., 1994, Raven Press, Ltd., New York, Chapter 32

G.A. Marko-Varga and P.L. Oroszlan (Editors)
Emerging Technologies in Protein and Genomic Material Analysis
Journal of Chromatography Library, Vol. 68

CHAPTER 11

Biological Single Molecule Applications and Advanced Biosensing

M. HEGNER,[a,*] Ch. GERBER,[a,b] Y. ARNTZ,[a] J. ZHANG,[a] P. BERTONCINI,[a] S. HUSALE,[a] H.P. LANG[a,b] and W. GRANGE[a]

[a]*Institute of Physics, NCCR Nanoscale Science, University of Basel, Klingelbergstrasse 82, CH-4056 Basel, Switzerland*

[b]*IBM Zurich Research Laboratory, Säumerstrasse 4, CH-8803 Rüschlikon, Switzerland*

11.1. INTRODUCTION

11.1.1. Macroscopic versus microscopic measurements in biology

Macroscopic experiments yield time and population averages of the individual characteristics of each molecule. At the level of the individual molecules, the picture is quite different: individual molecules are found in states far from the mean population, and their instantaneous dynamics are seemingly random. Whenever unusual states or the rapid random motions of a molecule are important, the macroscopic picture fails, and a microscopic description becomes necessary. Single molecule experiments differ from macroscopic measurements in two fundamental ways: first, in the importance of the fluctuations in both the system and in the measuring instrument, and second, in the relative importance of force and displacement as variables under experimental control and subject to direct experimental measurements. In single molecule experiments, the crucial parts of the measuring instruments themselves are small and subject to the same fluctuations as the system under study. Single molecule experiments thus give access to some of the microscopic dynamics that are hidden in the macroscopic experiments.

11.1.2. Force sensitive methods

During the last decade, new force measuring devices have been developed, which paved the way to explore the rich possibilities of mechanical measurements in the pico Newton (pN) range of force generating motor proteins and interacting biological macromolecules under physiological conditions.

* Corresponding author. *E-mail address:* martin.hegner@unibas.ch.

One of them is the optical trapping scheme, which consists of bringing a beam of laser light to a diffraction-limited focus using a 'good' lens, such as a microscope objective. Under the right conditions, the intense light gradient near the focal region can achieve stable three-dimensional trapping of dielectric objects, varying in size from a few tens of nanometers up to tens of micrometers. The term optical tweezers (OT) was defined to describe this so-called single-beam scheme.

Another force-measuring device is the scanning force microscope (SFM), which has evolved to a unique tool for the characterization of organic and biological molecules on surfaces. The SFM has proven its impact for biological applications, showing that it is possible to achieve sub nanometer lateral resolution on native membrane proteins in buffer solutions [1], to monitor enzymatic activity in situ [2] or to measure the unfolding of single proteins [3]. Up to now there is plenty of experimental information available regarding forces that arise in biological systems. Results from an overview of force-measuring experiments from as carried out during the last few years to get information on biological systems are shown in Fig. 11.1.

There is complementary biological information gathered by the various techniques. As visible in the graph, the force regime in which OT and scanning force microscopy are used overlap nicely. OT are applied preferably in experiments on molecular motors and entropic elasticity of molecules and conformational folding of proteins and rupturing of bonds are mainly investigated by SFM.

In addition, it has been shown over the last few years that the interaction of biomolecules on interfaces can be used as a tool for biosensing. The signal of the interaction of biomolecules on specific 'receptor' interfaces is transduced into a nanomechanical motion that is easily detected by the cantilever array technique, a specific method evolved from the scanning force microscopy. It is the purpose of this review to

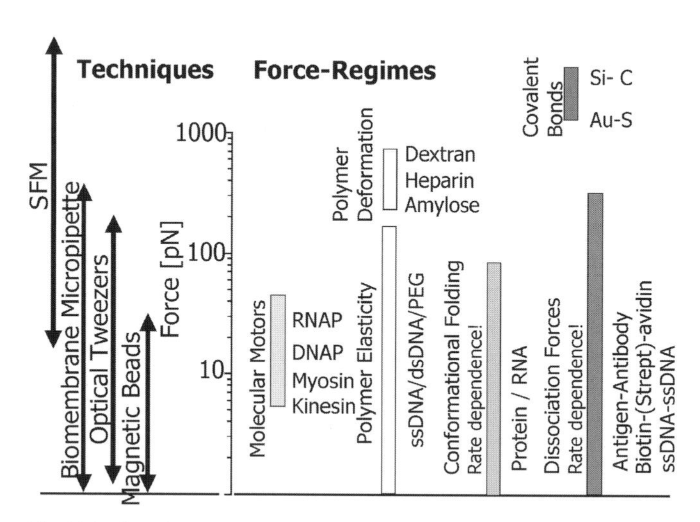

Fig. 11.1. Overview on techniques, which are applied to assess information upon forces in biological systems. On the right a series of experiments in the various areas of motors proteins or molecular mechanics are shown. Some of the experiments such as unfolding of individual proteins and dissociating (bio-molecular) bonds depend on the rate of the applied external force.

give an insight into these fields of our research activities in the Physics Institute at the University of Basel.

11.2. OPTICAL TWEEZERS

As mentioned previously, optical tweezers (OT) are instruments that allow the trapping or levitation of micron-sized dielectric particles using laser light. In addition, minute forces can be measured on the trapped particles with accuracy much better than what can be achieved with scanning force microscopy (in liquid and at room temperature). This explains why OT is nowadays considered as a technique of choice for the investigation of biomechanical forces. This section aims to give a basic introduction on this single molecule technique (a single molecule can be attached to the handle and therefore its mechanical properties can be studied), describing technical details and possible implementation and calibration of OT instruments. Moreover, typical experiments performed with OTs will be highlighted.

11.2.1. Origin of optical forces

It was first demonstrated in 1970 by Ashkin that light could be used to trap and accelerate dielectric micron-sized particles [4]. For this experiment, a stable optical potential well was formed using two slightly divergent counter-propagating laser beams. This pioneer study established the groundwork for the OT technique, where a single laser beam is focused by a high numerical aperture (NA) objective lens to a diffraction-limited spot [5]. At the focus, not only dielectric particle spheres can be trapped but also biological organisms such as cells, viruses, or bacteria [6,7]. Although it is still challenging for theory to calculate typical optical forces [8,9], the origin of optical forces can be understood easily (Fig. 11.2). Since a bundle of light rays is refracted when passing through a dielectric object, existing rays have different directions than incoming rays. This results in the change in light momentum. By conservation of momentum, the change in the light momentum causes a change ΔP in the momentum of the trapped particle. As a result and due to Newton's second Law, the particle will feel a force F:

$$F = \frac{\Delta P}{\Delta t} = \frac{nQW}{c} \tag{11.1}$$

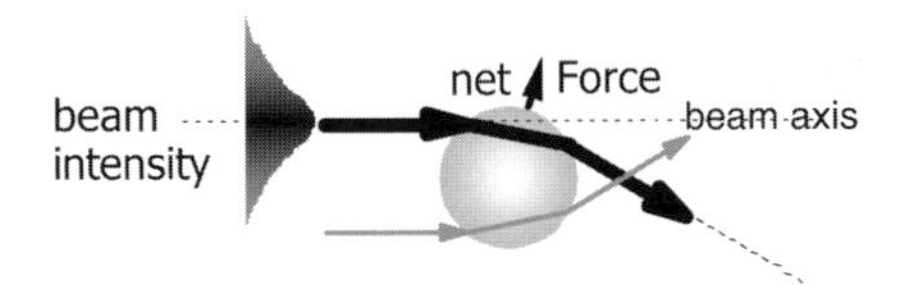

Fig. 11.2. A transparent dielectric micron-sized particle with an index of refraction larger than the surrounding medium is pushed towards the largest intensity of the light. All light rays are refracted when entering the particle. Due to Newton's second law, the change in light momentum flux (the force) causes a reaction force on the particle. Center rays contain more photons than outer rays and therefore exert more force. The resulting net force is shown.

where n is the index of refraction of the surrounding medium, W the power of the laser, c the celerity of the light, and Q is a dimensionless factor, known as the trapping efficiency.

Optical forces are however very small, since 100 mW of power at the focus (10^{+7} W/cm^2) produces forces of only a few tens of pN on a micron-sized particle.

For biological applications, it is therefore imperative to choose a laser excitation, which (i) does not raise the temperature of the surrounding medium (say water) and (ii) prevents biological damage. It has been shown that near infrared excitation is always best suited, although the wavelength region between 700 and 760 nm should be avoided in typical OT experiments [10,11].

As seen above, light exerts a pressure on dielectric particles. For practical applications, we need however to form a stable optical trap. In a simple picture, it can be shown that a stable trapping occurs when the scattering force is smaller than the gradient force. Basically, the scattering force (due to Fresnel reflections at the surface of the dielectric particle) is proportional to the power of the laser and acts only in the direction of propagation. As a consequence, this force does not trap. In contrast, the gradient force (proportional to the spatial gradient of the light) arises whenever the particle is out of the beam axis. This force acts therefore in three dimensions and tends to pull the particle towards the region of maximum spatial gradient (i.e. the focal point) if the index of refraction of the particle is larger than that of the buffer solution. This simple picture explains why a stable trapping is observed only when (i) high NA objective lenses are used and (ii) the back aperture of the objective lens has to be overfilled (to produce a diffraction limited spot and therefore a maximum spatial gradient). Moreover, it suggests that handles with high refractive index (at least larger than the surrounding medium) have to be used.

11.2.2. Experimental details

Knowing the origin of optical forces, we can build an optical trap. In principle, the design of such instruments should be easy since trapping requires only (i) a beam expander to overfill the back aperture of the microscope lens, (ii) a good microscope lens with a high NA to produce a steep spatial gradient and (iii) in addition when biological research is the focus of the experiment, lasers using wavelength in the near infrared should be used to avoid damage on the biological matter. Indeed, some rather simple modifications of a commercial inverted microscope are sufficient to build an OT [12]. Of course, whenever the following requirements have to be considered (beam steering, high mechanical stability, proper spatial filtering of the laser, reducing mode hopping of the laser…), it is best to build an OT on a conventional optical table with custom optics and electronics [8]. Note finally that oil immersion microscope lenses are less suited for OTs due to the difference in index of refraction between oil and water (inside the chamber). The major microscope providers have nowadays a high numerical water-immersion lens in their program, which circumvents this drawback.

11.2.2.1. Calibration procedure

One of the main difficulties in OT experiments is to correctly estimate the force that acts on the trapped particle. We already have mentioned that an object changes the direction of the refracted rays when it experiences a force. In principle, such a change in the light momentum flux can be easily monitored onto a position sensitive detector (PSD) if we place a condenser lens after the objective lens (Fig. 11.3).

To relate deflections observed on the detector to forces, we need to calibrate the instrument. Many different approaches have been proposed in the past to perform such a calibration [8]. However, the thermal fluctuation calibration method is certainly the most widely used. Having an object in the trap, the stiffness K of the light lever can be estimated from the power spectral density $S(f)$ of the displacement fluctuations, using:

$$S(f) = \frac{k_B T}{\pi^2 \gamma (f^2 + f_c^2)} \times A^2 \tag{11.2}$$

where γ denotes the viscous drag, $f_c = K(2\pi\gamma)^{-1}$ is the corner frequency (i.e. the frequency above which the particle does not feel the effect of the trap anymore), $k_B T = 4.1$ pN nm at room temperature, and A is a factor describing the sensitivity of the detector (V nm^{-1}). Fitting the measured power spectral density with Eq. (11.2) gives a robust estimate of the corner frequency and consequently of the trap stiffness if the viscous drag is computable (i.e. for an object of known shape).

In addition, the detector sensitivity can be obtained from the area under the spectral density curve, knowing that the mean square displacement of the particle is related to both $S(f)$ and K through:

$$\langle x^2 \rangle = \int S(f)\mathrm{d}f = \frac{k_B T}{K} \times A^2 \tag{11.3}$$

Such a procedure shows that both the trap stiffness and the detector sensitivity (which are the only parameters needed to estimate the force) can be determined from a detector that is not absolutely calibrated.

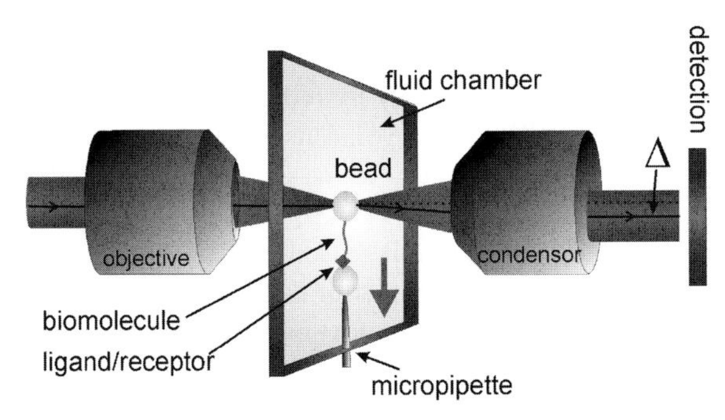

Fig. 11.3. Schematic representation of an OT experiment. A microscope objective lens with a high NA is used to focus the laser light to a diffraction-limited spot. At the focus, where the spatial gradient is maximal, particles such as beads can be trapped. Light is collected with a condenser lens, which converts angular deflections into transverse deflections that can be monitored on a PSD. A single molecule can be attached between the trapped bead and a bead on a micropipette through a receptor-ligand bridge. Mechanical properties of single molecules can be therefore investigated if the micropipette is placed onto a piezoelectric element. When the trapped bead experiences the force (arrow), it moves slightly away from its stable position. As in SFM, this leads to a deflection on the detector.

11.2.2.2. State of the art instrumentation

In the previous section, we have seen that single beam OTs can easily measure forces. However, an important drawback of such instruments is that a new calibration has to be performed for each new experiment (each time some local parameters such as the shape or size of the trapped particle, the local fluid viscosity, or power of the laser are changed). Indeed, single beam OTs do not directly measure the change in light momentum flux because of the scattering at the interface of the microscope lens that introduces marginal rays. As shown by Smith [13], dual beam OTs overcome such limitations. Such instruments, although more expensive than single beam OTs, need to be calibrated only once and have a very high trapping efficiency, which is of prime importance for biological investigations [14].

11.2.2.3. Thermal noise

At room temperature, Brownian motion of the trapped particle (OT) or the cantilever (SFM) limits the force resolution of this micromechanical experiment. As seen in the previous section, the power spectral density (i.e. thermal noise) is constant for frequency below the corner frequency and rolls off rapidly for frequency above f_c. However, $S(f)$ usually extents to high frequencies. When the bandwidth f_s of the measurement is much smaller than the corner frequency or when low-pass filters are used, the minimum detectable force F_{min} reads (see Eqs. (11.2) and (11.3)):

$$F_{min} = (K/A)\sqrt{\langle x^2 \rangle} = \sqrt{4\gamma k_B T f_S} \tag{11.4}$$

In this case, the force resolution is independent of the trap stiffness [15]. To improve the sensitivity of micromechanical experiments, we can decrease (i) the temperature, (ii) the bandwidth of our measurement, or (iii) the drag viscosity. Alternatively, the resonance frequency of the SFM cantilever can be increased to reduce the noise at a given bandwidth. Certainly the best approach is to decrease the drag viscosity by reducing the size of the force-sensing device (the trapped bead or the cantilever) [15,16]. This explains the need for small SFM cantilevers when high force sensitivity has to be achieved. Note that commercial SFM cantilevers have typical dimensions of 100 μm, whereas beads used for OT experiments have diameters of the order of 1 μm. For this reason, the force noise level of OT measurements (below 0.3 pN) is much smaller than that of SFM based techniques (10 pN, in liquid and at room temperature). However, the maximum force that can be measured with OTs is rather small compared with SFM based techniques. For instance, dual beam OTs can measure forces only up to 200 pN [14].

11.2.3. Recent experiments

We do not attempt to give an exhaustive review of all experiments that have been performed with OTs (see Refs. [17,18] for recent reviews). Rather, we would like to select and describe briefly typical applications of OTs in biology.

11.2.3.1. Molecular motors

Certainly, one of the most impressive applications of OTs is the study of molecular motors on a single molecule level. These molecular motors can be linear motors (Kinesin, Myosin) [19], DNA/RNA polymerase enzymes [20,21] or DNA packaging viruses (bacteriophage ϕ29) [22].

Kinesin and Myosin are two ATPase motor proteins. Kinesin, which is used for organelle transport or chromosome segregation, moves along microtubules. In contrast, Myosin interacts with actin filaments and is used not only for muscle contraction but is also involved in many forms of cell movement. For these studies, OTs are used to interact single Kinesin or Myosin in vitro with either a microtubule or an actin filament (Fig. 11.4(A)). These experiments have revealed how much ATP has to be hydrolyzed and the forces generated at each step, demonstrating possible mechanisms involved in the movement.

Other experiments investigated the function of motor enzymes used in DNA transcription or DNA polymerization. In this case a single DNA molecule is tethered between two beads (Fig. 11.4(B)), and the rate of transcription or polymerization can be followed in real time by applying a constant tension (force feedback) and allowing the distance between the beads to change accordingly. Such studies have direct implications for the mechanism of gene regulation or force-induced exonuclase activity.

11.2.3.2. Mechanical properties of single molecules

Due to the high force sensitivity, OTs have been used to study (i) mechanical properties of DNA [23], (ii) protein or RNA unfolding [24–26] and (iii) the polymerization of individual RecA-DNA filaments [27]. Again, these experiments provided new insights in biochemical processes on a single molecule and are of great relevance to biology. In a recent publication, Husale et al. [28] showed that OT can easily be applied to investigate the influence of small ligands directly interacting with DNA to elucidate the binding mechanism of the ligands on the DNA.

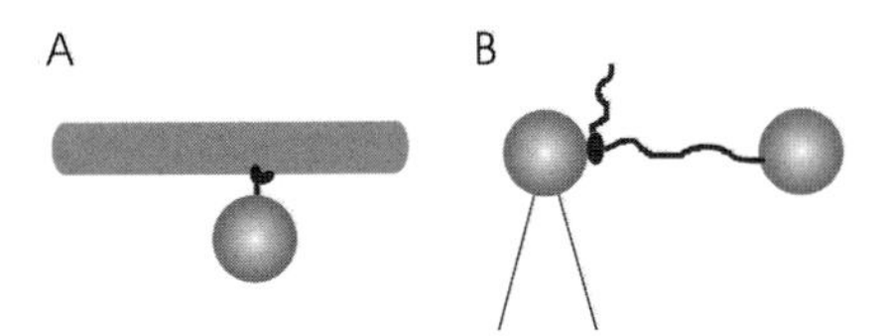

Fig. 11.4. Possible experimental setups used for single molecule observation of molecular motors. (A) Beads coated with either Kinesin or Myosin are approached in the vicinity of a microtubule or an actin filament. For Kinesin studies, the microtubule is placed on a surface and the bead is held by the OT. Kinesin can walk continuously for long distances (microns) before being released from the microtubule. In contrast, Myosin only binds weakly to actin and is released after a one-step movement. For these experiments, the bead has therefore to be immobilized onto a surface and an actin filament is stretched and held by two optical traps. (B) A bead coated with an enzyme (black) interacting with a DNA molecule is held by suction on top of a micropipette in a continuous flow of buffer solution. A bead coated with a receptor recognizing the ligand-modified DNA molecule end is approached till the end is connected through the biomolecular interaction of the receptor with the ligand using OT. The enzyme runs along the single DNA molecule and is used either to transcribe a dsDNA template into a messenger RNA (RNA polymerase) or to incorporate base pairs on ssDNA (DNA polymerase).

11.3. SCANNING FORCE SPECTROSCOPY

11.3.1. Introduction to dynamic force spectroscopy

It has long been known that only molecules with an excess of energy over the average energy of the population can participate in chemical reactions. Accordingly, reactions between ligands and receptors follow pathways (in a virtual energy landscape) that involve the formation of some type of high-energy transition states whose accessibility along a reaction coordinate ultimately controls the rate of the reaction. Until recently, chemists and biologists could only act on molecules if these were present in large quantities. Consequently, scientists could only access macroscopic thermodynamical quantities, e.g. the free energy of complex formation and/or dissociation.

Today, instruments offering a high spatial resolution and sensitivity down to the pico- or femto–Newton range allow one to study the adhesion of molecular bonds [29–41]. In particular, a novel type of force spectroscopy, the so-called dynamic force microscopy (DFS), has been developed. In Fig. 11.5 a setup of a SFM used for dynamic force spectroscopy is shown.

In a DFS experiment, the dependence of the rupture force on the loading rate is investigated using a SFM, a bio-membrane force probe (BFP), or eventually an OT setup. A rather detailed description of such experiments performed in our laboratory is given in this section [39–42]. For a typical DFS experiment using a SFM, a ligand is immobilized on a sharp tip attached to a micro-fabricated cantilever and the receptor is immobilized on a surface. When approaching the surface of the tip a bond may form between ligand and receptor. The bond is then loaded with an increasing force when retracting the surface from the tip. From these measurements, the energy landscape of a single bond can be mapped [43]. A typical force distance plot of these experiments is shown in Fig. 11.6.

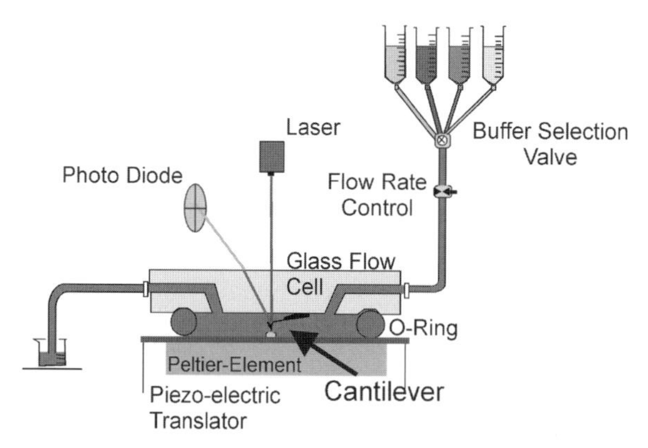

Fig. 11.5. Setup of a scanning force microscope. The instrument is working in buffer media under ambient conditions. A springboard type cantilever (dimensions $\sim 300 \times 20 \times 0.5\ \mu m$, spring constant 0.01 N/m) is deflected when adhesive forces between the ligand-modified tip and the receptor-modified interface arises during retraction of the sample surface. The motion of deflection of the cantilever is detected using a laser beam, which is mirrored on the levers backside and projected onto a four-quadrant photosensitive diode. Various liquids can be injected and a peltier-element allows precise varying of temperature.

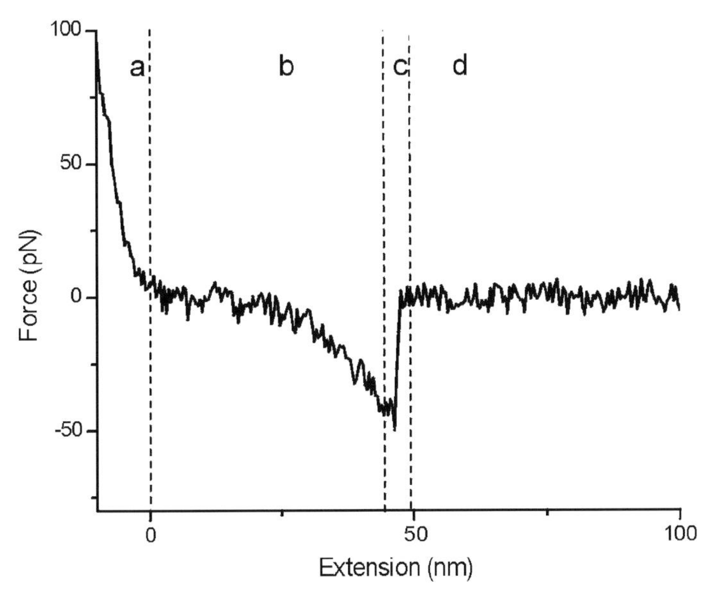

Fig. 11.6. A typical force–distance curve obtained in a stretching SFM experiment (retraction cycle). A DNA strand (TATTAATATCAAGTTG) is immobilized by its 5′-end via a PEG linker on the SFM tip and its complement is attached in a similar fashion by its 5′-end to the surface. When the tip is approached close to the surface a specific bond between the two strands is formed (**a**). The SFM tip is then retracted from the surface at constant loading rate (**b**–PEG stretching). The sudden drop in the force curve reflects unbinding of the duplex (**c**-[specific DNA unbinding]). The loading rate r (retract velocity v times the elasticity c) is determined from the slope of the force-displacement curve before the unbinding event occurs. d) Free cantilever, biomolecular bond is ruptured. The cantilever motion visible is induced by thermal motion.

This section is organized as follows. Part one introduces theoretical models that describe a chemical reaction when an external force is used to rupture a complex. Then, DFS experiments on complementary DNA strands are presented and illustrate the main ideas developed in Section 11.3.1.

11.3.2. Theoretical background

In this section, some thermo-dynamical models describing the rupture of a single bond will be briefly presented. More details can be found elsewhere [44–47].

Bell [45] first stated that the bond lifetime τ of an energy barrier reads:

$$\tau(F) = \tau_0 \exp[(E_0 - \Delta x F)/k_B T] \tag{11.5}$$

where T is the temperature, E_0 represents the bond energy (the height of the barrier), F is the external applied force per bond, Δx is the distance (projected along the direction of the applied force) between the ground state and the energy barrier (with energy E_0), and τ_0 is a pre-factor. Eq. (11.5) states that (i) a bond will rupture after a certain amount of time, thanks to thermal fluctuations and (ii) application of an external force dramatically changes the time it takes to overcome the energy barrier. Note finally that (Eq. (11.5)) can be re-written as:

$$k_{\text{off}}(F) = k_{\text{off}} \exp(F/F^0) \tag{11.6}$$

where k_{off} is the thermal off-rate of the barrier, and F^0 is a force-scale factor ($F^0 = k_B T/\Delta x$).

An important point is that the most probable force F^* needed to overcome an energy barrier should a priori depend on the loading rate, i.e. the velocity in a typical DFS experiment (typical values for velocities are in the range between 10 and 5000 nm/s). Indeed, when the loading rate decreases, F^* should decrease because of thermal fluctuations. In fact, a simple relation holds between F^* and the loading rate r ($r = k\nu$, where k is the stiffness of the DFS force sensor and ν is the retraction speed):

$$F^* = F^0 \ln(r/F^0 k_{\text{off}}) \tag{11.7}$$

By plotting F^* as a function of $\ln(r)$, one should therefore find different linear regimes, each of them corresponding to a specific region (a specific energy barrier) of the energy landscape. According to Evans [46], the kinetics runs as follows: application of an external force (i) selects a specific path (a reaction coordinate) in the energy landscape and (ii) suppresses outer barriers (Eq. (11.5)), and reveal inner barriers, which start to govern the process. For instance, recent BFP and SFM experiments have revealed an intermediate state for the streptavidin (or avidin)-biotin complex [38,41]. However, since each energy barrier defines a time-scale (a range of loading rate that has to be compatible with the time-scale of the experiment) only a specific part of the energy landscape can be mapped in a typical DFS experiment [47,48].

11.3.3. Experimental

DFS measurements were performed with a commercial SFM instrument using some external data acquisition and data output capabilities in addition. The spring constants of all cantilevers (ranging from 12 to 17 pN/nm) were calibrated by the thermal fluctuation method [49] with an absolute uncertainty of 20%. For the temperature measurements presented below, the temperature was controlled using a home built cell where the buffer solution that immersed both the probe surface and the SFM cantilever was in contact with a Peltier element, driven with a constant current source. Measurements at different points of the cell showed deviations of less than 2 °C.

The preparation and immobilization of all oligonucleotides follows the protocol described in Refs. [39,42].

11.3.4. Probability distribution and specificity of rupture forces

Unbinding events are caused by thermal fluctuations rather than by mechanical instability. Therefore unbinding forces show a distribution whose width σ is mainly determined by the force scale F^0, i.e. $\sigma = F^0(\Delta x)$.

When approaching the tip to the surface, many non-specific attachments may occur, even in the presence of treated surfaces or pure polymer samples. Therefore, it is imperative to test the specificity of the interaction (Fig. 11.7).

Unspecific interactions can be minimized using linkers [e.g. poly(ethylene)glycol (PEG) linkers] that shift the region where unbinding takes place away from the surface. Finally, to quantify the most probable value for the unbinding force of a single complex, one has to work under conditions in which the probability that two or more duplexes are attached to the tip is low.

These conditions are fulfilled for a low concentration and when the linkers have a length that is comparable with the diameter of the SFM-tip (about 50 nm). In this case, it is very unlikely that two or more linkers are extended to the same length when stretched. However, subsequent rupture events may be found. But still, the last rupture event will occur for an applied force equal to F^*.

11.3.5. Dynamic measurements

11.3.5.1. Base pair dependence

We now present DFS measurements performed on complementary DNA strands of different length [10, 20, and 30 base pairs (bp)] and pulled apart at their opposite 5′-ends. The base sequences of the oligonucleotides were designed to favor the binding to its complementary oligonucleotides in the ground state with respect to intermediate duplexes in which the strand is shifted relative to its complement. We have chosen the oligomer **a** (5′-G-G-C-T-C-C-C-T-T-C-T-A-C-C-A-C-T-G-A-C-A-T-C-G-C-A-A-C-G-G-3′), which contains 30 bases and in which every three base motive occurs only once in the sequence. For this sequence, self-complementarities are avoided because the complement of each three-base motive is not contained in the sequence. **a** was tested against its complement **b** (30 bp) and against truncated components **c** (20 bp) and **d** (10 bp), respectively.

As expected, a F^* versus $\ln(\nu)$ plot shows a linear behavior for each duplex (Eq. (11.7), see Fig. 11.8).

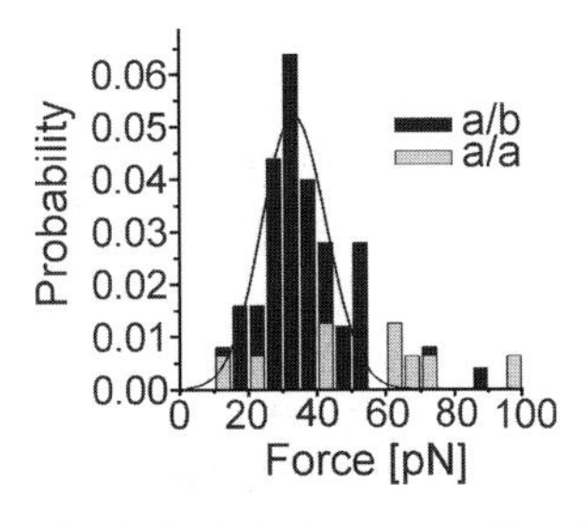

Fig. 11.7. A typical probability distribution for the rupture force (about 500 approach/retract cycles, retract velocity 100 nm/s) [39]. For this experiment, an oligomer **a** (see text) was attached to the tip of the SFM-cantilever and its complement **b** was immobilized on the surface (complements were pulling apart at their opposite 5′-ends). Gray rectangles (**a** against **a**), black rectangles (**a** against **b**). To minimize unspecific interactions (e.g. **a** against **a**) and multiple unbinding events, 30-nm-long PEG linkers were attached to the 5′-ends. Note that the scale-force F^0 can be in principle determined from the width of the distribution.

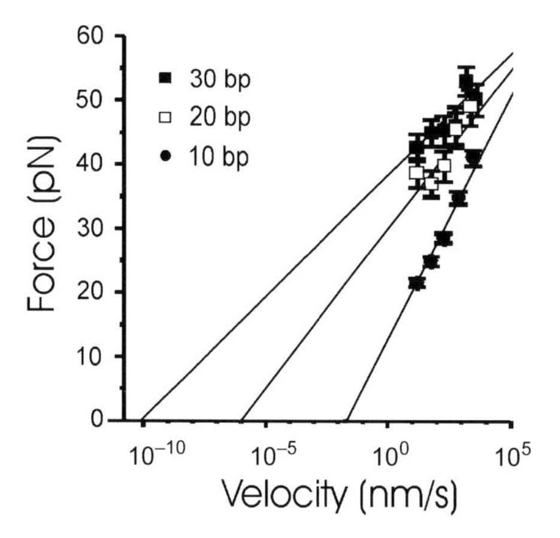

Fig. 11.8. Velocity dependence of the most probable unbinding force [39]. Back squares (**a**–tip/**b**–surface, 30 bp), empty squares (**a**–tip/**c**-surface, 20 bp), circles (**a**–tip/**d**–surface, 10 bp). From a linear fit, both the force-scales $F^0 = k_B T/\Delta x$ and thermal off-rates can be determined.

For each duplex, the distance Δx from the ground state to the energy barrier and the thermal off rate k_{off} were determined according to Eq. (11.7). The Δx distance was found to follow the linear relation: $\Delta x = [(0.7 \pm 0.3) + (0.07 \pm 0.03) \times n]$ nm, where n is the number of bp. This increase of Δx with n clearly indicates cooperativity in the unbinding process. Measurements of k_{off} can be described by: $k_{\text{off}} \approx 10^{\alpha - \beta n}\ \text{s}^{-1}$, where $\alpha = 3 \pm 1$ and $\beta = 0.5 \pm 0.1$. The obtained k_{off} values are in good agreement with thermodynamical data [50]. Let us finally point out that an exponential decrease of the thermal off-rate with the number of bp is expected because of the increase of the activation energy for dissociation (Eq. (11.5)). However, the pre-factor τ_0 in Eq. (11.5) also strongly decreases with the number of bp because of the increasing number of degrees of freedom of the system.

11.3.5.2. Temperature dependence

In this section, temperature dependent DFS measurements are briefly discussed. The sequence **e** (5′-T-A-T-T-A-A-T-A-T-C-A-A-G-T-T-G-3′) [51] attached to the tip and its complement **f** was immobilized on the surface. As previously, PEG linkers were used and DNA strands were pulled apart at their opposite 5′-ends. The specificity of the interaction was comparable with the one obtained in base-pair dependent measurements (Fig. 11.7).

As seen in Fig. 11.9, the slope of the F^* versus $\ln(r)$ plots changes as a function of temperature, which evidences for a strong temperature dependence of Δx. This result emphasizes the fact that for the DNA-duplex, the energy landscape is much more complicated than that of ligand-receptor bonds.

As a consequence, the unbinding process may involve many different reaction paths. In this case, thermal fluctuations are expected to play a key role.

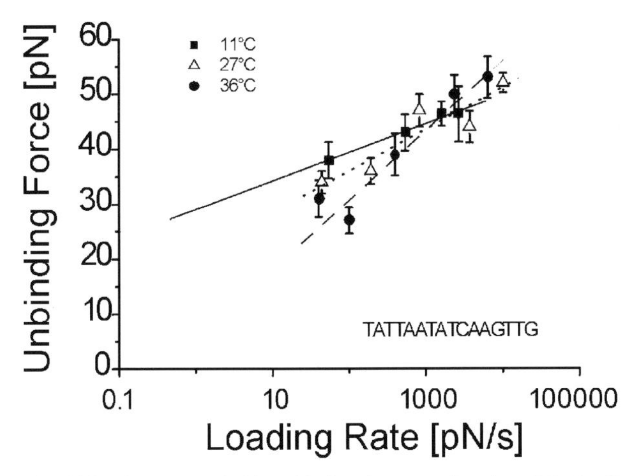

Fig. 11.9. The most probable unbinding force as a function of the loading rate (**e**–tip/**f**–surface, 16 bp) obtained at different temperatures. Squares (11 °C), triangles (27 °C), circles (36 °C).

11.3.6. Future of dynamic force spectroscopy

Using DFS measurements, the energy landscape of molecular bonds can be mapped. Moreover, relevant parameters such as the location and height of the barriers and the thermal off-rates can be determined. Our measurements confirm that the most probable force for unbinding scales is the logarithm of the loading rate. From this dependence, both the natural thermal off-rate for dissociation k_{off} and the bond length x along the reaction coordinate were determined. Our measured k_{off} values are in agreement with bulk temperature measurements indicating the validity of our measurements. The bp dependent measurements indicate that unbinding of DNA strands is a cooperative process. Temperature dependent measurements are evidence for a decrease of Δx as the temperature increases [42]. This behavior, which is not expected in the case of one-dimensional energy landscape with a sharp energy barrier, indicates the role played by entropic contributions when unbinding DNA and unfolding RNA or proteins. However, the linear decrease of x with the temperature is still an open question. It is obvious that the exact relationship between the bond length and the temperature is not straightforward and calculations are needed to explain the observed properties. Since the limited range of loading rates available in a SFM experiment does not allow one to map the whole energy landscape, such experiments should be combined in the future with other DFS setups such as BFP or OT setups. An additional solution is to apply small cantilevers, which allow faster pulling and exhibit less thermal noise; so smaller unbinding forces can be detected. These small cantilevers are still experimental [52] and great efforts are being made to commercialize them in the future. These developments will also ask for instrument development so it is expected that it will be a few years before they are widely used. One could envision that the dynamic force spectroscopy will be applied in the future to assess the binding affinity of biomolecules on bio-arrays but the experimentalists have to catch up to step in this direction.

11.4. ADVANCED BIO-SENSING USING MICRO-MECHANICAL CANTILEVER ARRAYS

11.4.1. Introduction to micro-mechanical bio-sensors

Over the last few years a series of new detection methods in the field of biosensors have been developed. Biosensors are analytical devices, which combine a biologically sensitive element with a physical or chemical transducer to selectively and quantitatively detect the presence of specific compounds in a given external environment.

These new biosensor devices allow sensitive, fast and real-time measurements. The interaction of biomolecules with the biosensor interface can be investigated by transduction of the signal into a magnetic [53], an impedance [54] or a nanomechanical [55] signal. In the field of nanomechanical transduction, a promising area is the use of cantilever arrays for biomolecular recognition of nucleic acids and proteins. One of the advantages of the cantilever array detection is the possibility to detect interacting compounds without the need of introducing an optically detectable label on the binding partners. For biomolecule detection the liquid phase is the preferred one but it has been shown that the cantilever array technique is also very appropriate for use as a sensor for stress [56], heat [57] and mass [58]. Recent experiments have shown that this technique can also be applied as an artificial nose for analyte vapors (e.g. flavors) in the gas phase [59].

11.4.2. Nanomechanical cantilevers as detectors

The principle of detection is based on the functionalization of the complete cantilever surface with a layer, which is sensitive to the compound to be investigated. The detection is feasible in different media (e.g. liquids or gas phase). The interaction of the analyte with the sensitive layer is transduced into a static deflection by inducing stress on one surface of the cantilever due to denser packing of the molecules [60] or a frequency shift in the case of dynamic detection mode [61] due to changes in mass.

11.4.3. Overview of the two detection modes

11.4.3.1. Static mode

In static mode detection, the deflection of the individual cantilever depends on the stress induced by the binding reaction of the specific compounds to the interface. The interface has to be activated in an asymmetrical manner, as shown in Fig. 11.10. Most often one of the cantilever surfaces is coated with a metallic layer (e.g. gold) by vacuum deposition techniques and subsequently activated by binding a receptor molecule directly via a thiol group to the interface (e.g. thiol modified DNA oligonucleotides) or as in case of protein recognition by activating the fresh gold interface with a self-assembling bi-functional bio-reactive alky-thiol molecule to which the protein moiety is covalently coupled [62].

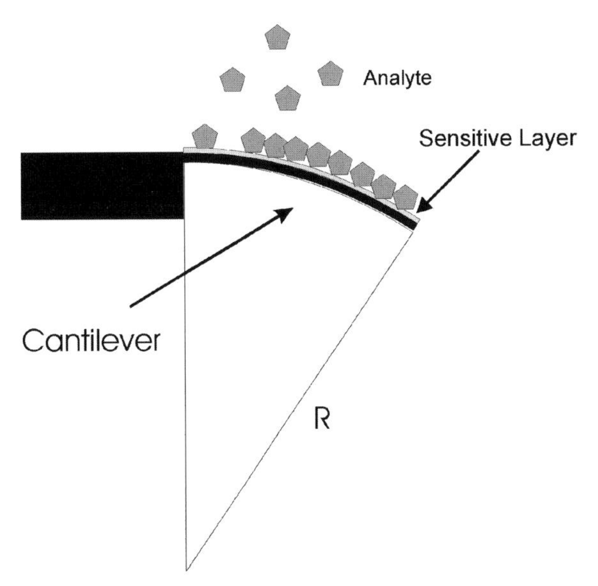

Fig. 11.10. Interaction of the analyte (light gray pentagons) with the sensitive layer induces a stress on the interface and bends the cantilever (note the asymmetric coating of the individual cantilever surface).

The radius R of the curvature of the cantilever is given by Stoney's law [63]:

$$\sigma = Et_{\text{cant}}^2(6R(1-\gamma))^{-1} \tag{11.8}$$

where σ is the stress, γ the Poisson ratio, E the Young's modulus and t_{cant} the thickness of cantilever. The thickness of the lever is an important parameter, which can be varied to increase or decrease the sensitivity of the device. By reducing the thickness a larger deflection due to stress change at the interface is possible. Note, that the interaction of the ligand with the receptor molecule has to occur in the vicinity of the interface. No flexible linking of the receptor molecule is allowed due to the fact that the induced stress will be diminished. In addition the receptor molecules should be immobilized, natively tightly packed on the interface, to interact with the substances to be analyzed.

11.4.3.2. Dynamic mode

In the case of dynamic mode detection, the resonance frequency of the individual cantilever, which has to be excited, depends on the mass. The binding reaction of the analyte to the interfaces is increasing the mass and the resonance frequency is normally decreased. In Fig. 11.11 the scheme of dynamic cantilever detection is shown.

The cantilever is excited by a piezo element. The change in mass (Δm) during the experiment due to an uptake of interacting biomolecules induces a change in the resonance frequency of cantilever, which can be described by the following formula [61]:

$$\Delta m = K(4n\pi^2)^{-1}(f_1^{-2} - f_0^{-2}) \tag{11.9}$$

where the resonance frequency prior and during experiment are f_0 and f_1, respectively, K is the spring constant of cantilever and n is a factor depending on the geometry of the

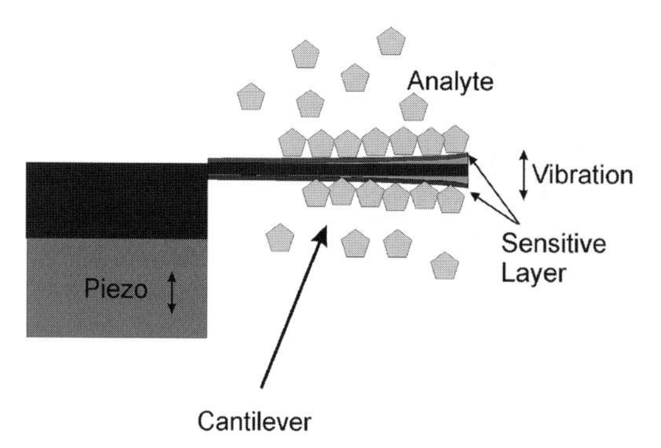

Fig. 11.11. Interaction of analyte (light gray pentagons) with sensitive layers induces a change in the resonance frequency of the cantilever.

cantilever. The uptake of mass due to specifically interacting molecules is doubled in this manner and the cantilever does not respond to temperature changes via a bimetallic effect. Additionally the preparation involves fewer steps than in the case of the static detection mode [57].

11.4.4. Setups

At the Institute of Physics at the University of Basel in collaboration with the IBM Research Laboratory Zurich, we developed cantilever array setups both for static and dynamic mode operations in liquids and in the gas phase.

The principal part of the setup is an array of eight cantilevers, produced by classical lithography technology with wet etching. A typical picture of such a cantilever array is shown in Fig. 11.12.

The structure of an array is composed of eight cantilevers. The etching process provides cantilever thicknesses ranging from 250 nm to 7 μm adapted for the individual application (i.e. static or dynamic mode).

A classical laser beam deflection optical detection for both the static and dynamic mode setups is used (see Fig. 11.13).

The laser source consists of an array of eight vertical cavity surface emitting lasers (VCELs, 760 nm wavelength, pitch 250 μm) and the position detection is obtained through a linear PSD. The array is mounted in a cell, which can be used for measurements in gas or liquid environments.

A scheme showing the setup is displayed in Fig. 11.14. The operation of our instruments is fully automatic. During the time course of a few hours up to eight different samples can be probed using the automatic fluid delivery. The instrumental noise of the static setup lies in the sub-nanometer range and the dynamic setup is able to detect mass changes in the order of picograms.

The key advantage to using cantilever arrays is to offer the possibility of in situ reference and the simultaneous detection of different substances. The in situ reference is needed to avoid the thermo-mechanical noise, especially in fluid-phase detection. Changes in

Fig. 11.12. SEM picture of an array of eight cantilevers. Dimensions: width 100 μm, length 500 μm, 0.5 μm with a pitch of 250 μm in-between.

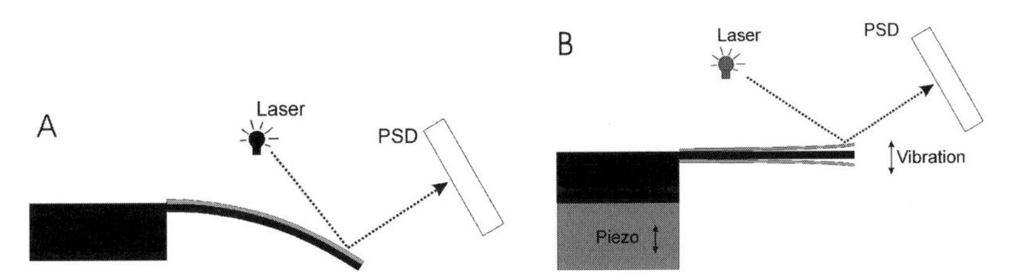

Fig. 11.13. Optical detection used for static (A) or dynamic (B) mode detection of average cantilever position using a multiple laser source VCELs and a position sensitive device (PSD).

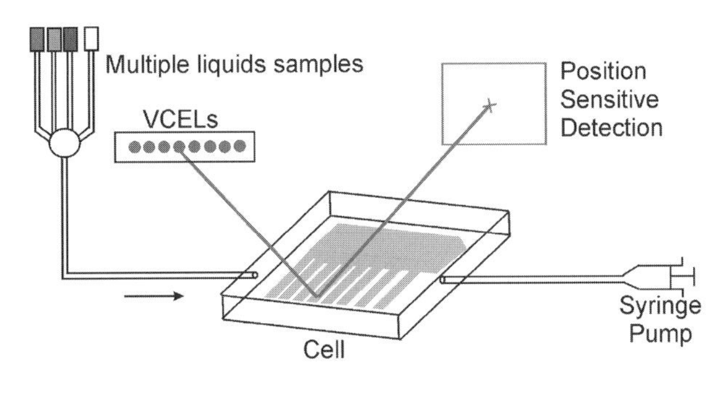

Fig. 11.14. General structure of cantilever array setups for gas/liquid samples.

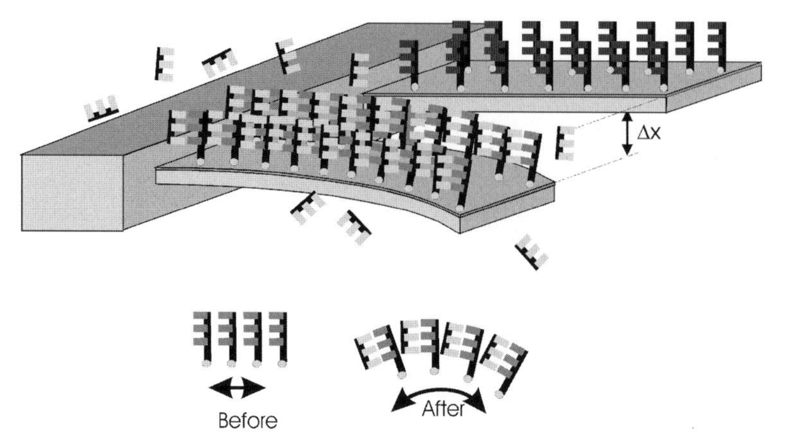

Fig. 11.15. Static detection of biomolecular interaction. The cantilevers have to be equilibrated and then the biomolecule of interest is injected. Due to the specific interaction of the injected biomolecules (light gray) with thed biomolecules on the cantilever shown in front stress builds up. A scheme is shown below. The interaction of the biomolecules with the receptor molecules induces stress at the interface, which deflects the individual cantilever specifically.

refractive index when the buffer changes will also contribute to a 'virtual' motion of the cantilever. As visible in Fig. 11.15 only the 'real' motion, which is the difference between the cantilevers on the same chip, is originating from the specific biomolecular interaction.

In Fig. 11.16(A) a raw signal of the cantilever array is displayed. Since there will always be instrumental or thermal drift, the differential signal detection is mandatory. The figure shows an experiment with a set of three cantilevers (thickness ~500 nm). In this experiment we used two reference cantilevers with different coatings and one specific biorecognition cantilever. This cantilever is being stressed upon binding of the corresponding interacting biomolecule.

As visible in Fig. 11.16(B) the differential signal lacks any external influences except for the specific biomolecular interaction that induces a differential signal of ~90 nm relative to the in situ reference. The experiment is reversible and can be repeated using

---

Fig. 11.16. (A) Raw data of a three-lever bio-array experiment. In the top traces (light gray, gray) the motion of the reference cantilevers is shown. In black color the motion of the biologically specific cantilever is displayed. Upon injection of interacting biomolecules (~1000 s) turbulences of the liquid cause all levers to undergo some motion, which is stabilized immediately when the flow is stopped (~1200 s). The specific binding signal quickly builds up and remains stable. The interaction is fully reversible and can be broken by shifting the equilibrium of the binding reaction by injecting pure buffer solution (~6500 s) into the fluid chamber. During the time course of 2–3 h, we regularly see a drift of the cantilever arrays on the order of tens of nanometers even though the setup is temperature-stabilized (0.05 °C). (B) Differential data of the experimental set of (A). The difference between the two reference cantilevers is shown (light grey). Except for some small motions, no differential bending is observed, whereas in the dark gray and black colors the difference of the specifically reacting cantilever with respect to the reference cantilevers is shown. As shown, after ~6500 s pure buffer solution is injected and the differential signal collapses to values close to the starting point where no interacting biomolecules were present in the experiment.

different concentrations of analytes. In a recent work we presented data which allow extracting of the thermodynamics of the interacting biomolecules (i.e. DNA) [64]. Deflection signals as small as a few nanometers are easily detected. Currently, the detection limit in static experiments lies in the range of nanomolar concentration [64] but can be significantly lowered by using cantilever arrays of thickness in the range of 250–500 nm in the future.

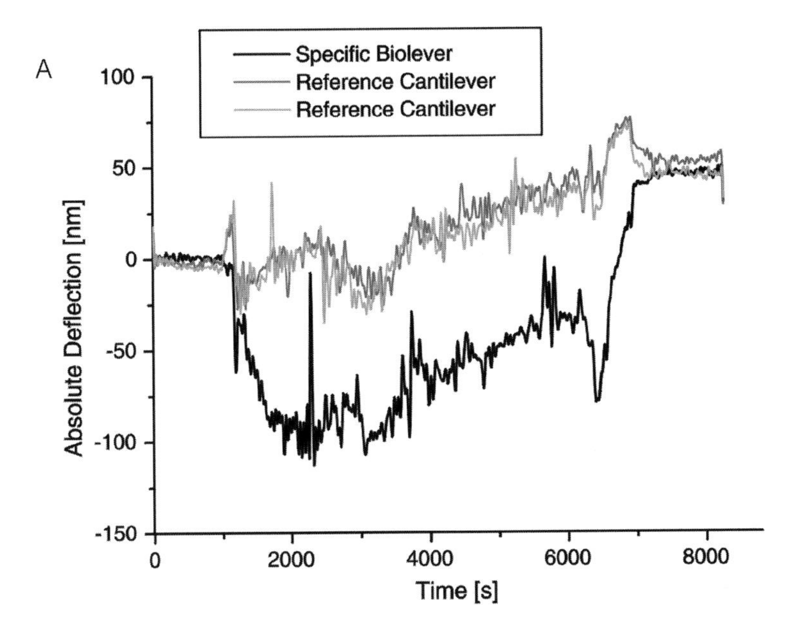

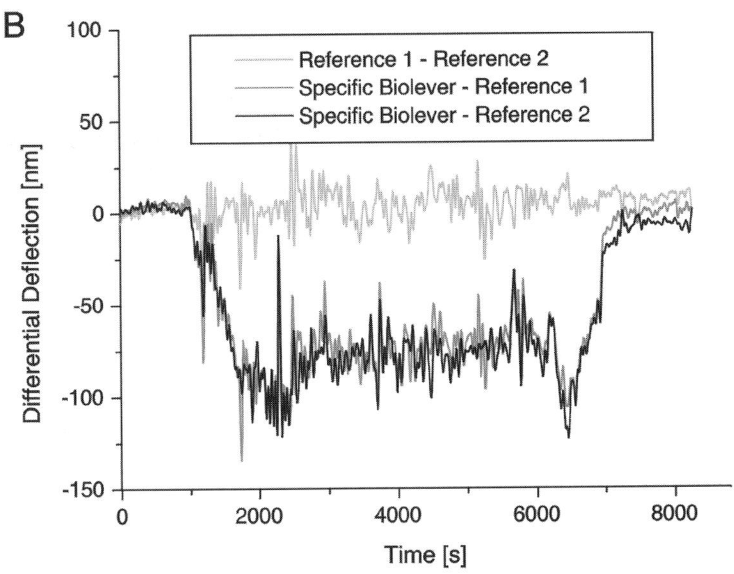

References pp. 262–263

Great care has to be taken in the selection of the internal reference lever. In the case of DNA detection, an oligonucleotide displaying a sequence, which does not induce cross talk binding reactions to the sequences to be detected is chosen. Coating with thin layers of titanium and gold using vacuum deposition modifies one side of the cantilever array. Onto this metallic interface a thiol-modified oligonucleotide self-assembles in a high-density layer. Complementary and unknown oligonucleotide sequences are then injected and the specific interaction is directly visible within minutes. Stress at the interface is built up due to a higher density of packing (see Fig. 11.15). In protein detection, a protection of the asymmetrically coated cantilever has to be considered [65, 66]. Preparation of protein detecting cantilevers is a multi-step procedure and requires surface chemistry knowledge. The side opposite to the biomolecular-modified side is generally protected by a poly-ethylene-glycol (PEG) layer. The bio-reference surface can be coated by using unspecifically interacting proteins (e.g. bovine serum albumin). In protein detection experiments larger 'fluctuations' of the cantilevers are observed (e.g. Fig. 11.16) than in the ssDNA-ssDNA experiments. A possible interpretation of this difference might be due to the fact that the proteins used [μM] absorb light within the visible spectrum and therefore induce some local changes in index of refraction. Specific signals are normally measured within minutes without problems. Usually some drift of a few tens of nanometers is observed from the complete set of cantilevers during the time course of the experiment even though the instruments' temperature is stabilized within ± 0.05 °C. However, these effects are completely eliminated by using a differential read out on the very same cantilever array.

Cantilevers arrays are already applied as detectors both in static and dynamic mode [60, 61]. Recent articles show the potential for detection of DNA hybridization [55,64], cell capture or toxin detection [53]. Integrating cantilever arrays into micro-fluidic channels will significantly reduce the amount of sample required [67]. Attempts have been made to get data from single cantilever experiments for DNA [68] or antibody antigen reactions [69] or from a two-cantilever setup using different stiffness for the individual cantilevers [70]. We would like to point out that these approaches have serious drawbacks. Information extracted from these experiments, which often last for multiple hours, cannot exclude unspecific drift of any kind. The signal in these experiments is interpreted as specificity on the biomolecular level but no correlation from one lever to the next is applicable if only one lever is used at a time. In a second approach, cantilevers with different stiffness are used to monitor the nanometer motions. Since the individual cantilever used shows a difference of factor four in terms of stiffness, the response, which originates from specific interaction, is difficult to extract. The sensitivity of this approach is hampered due to the differences in stiffness, which is directly correlated to the thickness of the cantilever used (see Eq. (11.8)). An interaction of the biomolecule with the stiffer reference cantilever might not be detectable if the stress signal lies within the thermal noise of that lever.

11.4.5. Future applications of cantilever arrays

The cantilever array technology explores a wide area of applications; all biomolecular interactions are in principle able to be experimentally detected using cantilever array as long as mass change or surface stress is induced due to the specific interaction. A few applications have so far demonstrated promising results in the field of biological detection.

The cantilever based sensor platform might fill the gap between the very expensive analytical instrumentation (e.g. mass-spectroscopy, HPLC, SPR), which are sensitive but costly and relatively slow and the chip technologies (e.g. gene-arrays) with its advantage of easy multiplexing capabilities, but there is a need for fluorescence labeling and it is restricted to higher molecular weight compounds like proteins and nucleic acids.

In comparison with the methods described above, the cantilever technology is cheap, fast, sensitive and applicable to a broad range of compounds. The cantilever arrays can be used repeatedly for successive experiments. The lack of multiplexing could be overcome by the application of large cantilever arrays with > 1000 cantilevers/chip. At present there are projects launched to introduce commercial platforms providing arrays of eight cantilevers for applications in the liquid or gas phases. A critical point for future developments in this field will be the access to the cantilevers arrays as it is in the 'normal' field biological applications using single cantilever scanning force microscopy. At the moment, there are no biological experiments published that use the dynamic mode detection. But as we believe the ease of preparation (symmetrically as pointed above) and the fact that the sensitivity towards environmental changes is reduced, this might be the instrumental approach of choice for the future biological detection using cantilever arrays.

11.5. CONCLUSION

During the last decade, single molecule experiments provided ample information in the field of biological basic research. We would like to point out again that these kinds of experiments do not probe an ensemble of molecules and therefore give access to information or properties of sub-populations of biomolecules. These experiments do not have to be synchronized and therefore no averaging occurs. In nature, many cases that define the status of an organism depend on properties or activities of individual sub-populations (e.g. start of cancer in an individual cell). There is a long way to go before having real implications of single molecule manipulation experiments on daily life, but the information revealed so far shows that the clues to some specific biological problems might lie in detail (e.g. Ref. [22]). In addition, it is important to mention that single molecule experiments are always technically most demanding and future results obtained on single molecules will mainly depend on instrumentation capabilities. One 'drawback' of single molecule experiments is that, one experiment is not sufficient to elucidate the properties of a subpopulation. Enough experimental data have to be gathered, which is time consuming, to allow applying statistics.

A way out might be to combine the high sensitivity of these force-measuring devices and sample a few thousand molecules at a given time as is done by using cantilever arrays. This new array technology is not limited to genomic studies but can also detect protein–protein interactions [55], and will thus find applications in the fields of proteomics, biodiagnostics and combinatorial drug discovery where rapid, quantitative binding measurements are vital. The ability to directly translate biochemical recognition into nanomechanical motion might have wide ranging implications, for example in DNA computing applications or nanorobotics. The nano–Newton forces generated are sufficient to operate micromechanical valves or microfluidic devices and in situ delivery devices could be triggered directly by signals from gene expression, immune response or single cells.

References pp. 262–263

11.6. ACKNOWLEDGEMENTS

Financial support of the NCCR 'Nanoscale Science', the Swiss National Science Foundation and the ELTEM Regio Project Nanotechnology is gratefully acknowledged. We would like to thank our colleagues from Basel and IBM Rüschlikon for the great collaborative effort and for their valuable contributions to make progress in the field of single molecules manipulation using OT and SFM and in the field of cantilever arrays. We are especially grateful to Ernst Meyer, Peter Vettiger, Felice Battiston, Torsten Strunz, Irina Schumakovitch and Hans-Joachim Güntherodt.

11.7. REFERENCES

1. A. Engel and D.J. Müller, Nat. Struct. Biol., 7 (2000) 715
2. M. Guthold, X.S. Zhu, C. Rivetti, G.L. Yang, N.H. Thomson, S. Kasas, H.G. Hansma, B. Smith, P.K. Hansma and C. Bustamante, Biophys. J., 77 (1999) 2284
3. M. Rief, M. Gautel, F. Oesterhelt, J.M. Fernandez and H.E. Gaub, Science, 276 (1997) 1109
4. A. Ashkin, Phys. Rev. Lett., 24 (1970) 156
5. A. Ashkin, J.M. Dziedzic, J.C. Bjorkholm and S. Chu, Opt. Lett., 11 (1986) 288
6. A. Ashkin and J.M. Dziedzic, Science, 235 (1987) 1517
7. A. Ashkin, J.M. Dziedzic and T. Yamane, Nature, 330 (1987) 769
8. K. Svoboda and S.M. Block, Annu. Rev. Biophys. Biomol. Struct., 23 (1994) 247
9. N. Pam and H.M. Nussenzveig, Eur. Phys. Lett., 50 (2000) 702
10. K.C. Neuman, E.H. Chadd, G.F. Lion, K. Bergman and S.M. Block, Biophys. J., 77 (1999) 2856
11. G. Leitz, E. Fällman, S. Tuck and O. Axner, Biophys. J., 82 (2002) 2224
12. R.S. Afzal and E.B. Treacy, Rev. Sci. Instrum., 63 (1992) 2157
13. S.B. Smith, PhD Thesis, University of Twente, The Netherlands, 1998
14. W. Grange, S. Husale, H.J. Güntherodt and M. Hegner, Rev. Sci. Instrum., June (2002), in press
15. F. Gittes and C.F. Schmidt, Eur. Biophys. J., 27 (1998) 75
16. M.B. Viani, T.E. Schaffer, A. Chand, M. Rief, H.E. Gaub and P.K. Hansma, J. Appl. Phys., 86 (1999) 2258
17. A.D. Metha, M. Rief, J.A. Spudich, D.A. Smith and R.M. Simmons, Science, 283 (1999) 1689
18. C. Bustamante, J.C. Macosko and G.J.L. Wuite, Nat. Rev., Mol. Cell Biol., 1 (2000) 130
19. G. Woehlke and M. Schliwa, Nat. Rev., Mol. Cell Biol., 1 (2000) 50
20. G.J. Wuite, S.B. Smith, M. Young, D. Keller and C. Bustamante, Nature, 404 (2000) 103
21. R.J. Davenport, G.J. Wuite, R. Landick and C. Bustamante, Science, 287 (2000) 2497
22. D.E. Smith, S.J. Tans, S.B. Smith, S. Grimes, D.L. Anderson and C. Bustamante, Nature, 413 (2001) 748
23. S.B. Smith, Y. Cui and C. Bustamante, Science, 271 (1996) 795
24. M.S. Kellermayer, S.B. Smith, H.L. Granzier and C. Bustamante, Science, 276 (1997) 1112
25. L. Tskhovrebova, J. Trinick, J.A. Sleep and R.M. Simmons, Nature, 387 (1997) 308
26. J. Liphardt, B. Onoa, S.B. Smith, I. TinocoJr. and C. Bustamante, Science, 292 (2001) 733
27. M. Hegner, S.B. Smith and C. Bustamante, Proc. Natl. Acad. Sci. USA, 96 (1999) 10109
28. S. Husale, W. Grange and M. Hegner, Single Mol., June (2002), in press
29. G.U. Lee, L.A. Chrisley and R.J. Colton, Science, 266 (1994) 771
30. G.U. Lee, D.A. Kidwell and R.J. Colton, Langmuir, 10 (1994) 354
31. V.T. Moy, E.L. Florin and H.E. Gaub, Science, 266 (1994) 257
32. U. Dammer, O. Popescu, P. Wagner, D. Anselmetti, H.-J. Güntherodt and G.N. Misevic, Science, 267 (1995) 1173
33. P. Hinterdorfer, W. Baumgartner, H.J. Gruber, K. Schilcher and H. Schindler, Proc. Natl. Acad. Sci. USA, 93 (1996) 3477
34. U. Dammer, M. Hegner, D. Anselmetti, P. Wagner, M. Dreier, W. Huber and H.-J. Güntherodt, Biophys. J., 70 (1996) 2437

35. S. Allen, X.Y. Chen, J. Davies, M.C. Davies, A.C. Dawkes, J.C. Edwards, C.J. Roberts, J. Sefton, S.J.B. Tendler and P.M. Williams, Biochemistry, 36 (1997) 7457
36. P.E. Marszalek, H. Lu, H.B. Li, M. Carrion-Vazquez, A.F. Oberhauser, K. Schulten and J.M. Fernandez, Nature, 402 (1999) 100
37. R. Ros, F. Schwesinger, D. Anselmetti, M. Kubon, R. Schäfer, A. Plückthun and L. Tiefenauer, Proc. Natl. Acad. Sci. USA, 95 (1998) 7402
38. R. Merkel, P. Nassoy, A. Lueng, K. Ritchie and E. Evans, Nature, 397 (1999) 50
39. T. Strunz, K. Oroszlan, R. Schäfer and H.-J. Güntherodt, Proc. Natl. Acad. Sci. USA, 96 (1999) 11277
40. F. Schwesinger, R. Ros, T. Strunz, D. Anselmetti, H.-J. Güntherodt, A. Honegger, L. Jermutus, L. Tiefenauer and A. Plückthun, Proc. Natl. Acad. Sci. USA, 97 (2000) 7402
41. R. De Paris, T. Strunz, K. Oroszlan, H.-J. Güntherodt and M. Hegner, Single Mol., 1 (2000) 285
42. I. Schumakovitch, W. Grange, T. Strunz, P. Bertoncini, H.-J. Güntherodt and M. Hegner, Biophys. J., 82 (2002) 517
43. E. Evans, A. Leung, D. Hammer and S. Simon, Proc. Natl. Acad., 98 (2001) 3784
44. H.A. Kramers, Physica Utrecht, 7 (1940) 284
45. I. Bell, Science, 200 (1978) 618
46. E. Evans, Annu. Rev. Biophys. Biomol. Struct., 30 (2001) 105
47. E. Galligan, C.J. Roberts, M.C. Davies, S.J.B. Tendler and P.M. Williams, J. Chem. Phys., 114 (2001) 3208
48. B. Heymann and H. Grubmüller, Phys. Rev. Lett., 84 (2000) 6126
49. J.L. Hutter and J. Bechhoefer, Rev. Sci. Instrum., 64 (1993) 1868
50. D. Pörschke and M. Eigen, J. Mol. Biol., 62 (1971) 361
51. N. Tibanyenda, S.H. Debruin, C.A.G. Haasnoot, G.A. Vandermarel, J.H. Vanboom and C.W. Hilbers, Eur. J. Biochem., 139 (1984) 19
52. M.B. Viani, T.E. Schaffer, G.T. Paloczi, L.I. Pietrasanta, B.L. Smith, J.B. Thompson, M. Richter, M. Rief, H.E. Gaub, K.W. Plaxco, A.N. Cleland, H.G. Hansma and P.K. Hansma, Rev. Sci. Instrum., 70 (1999) 4300
53. D.R. Baselt, G.U. Lee and R.J. Colton, Vac. Sci. Technol. B, 14(2) (1996) 789
54. H. Masayuki, Y. Yoshiaki, T. Hideki, O. Hideo, N. Tsunenori and M. Jun, Biosens. Bioelectron., 17 (2002) 173
55. J. Fritz, M.K. Baller, H.P. Lang, H. Rothuizen, P. Vettiger, E. Meyer, H.-J. Güntherodt, Ch. Gerber and J.K. Gimzewski, Science, 288 (2000) 316
56. R. Berger, E. Delamarche, H.P. Lang, Ch. Gerber, J.K. Gimzewski, E. Meyer and H.-J. Güntherodt, Science, 276 (1997) 2021
57. R. Berger, Ch. Gerber, J.K. Gimzewski, E. Meyer and H.-J. Güntherodt, Appl. Phys. Lett., 69 (1996) 40
58. T. Bachels, R. Schäfer and H.-J. Güntherodt, Phys. Rev. Lett., 84 (2000) 4890
59. M.K. Baller, H.P. Lang, J. Fritz, Ch. Gerber, J.K. Gimzewski, U. Drechsler, H. Rothuizen, M. Despont, P. Vettiger, F.M. Battiston, J.P. Ramseyer, P. Fornaro, E. Meyer and H.-J. Güntherodt, Ultramicroscopy, 82 (2000) 1
60. J. Fritz, M.K. Baller, H.P. Lang, T. Strunz, E. Meyer, H.-J. Güntherodt, E. Delamarche, Ch. Gerber and J.K. Gimzewski, Langmuir, 16 (2000) 9694
61. F.M. Battiston, J.P. Ramseyer, H.P. Lang, M.K. Baller, Ch. Gerber, J.K. Gimzewski, E. Meyer and H.-J. Güntherodt, Sensors Actuators B, 77 (2001) 122
62. P. Wagner, M. Hegner, P. Kernen, F. Zaugg and G. Semenza, Biophys. J., 70 (1996) 2052
63. G.G. Stoney, Proc. R. Soc. London Soc. A, 82 (1909) 172
64. R. McKendry, T. Strunz, J. Zhang, Y. Arntz, M. Hegner, H.P. Lang, M. Baller, U. Certa, E. Meyer, H.-J. Güntherodt and Ch. Gerber, Proc. Natl. Acad. Sci. USA, 99 (2002) 9783
65. H.P. Lang, M. Hegner, E. Meyer, and Ch. Gerber, Nanotechnology, 13 (2002) R29
66. Y. Arntz, J.E. Seelig, J. Zhang, H.P. Lang, P. Hunziker, J.P. Ramseyer, E. Meyer, M. Hegner, and Ch. Gerber, Nanotechnology, 14 (2003) 86
67. J. Thaysen, R. Marie, A. Boisen, IEEE Int. Conf. Micro. Electro. Mech. Syst., Tech. Dig., Vol. 14, Institute of Electrical and Electronic Engineers, New York, 2001, p. 401
68. K.M. Hansen, H.F. Ji, G.H. Wu, R. Datar, R. Cote, A. Majumdar and T. Thundat, Anal. Chem., 73 (2001) 1567
69. G.H. Wu, R.H. Datar, K.M. Hansen, T. Thundat, R.J. Cote and A. Majumdar, Nat. Biotech., 19 (2001) 856
70. C. Grogan, R. Raiteri, G.M. O'Connor, T.J. Glynn, V. Cunningham, M. Kane, M. Charlton and D. Leech, Biosens. Bioelectron., 17 (2001) 201

Subject Index

JOURNAL OF CHROMATOGRAPHY LIBRARY

A Series of Books Devoted to Chromatographic and Electrophoretic Techniques and their Applications

Although complementary to the Journal of Chromatography, each volume in the library series is an important and independent contribution in the field of chromatography and electrophoresis. The library contains no material reprinted from the journal itself.

Recent volumes in this series

Volume 60	**Advanced Chromatographic and Electromigration Methods in BioSciences** edited by Z. Deyl, I. Mikšík, F. Tagliaro and E. Tesařová
Volume 61	**Protein Liquid Chromatography** edited by M. Kastner
Volume 62	**Capillary Electrochromatography** edited by Z. Deyl and F. Svec
Volume 63	**The Proteome Revisited: theory and practice of all relevant electrophoretic steps** by P.G. Righetti, A. Stoyanov and M.Y. Zhukov
Volume 64	**Chromatography - a Century of Discovery 1900-2000: the bridge to the sciences/technology** edited by C.W. Gehrke, R.L. Wixom and E. Bayer
Volume 65	**Sample Preparation in Chromatography** by S.C. Moldoveanu and V. David
Volume 66	**Carbohydrate Analysis by Modern Chromatography and Electrophoresis** edited by Z. El Rassi
Volume 67	**Monolithic Materials: Preparation, Properties and Applications** by F. Svec, T. Tennikova and Z. Deyl